Springer Tracts in Civil Engineering

Series Editors

Sheng-Hong Chen, School of Water Resources and Hydropower Engineering, Wuhan University, Wuhan, China

Marco di Prisco, Politecnico di Milano, Milano, Italy

Ioannis Vayas, Institute of Steel Structures, National Technical University of Athens, Athens, Greece

Springer Tracts in Civil Engineering (STCE) publishes the latest developments in Civil Engineering - quickly, informally and in top quality. The series scope includes monographs, professional books, graduate textbooks and edited volumes, as well as outstanding PhD theses. Its goal is to cover all the main branches of civil engineering, both theoretical and applied, including:

- Construction and Structural Mechanics
- Building Materials
- Concrete, Steel and Timber Structures
- Geotechnical Engineering
- Earthquake Engineering
- Coastal Engineering; Ocean and Offshore Engineering
- Hydraulics, Hydrology and Water Resources Engineering
- Environmental Engineering and Sustainability
- Structural Health and Monitoring
- Surveying and Geographical Information Systems
- Heating, Ventilation and Air Conditioning (HVAC)
- Transportation and Traffic
- Risk Analysis
- Safety and Security

Indexed by Scopus

To submit a proposal or request further information, please contact:
Pierpaolo Riva at Pierpaolo.Riva@springer.com (Europe and Americas) Wayne Hu at wayne.hu@springer.com (China)

Yu Bai
Editor

Composites for Building Assembly

Connections, Members and Structures

 Springer

Editor
Yu Bai
Department of Civil Engineering
Monash University
Clayton, Australia

ISSN 2366-259X ISSN 2366-2603 (electronic)
Springer Tracts in Civil Engineering
ISBN 978-981-19-4277-8 ISBN 978-981-19-4278-5 (eBook)
https://doi.org/10.1007/978-981-19-4278-5

© The Editor(s) (if applicable) and The Author(s), under exclusive license to Springer Nature Singapore Pte Ltd. 2023
This work is subject to copyright. All rights are solely and exclusively licensed by the Publisher, whether the whole or part of the material is concerned, specifically the rights of translation, reprinting, reuse of illustrations, recitation, broadcasting, reproduction on microfilms or in any other physical way, and transmission or information storage and retrieval, electronic adaptation, computer software, or by similar or dissimilar methodology now known or hereafter developed.
The use of general descriptive names, registered names, trademarks, service marks, etc. in this publication does not imply, even in the absence of a specific statement, that such names are exempt from the relevant protective laws and regulations and therefore free for general use.
The publisher, the authors, and the editors are safe to assume that the advice and information in this book are believed to be true and accurate at the date of publication. Neither the publisher nor the authors or the editors give a warranty, expressed or implied, with respect to the material contained herein or for any errors or omissions that may have been made. The publisher remains neutral with regard to jurisdictional claims in published maps and institutional affiliations.

This Springer imprint is published by the registered company Springer Nature Singapore Pte Ltd.
The registered company address is: 152 Beach Road, #21-01/04 Gateway East, Singapore 189721, Singapore

Preface

Fibre reinforced polymer (FRP) composites are materials consisting of polymer matrix reinforced by fibres. FRP composites have been considered in the construction of a range of structures in civil engineering because of their light weight in combination with high strength, and especially, corrosion resistance. Pultrusion process, as an automated and continuous manufacturing technique, has been well developed to produce FRP composites, with various sections and shapes, at affordable costs (especially when glass fibres are used, i.e. GFRP) and a fast production rate. Pultruded FRP composites, when used as structural members in building construction, are mostly welcomed in the approaches of design for manufacturing and assembly (DfMA) and modular construction methods.

Pultruded FRP composites show their higher stiffness and strength in the longitudinal pultrusion direction where the majority of continuous fibres runs in this direction. In this direction, pultruded GFRP composites can be with tensile strength similar to steel but much lower elastic modulus and shear strength, and lack of ductility at the material level. Such material characteristics need to be properly addressed in the designs and applications for pultruded GFRP composites as load-carrying structural members in construction. This work aims to introducing a series of design concepts and structural forms in the use of pultruded GFRP composites with considerations of their distinct material properties. Such concepts and forms include beam or slab and column or wall structures with web-flange sandwich configurations assembled by GFRP components, and various connection details for GFRP beams to columns and columns to columns.

The web-flange sandwich sections may achieve improved structural stiffness through the placement of sectional areas away from the neutral axis. To address the material brittleness of GFRP composites, steel is incorporated in some cases such as the beam-to-column and column-to-column connections and the steel-GFRP composite beam system, where steel is designed to yield prior to the failure of GFRP composites and therefore ductility can be seen; or load-dependent composite action may be considered for a composite section such as in composite beams and sandwich structures. Fire performance of GFRP structures is also of importance because of the glass transition and decomposition of polymer matrix at elevated and high temperatures. Therefore, this work presents improved fire endurance time for the proposed GFRP web-flange sandwich structures, subjected to mechanical loading and ISO fire curve simultaneously, where proper fire-resistant panels are installed.

With the support of the results from experimental investigations and numerical studies for the structural concepts and forms, large-scale structures assembled using the developed GFRP members and connections are also developed and introduced in this work. These include demonstrations for pedestrian bridge structures, space frame structures, house frame structures and modular retaining walls as examples. These examples may help illustrating the developed structural concepts and also the practices of the DfMA and modular construction methodology.

This work could not be accomplished without the contributions from the Ph.D. students and research associates and visiting scholars and the collaborations with a number of colleagues in the past ten years resulting in relevant research papers for the main content of this work. The support received from the Australian Research Council including the inaugural Discovery Early Career Research Award fellowship and the Discovery and Linkage projects, as well as the support from a range of industries especially the Multiplex building innovation project and iBuild building solutions, is gratefully acknowledged.

During the course of this work, support and assistance are received from a large group of colleagues and friends. Thanks should be given to Prof. Thomas Keller at Composite Construction Laboratory of École Polytechnique Fédérale de Lausanne, Prof. João Ramôa Correia at the Instituto Superior Técnico of University of Lisbon, Prof. Toby Mottram in the School of Engineering of the University of Warwick, and Prof. Lawrence Bank in the School of Architecture of Georgia Tech, for sharing their research experiences and outcomes on the developments of FRP building structures. Thanks should also go to Strongwell, Exel Composites, Holland Composites, Valeria Carullo, Foster + Partners, TOP GLASS Industries S.p.A and IUAV University of Venice, and Armourwall Group Australia for providing relevant information about FRP composites in construction introduced in this work.

The editorial assistance from Fengyi Huang, Dr. Jie Yu, Dr. Yiming Zhang and Joan Rosenthal are very much appreciated. It is also a great enjoyment in working with the team of the publisher Springer Nature, and thanks for their professional support for improving the quality of this work.

Melbourne, Australia   Yu Bai

Disclaimer

While the editor, authors and the contributors have used reasonable endeavours to ensure that material contained in this book was correct to their best understanding at the time of publication, they provide no express or implied warranties as to the content of the book. The book is provided as its version without any warranty or guarantee as to its accuracy, reliability, currency or completeness. The book should not be construed as an endorsement of any entity, product or technique in it.

The content of the book is provided for information only and does not constitute professional advice. Use of the book is strictly at your own risk and users should seek appropriate independent professional advice prior to relying on, or entering into any commitment based upon, the material in the book.

The editor, authors and the contributors accept no responsibility for any loss or damage arising from the use of or reliance upon the book by any person. The book does not override the approval processes in any jurisdiction. The right to make changes to the book is reserved at any time without notice.

Contents

1 **Introduction** 1
   Yu Bai and Chengyu Qiu

2 **Fibre Reinforced Polymer Built-Up Beams and One-Way Slabs** 27
   Sindu Satasivam, Yu Bai, and Xiao-Ling Zhao

3 **Fibre Reinforced Polymer Composites Two-Way Slabs** 49
   Sindu Satasivam, Yu Bai, Yue Yang, Lei Zhu, and Xiao-Ling Zhao

4 **Steel- Fibre Reinforced Polymer Composite Beams** 75
   Sindu Satasivam and Yu Bai

5 **Composite Actions of Steel-Fibre Reinforced Polymer Composite Beams** 99
   Sindu Satasivam, Peng Feng, Yu Bai, and Colin Caprani

6 **Fibre Reinforced Polymer Columns in Axial Compression** 121
   Lei Xie, Yu Bai, Yujun Qi, Colin Caprani, and Hao Wang

7 **Fibre Reinforced Polymer Wall Assemblies in Axial Compression** 143
   Lei Xie, Yujun Qi, Yu Bai, Chengyu Qiu, Hao Wang, Hai Fang, and Xiao-Ling Zhao

8 **Fibre Reinforced Polymer Columns with Bolted Sleeve Joints under Eccentric Compression** 167
   Lei Xie, Yu Bai, Yujun Qi, and Hao Wang

9 **Connections of Fibre Reinforced Polymer to Steel Members: Experiments** 187
   Chao Wu, Zhujing Zhang, and Yu Bai

10 **Connections of Fibre Reinforced Polymer to Steel Members: Numerical Modelling** 211
    Zhujing Zhang, Chao Wu, Xin Nie, Yu Bai, and Lei Zhu

11	**Cyclic Performance of Bonded Sleeve Beam-Column Connections** 231
	Zhujing Zhang, Yu Bai, Xuhui He, Li Jin, and Lei Zhu
12	**Joint Capacity of Bonded Sleeve Connections for Tubular Fibre Reinforced Polymer Members** 255
	Chengyu Qiu, Peng Feng, Yue Yang, Lei Zhu, and Yu Bai
13	**Axial Performance of Splice Connections for Fibre Reinforced Polymer Columns** 285
	Chengyu Qiu, Chenting Ding, Xuhui He, Lei Zhang, and Yu Bai
14	**Cyclic Performance of Splice Connections for Fibre Reinforced Polymer Members** 309
	Chengyu Qiu, Yu Bai, Zhenqi Cai, and Zhujing Zhang
15	**Fire Performance of Loaded Fibre Reinforced Polymer Multicellular Composite Structures** 333
	Lei Zhang, Yiqing Dai, Yu Bai, Wei Chen, and Jihong Ye
16	**Large Scale Structural Applications** 357
	Yu Bai, Sindu Satasivam, Xiao Yang, Ahmed Almutairi, Hosea Ivan Christofer, and Chenting Ding

Chapter 1
Introduction

Yu Bai and Chengyu Qiu

Abstract Structural members made from fibre reinforced polymer (FRP) composites are alternatives for construction of durable and lightweight structures. Generally, FRP composites consist of a polymer matrix reinforced by high-strength and stiffness fibres. The way that the fibres are arranged in the polymer matrix forms an anisotropic composite. Compared to conventional construction materials, the key advantages of FRP composites are lightness in weight, resistance to corrosion, and embrace to design for manufacturing and assembly, with possible other features from specific constituents and tailored formations. The success of FRP composites as the primary structural members in building construction requires proper designs of such structural members and connections, in consideration of the anisotropic and brittle characteristics of the composite materials. This work therefore aims to introduce a series of FRP structures and connections developed in this way dedicated for prefabrication and design for manufacturing and assembly. It is supported by a range of research activities, in order to advance member and connection solutions and to provide structural stiffness in compensation of low elastic modulus for glass fibre reinforced polymer (GFRP) composites and to provide ductility in contrast to their material brittleness. With the experience gained from these researches, large scale FRP structures built using the developed structural components and connections are also introduced.

1.1 Background

Many commonly used construction materials in civil engineering, such as steel and reinforced concrete, are subjected to corrosion in aggressive environments, resulting in increased cost for maintenance and rehabilitation. Furthermore, the transportation, forming and erection of structural members made from heavy construction materials require intensive labouring and the use of substantial equipment. Under the desire to produce durable structures that are lighter to transport and more convenient to

---

Y. Bai (✉) · C. Qiu
Department of Civil Engineering, Monash University, Clayton, Australia
e-mail: yu.bai@monash.edu

© The Author(s), under exclusive license to Springer Nature Singapore Pte Ltd. 2023
Y. Bai (ed.), *Composites for Building Assembly*, Springer Tracts in Civil Engineering,
https://doi.org/10.1007/978-981-19-4278-5_1

build, the construction industry constantly show interest in innovative construction materials. Fibre reinforced polymer (FRP) composites, which have been routinely used in the aerospace and marine industries since the 1970s [1, 2], come up as a promising option. The matrix of FRPs usually is one of the thermosetting polymers of polyester, epoxy, vinylester and phenolic [3]. Carbon, aramid, basalt and glass are the common types of fibres used in FRPs. Among them glass fibres are the least expensive while exhibit relatively lower mechanical strength and stiffness.

In civil engineering, the early use of FRPs mainly focuses on the strengthening and rehabilitation of existing structures [4]. It was typically done through manually laying layers of fibre sheets and impregnating them with a polymer resin that will later cure in place. The potential use of FRPs as the primary load-carrying members is prompted by the advance of the pultrusion manufacturing technique [5]. Pultrusion, as shown in Fig. 1.1a, is an automated process that entails impregnated fibres pulled continuously through a heated die where a thermosetting polymer resin is fed and binds the fibres together as it cures. The fibre system of a pultruded FRP section usually includes unidirectional fibre rovings and multi-directional fibre mats in symmetric alternating layers as shown in Fig. 1.1b; a layer of surfacing veils is also used when a smooth surface is desired [3].

Due to the one-dimensional 'pulling' process, pultruded FRP composites mostly are characterised by constant cross-sections and majority of the continuous fibres aligned in the longitudinal pultrusion direction (thus majority of strength and stiffness in this direction). Glass fibres are often used when FRP composites are pultruded in large quantities. In the longitudinal direction, pultruded GFRP (with glass fibres) composites typically possess strengths comparable to mild steels, but elastic moduli from 1/8 to 1/5 [8]. GFRP composites can be pultruded in a range of cross-sections similar to structural steel. Researches have since been carried out in attempt to bring these sections from factory for implementation in structural construction. Also, the publication of design guidelines and standards specially for pultruded FRP composites [9–11] are developed to envisage their prospect in construction.

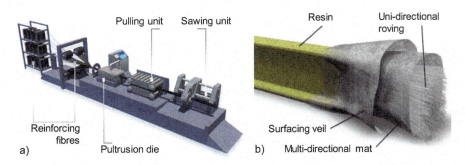

Fig. 1.1 Pultruded FRP composites: **a** pultrusion process (image courtesy of Exel Composites) [6]; **b** fibre system (image courtesy of Strongwell) [7]

1.2 Development of FRP Building Structures

1.2.1 Free-Form Structures by Lay-Up and Moulding Process

In the 1950s the use of FRP composites began to extend from the aerospace and marine industries to the building industry. Similar as in the former two industries, components or entirety of the early-generation FRP buildings were manufactured from the fibre lay-up and resin moulding processes. This type of structures featured a free-form geometry owing to the flexibility in fibre architecture and in overall moulded shape.

The year of 1957 witnessed the erection of an early FRP building, the Monsanto House of the Future [12, 13]. The structure, as shown in Fig. 1.2a, featured four symmetric wings cantilevered off a 5 × 5 m central concrete core. The wings consisted of curved GFRP shells made of glass fibres and polyester resin, and a honeycomb core made of kraft paper. The components of the house were assembled on-site using bolts and epoxy. The tourist attraction, reported as able to withstand earthquakes and strong wind loads, was demolished in 1968.

Another example of early FRP buildings is the Futuro. Originally intended as a ski cabin, the Futuro house, shown in Fig. 1.2b, was 4 m in height and 8 m in diameter, and consisted of 16 elements that could be bolted together to form the floor and the roof [14, 15]. The elements were composed of GFRP skins (glass fibres and a polyester resin) sandwiching a polyurethane foam core. Because of the light weight of the FRP composites, the house could be assembled and disassembled on site in two days, or even air-lifted in one piece. The house may be placed in a variety of rough terrains as only four concrete piers were required. It was also reported that the room temperature could be increased from −29 to 15 °C in half an hour, thanks to the integrated polyurethane insulation and the electric heating system.

In the mid-1970s there were several other free-form types of FRP buildings, such as the Rudolf Doernach plastic house (1958) in Germany and the Bakelite Telephone Exchange Room (1962) in the UK [16]. The period from mid-1970s

Fig. 1.2 Early FRP buildings: **a** The Mosanto House of the Future (photo by Tullio Saba, used under the *Public Domain Mark 1.0* license) [16]; **b** the Futuro (photo by J.-P. Kärnä, used under the *Creative Commons Attribution-ShareAlike 3.0 Unported* license) [17]

to early-2000s saw a decline of such FRP structures, perhaps due to the mature of automatic manufacturing methods for FRPs, in particular the pultrusion method [18]. Nevertheless, the lay-up and moulding method retains its unique appeal to architects due to its capability to produce complex architectural features and free-forming geometries.

In 1999 the Millennium Dome was constructed in London UK to house the Millennium Experience exhibition. Being the ninth largest building then in the world by usable volume, the Millennium Dome (see Fig. 1.3a) comprised of roof skins supported by a web of 2,600 cables which were suspended by a circle of 12 slightly inclined, almost 100 m-high steel support towers [19]. The roof skins were made of polytetrafluoroethylene coated GFRP for durability and weather resistance. Several complete FRP structures were also housed in the Millennium Dome, including the Rest Zone (Fig. 1.3b) and the Home Planet (Fig. 1.3c). Both were assembled on site from FRP sandwich shell panels [20].

FRP modular classroom buildings were developed in UK [22, 23]. In 2004 such a classroom system was pioneered in Grey Court Secondary School and Meadlands Primary School, Richmond, Surrey UK. The modular building was a three-dimensional shell structure (see Fig. 1.4a) without internal frames. The complete structure was moulded in factory including internal linings, thereby enabling rapid on-site assembly. Another modular FRP system was the SpaceBox [24]. A demonstration of the system was the Student Housing Service in Delft University. Figure 1.4b shows the erection of such a student residential building in the university using the SpaceBox modules. The modules could be stacked up to three storeys

Fig. 1.3 **a** The Millennium Dome (adapted from photo by Matt Buck, used under the *Creative Commons Attribution-ShareAlike 3.0 Unported* license) [21]; **b** the Rest Zone (with permission from Elsevier) [20]; and **c** the Home Planet (with permission from Elsevier) [20]

Fig. 1.4 FRP modular buildings: **a** the future classroom (image courtesy of Valeria Carullo); **b** the SpaceBox (image courtesy of Holland composites http://www.hollandcomposites.nl/en)

and, due to lightness in weight, did not require special foundation. Each individual module was completely prefabricated, plumbed, wired, and furbished, such that it could be inhabited within hours of deployment.

In 2006, the Novartis campus entrance building in Basel Switzerland was constructed with a GFRP sandwich roof structure that integrated load-carrying, building physical and architectural functions [25]. The entire roof structure, measured 21.6 × 18.5 m and weighted 28 ton, was supported by glass walls with stiffeners as shown in Fig. 1.5a that arranged in a rectangular floor plan of 17.6 × 12.5 m. In the shape of a wing, the roof structure had a maximum thickness of 620 mm at the middle which tapered off to 70 mm at the overhang edges. The roof structure consisted of a polyurethane foam core sandwiched by GFRP face sheets and stiffened by GFRP webs. The GFRP face sheets and webs, 6–10.5 mm in thickness, were hand-lay using several layers of glass fabric and a polyester resin that showed low flammability. The entire roof structure was manufactured into four elements, which were transported to site, lifted and assembled as shown in Fig. 1.5b. On-site assembly of the four elements was completed by adhesive bonding of the webs and additional GFRP laminate strips.

Another recent example of free-form FRPs in building applications is the CFRP roof of the Steve Jobs Theatre constructed in 2017 at the Apple Park in Cupertino,

Fig. 1.5 GFRP roof structure of the Novartis campus main gate building: **a** in service and **b** on-site assembly (Image courtesy Prof. Thomas Keller, Composite Construction Laboratory, École Polytechnique Fédérale de Lausanne)

Fig. 1.6 CFRP roof of the Steve Jobs Theatre [26]: **a** in service (image courtesy of Nigel Young/Foster+Partners); **b** overall view (image courtesy of Nigel Young/Foster+Partners); **c** assembly (image courtesy of Foster+Partners); and **d** installation (image courtesy of Foster+Partners)

California, USA. Shown in Fig. 1.6a, the lens-shaped roof, 47 m in diameter and 73.2 ton in weight, was supported by a glass wall at the perimeter of 41.1 m in diameter [26, 27]. The circular roof consisted of 44 identical radial CFRP panels, each shaped from four layers of 12 mm-thick plies, and a central circular panel. It was reported that these CFRP panels were manufactured and examined before shipping to the construction site [27]. On-site, the panels were assembled together (see Fig. 1.6b) in an adjacent area before lifted in one piece as shown in Fig. 1.6c and installed onto the glass supporting structure.

In most of the free-form FRP structures, a high capability of thermal insulation was integrated into the structural panels through the sandwich construction. Besides the architectural appeal, the geometries of this type of structures often had been optimized with respect to the design loadings. The development of such free-form structures involved a specific manufacturing process and was mostly dedicated for highly customised projects, directing the application of such type of FRP structures for particular needs.

1.2.2 Snap-Fit Panelised Structures

As the pultrusion manufacturing technique developed and matured, FRP sections made from this technique started to show potential as an alternative to conventional structural members. The main motivations were their lightness thus cost saving in transportation and erection, and chemical resistance (including corrosion resistance) thus reduced maintenance in service. To further ease the erection on-site, FRP building systems was developed based on standardised building "panels" or "blocks" with new features of "snap-fit" interlocking connections.

The advanced composite construction system (ACCS), developed in the 1980s, represents one of the earliest versions of this kind of systems [28, 29]. Figure 1.7a and b shows examples of the pultruded multi-cell wall or floor panels, including the pultruded three-way connectors, end-caps and toggle for joining the panels. Adhesive bonding could be applied on the contact surfaces if weatherproof sealing was required. A temporary two-storey site office at the Severn Crossing road bridge in Bristol UK was constructed of the ACCS and was later converted to a visitor centre (see Fig. 1.7c). In this application the multi-cell panels were infilled with polyurethane foam to enhance thermal insulation [28]. Because of the corrosion resistance of the FRP material, the ACCS was also used in a series of industrial applications, including chemical plants walkways, equipment wash houses, odour covers and the outfitting of cooling towers (see Fig. 1.7d) [29].

A more recent development was the Startlink modular FRP building system in 2004. For design flexibility the system offered a range of open and closed pultruded sections as shown in Fig. 1.8a) that could be snap-fit together with rubber seals in consideration of weatherproofing [22, 30]. A two-storey demonstration house was built out of the Startlink system in Lincolnshire UK in 2012. The reduced weight of the FRP composites offered noted saving in transportation and enabled the use of FRP-steel composite piles instead of a concrete foundation [31]. Another example of the snap-fit system was the Hambleside-Danelaw modular building unit developed as a temporary shelter [22]. The building panels, comprised of pultruded FRP-steel

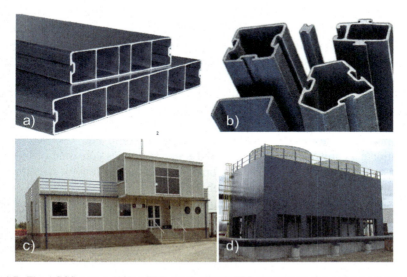

Fig. 1.7 The ACCS system [28]: **a** FRP panel units; **b** FRP connector units; and **c** application for site office building; and **d** application for cooling towers. (image courtesy Strongwell)

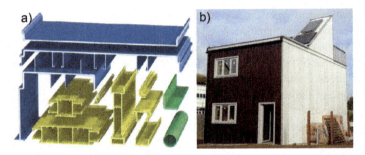

Fig. 1.8 The Startlink system: **a** units (with permission from Elsevier) [20] and **b** demonstration house (image courtesy Prof. Toby Mottram, University of Warwick) [32]

hybrid tubular frames and FRP panels with foam core, could be slided together without the need of mechanical fastener and adhesive bonding.

Such snap-fit panel systems feature an open-building style that may offer certain flexibility in two dimensions for their applications of low-rise structures. The panelised construction and the snap-fit assembly feature largely ease the erection of emergency housing and temporary shelter where assembly may have to be quickly carried out by semi-skilled operators.

1.2.3 Frame Structures Assembled from Pultruded FRPs

Through the pultrusion manufacturing process, FRP profiles were produced with standardized sections similar to the structural steel beams and columns. A number of frame structures have been designed and built using such FRP beam and column members with open sections. Compared to the aforementioned free-form and snap-fit panel systems, this kind of FRP building systems has its popularity because of the familiarity with existing steel sections.

The computer and electronics industry might be the early customer of the FRP frame systems, with the consideration to minimize electromagnetic interference in such applications [33, 34]. In the 1980s, portal frame structures made of pultruded GFRP profiles were designed and constructed as shown in Fig. 1.9 [34]. Like in conventional steel works, these FRP portal frame structures consisted of I-section columns and rafters, diaphragms, and bracings; the connections of columns and rafters were mainly accomplished by web gussets with bolt fasteners.

Another major application of FRP frame systems was the cooling tower industry in need of a corrosion-proof building solution. Figure 1.10 shows a light-frame cooling tower structure developed in the late 1980s [35]. The braced frame structure was constructed of pin-connected FRP tubular and channel sections with an FRP or nonreinforced polymer cladding and flooring system.

FRP frame systems were also used in civil facilities. A noted example was the 19.2 m-tall stair-tower at the US army training base in Fort Story, Virginia (Fig. 1.11a).

Fig. 1.9 FRP building frame structure (image courtesy Strongwell)

Fig. 1.10 FRP light-frame cooling tower: **a** overview at erection and **b** close view of structural elements (Image courtesy Strongwell)

Fig. 1.11 FRP stair-tower: **a** overview and **b** close view of connections (Image courtesy Strongwell)

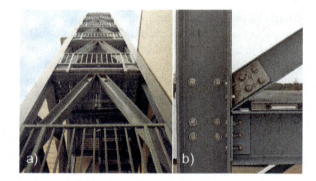

The stair-tower was reported capable of carrying a vertical load of approximately 408 kN and a hurricane wind up to 62 m/s [36]. GFRP profiles were selected as the structural members for quick erection and to resist salt water corrosion from the nearby Chesapeake Bay. The I-section beams, columns and bracings were connected through stainless-steel web angles and bolts as shown in Fig. 1.11b. Such beam-to-column connections transferred little bending moments and the lateral resistance therefore was provided through the diagonal bracing members [37].

Fig. 1.12 The eyecatcher building: **a** overview; **b** structural frame. (image courtesy Prof. Thomas Keller, Composite Construction Laboratory, École Polytechnique Fédérale de Lausanne)

In 1999 the Eyecatcher building with five-storey and 15 m-high as shown in Fig. 1.12a, was constructed of pultruded FRP members. Originally built for the Swissbau exhibition, the structure, with steel bolted connections, was disassembled, transported, and reassembled for the use as an office building [38]. It was formed by several building frames (see Fig. 1.12b) where built-up FRP members were developed from standard I- and channel sections and flat plates through adhesive bonding [39, 40]. To enclose the building frame, sandwich walls and facades were attached to the structural members for thermal insulation.

A series of FRP frames were constructed to support cooling containers in Yangshan deepwater port in the Shanghai international shipping centre [41]. FRP composites were selected to be the option in this application because of its durability in the coastal corrosive environment and the consideration to minimize electrical interference generated by the automated container terminal. The three-storey frame structures, 8.1 m in height and 6.1 × 1.5 m in plan, consisted of three bays in the longitudinal direction and one bay in the transverse direction (Fig. 1.13a). The beams and columns, in rectangular hollow sections of 184 × 184 × 10 mm and 210 × 110 × 10 mm respectively, were pultruded with a wood core to prevent local indentation, splitting and buckling of the FRP skins. The beams were connected to the column

Fig. 1.13 FRP frames for cooling containers at Yangshan Port: **a** overview; **b** FRP grid decks [41]

through embedded steel plates and stainless steel bolts. FRP grids manufactured from a resin infusion process were used as the floor decks shown in Fig. 1.13b.

After the partial collapse of the Santa Maria Paganica church in Italy due to the L'Aquila earthquake in 2009, a series of FRP temporary shelter structures were constructed to accommodate restoration activities [42]. Figure 1.14a shows one of the shelter structures that measured up to 29.4 m high and covered an area of 266 m^2. The shelter structures were rectilinear space frames assembled from pultruded FRP elements. Typical primary members consisted of channel or angle pultruded sections joined together using steel bolts. The frame members were connected through bolting to a moulded FRP gusset web plate as shown in Fig. 1.14b. The base of the column members was embedded into reinforced concrete pedestals on the ground.

Being lightness in weight and integrated with architectural functions, pultruded FRP profiles are ideal structural member used in prefabricated systems. In 2014 an FRP prefabricated modular housing system was developed and introduced in [43–45]. It was aimed to provide prefabricated modular housing system for disaster zones, emergency situations and construction sites. The resulting prototype, shown in Fig. 1.15, consisted of two 3 × 3 × 3 m building modules sharing adjacent beams and columns. The beams and columns were pultruded GFRP members with 120 × 120 × 10 mm square hollow section. Connections between the beams and columns were realized by stainless-steel bolted tube connectors inside the hollow section members. To avoid positioning of the fasteners in the closed sections, bolt nuts were welded onto the inside face of the steel connectors. Sandwich panels with GFRP skins and polyurethane core were adopted for roofs, floors and walls. Water, electricity networks and sanitation facilities etc. may be pre-installed before the house is transported to site.

Recently a hotel building was constructed near the Nanjing Foshou lake using a FRP-wood composite solution. The three-storey building (see Fig. 1.16a) was 10.3–11.1 m in height and covered a plan area of approximately 800 m^2. As shown in Fig. 1.16a, the composite column, 120 × 90 mm in cross-section and continuous through the three storeys, were pultruded with GFRP sections confining southern

Fig. 1.14 FRP shelter structure for the Santa Maria Paganica church [42]: **a** overall view; **b** connection between members. (image courtesy of TOP GLASS Industries S.p.A and IUAV University of Venice)

Fig. 1.15 The ClickHouse FRP modular housing. (image courtesy Prof. João Ramôa Correia, University of Lisbon)

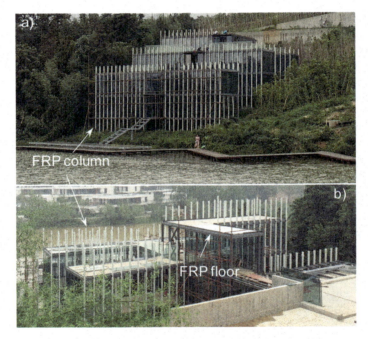

Fig. 1.16 A hotel building in Nanjing China **a** in an overview and **b** with FRP wood composite columns and FRP wood sandwich floor

pine wood core [46]. Composite sandwich panels were used as the floor slabs in the building as shown in Fig. 1.16b. The sandwich panels, manufactured from a resin infusion process, were comprised of GFRP face sheets and a paulownia wood core strengthened with GFRP webs [47].

1.3 Scope of the Work

Buildings using FRP composites as the load-carrying system have emerged since the 1950s, motivated by their reduced weight, corrosion resistance, low thermal and electro-magnetic conductivity and tailored design options. In fact, GFRP composites have been reported to offer further benefits such as saving in energy consumption and reduction in water and air pollution during their manufacture compared to reinforced concrete, steel and aluminium [48]. The mature of cost-effective automatic manufacturing techniques, in particular pultrusion, expands the application of GFRPs in building construction. Through the pultrusion method, GFRP profiles were produced with a range of standardized sections similar to structural steel.

However, the elastic modulus of a typical pultruded GFRP profile ranges from about 20–40 GPa [8]. This is approximately 10–20% that of steel, and hence serviceability requirements such as deflection limits become more critical. The sandwich concept may provide a way to enhance the structural stiffness by improvement in inertial moment of the section rather than material elastic modulus, as sectional areas are placed away from the neutral axis. The design of pultruded GFRP structural members under flexure also requires the consideration of shear deformations because GFRP has a relatively low shear modulus compared to its longitudinal elastic modulus. Furthermore, due to their laminate and orthotropic nature in comparison to isotropic steel, GFRP composites are often associated with lower shear strength although tensile and compressive strengths are comparable [49]. This reminds particular attentions on potential shear failure of GFRP composites in bending or even compression considering second order effect [50].

The constituent components of GFRP composites are brittle and hence catastrophic failure may occur suddenly without pre-failure warning. In civil engineering, ductile failure modes, where structures undergo large deformations while still carrying loads, are preferred in structural design, as warning can be provided prior to catastrophic failure. This may be tackled for structures made from GFRP composites through progressive failure to provide pseudo-ductility [51], or in combination with ductile materials such as steel in dedicated locations [52]. Large nonlinear deformation can also be achieved by taking advantage of the post-buckling process of certain structural components in compression at the cost of loss in structural redundancy [53]. Recently a new structural concept of load-dependent composite action is further proposed and demonstrated for layered beam structures to achieve nonlinear load-deformation responses and structural ductility [54].

In steel structures, open sections such as I sections are prevailingly used over closed shapes since they are more convenient for assembly in the field with the use of bolted connections [55]. However, for pultruded GFRP composites, I sections may be susceptible to lateral-torsional buckling and likely critical in web-flange junctions due to the presence of a roving-rich core within the junction [56–58]. On the other hand, closed GFRP sections have improved torsional rigidities, are more resistant to local flange buckling and are stronger and stiffer in the transverse direction compared to open sections [55]. However, connection methods for closed section may be more challenging than open sections.

In face of the challenges aforementioned, this work presents a range of research activities conducted in recent years by its contributors in order to develop GFRP members with web-flange sandwiches as built-up closed sections for structural applications including beams, slabs, composite beams, and column or wall units. Connection solutions are also proposed in form of a sleeve configuration in corporation of steel elements, as introduced in this work, to improve joint rotational stiffness and to provide structural ductile performance. Built upon such research experience and outcomes, large scale structures are developed and demonstrated as further presented in this work, with inspiration of the proposed structural members and connections.

1.3.1 Beam and Slab Members

Modular GFRP web-flange sandwich system was introduced for building floor construction first. This modular system consists of built-up sections using standard pultruded GFRP box or I-sections that are incorporated between two pultruded flat panels as web core to form a web-flange sandwich assembly. Such an sandwich section has an improved second moment of area compared to individual pultruded sections. Moreover, it permits adjustment of various geometric parameters such as the width and depth of the section, the spacing between the web-profiles and the thickness of the flat panels and webs, and thus offers more design flexibility. In addition, the modular nature of this system allows for easy integration of additional core materials such as foam and lightweight timber blocks to enhance structural and building performance.

Chapter 2 investigates the mechanical performance of modular GFRP sandwich beams or one-way slabs under monotonic loading, in terms of both load-carrying capacity and stiffness. The connection type is also under investigation, particularly in regards to the degree of composite action and how this affects strength and stiffness. The addition of lightweight foam materials as the core of the structure and the effect of core addition on the structural performance is also examined. Finally, it further involves the development of analytical and numerical techniques to understand and evaluate the structural behaviour.

Chapter 3 focuses on such sandwich structures under two-way bending for slab applications. The slab structures are designed based on the major findings from Chap. 2. In addition to investigation of strength, stiffness, the connection detail and the insertion of lightweight foam core, the effect of incorporating two different pultrusion directions perpendicular to each other is also analysed. This orientation configuration is used to prevent localised cracking between fibres along the pultrusion direction of the flat panels when subjected to two-way bending, as observed in existing unidirectional GFRP members and structures.

Built on the design concepts and mechanical understanding that were formed in Chaps. 2 and 3, Chaps. 4 and 5 present the GFRP-steel composite beams and their innovative shear connection approach. Mechanical performance of the composite beams in bending is focused in Chap. 4 while the shear connection is focused in Chap. 5 to understand the composite action provided. The composite beam structures consist of the modular GFRP sandwich assembled as the top flange and the supporting steel girder. Again, a bidirectional pultrusion orientation can be achieved for the flange. Such a configuration is made with the intention to prevent localised failures that have been observed in existing FRP bridge decks where the pultrusion direction of the deck runs transverse to the steel beam. The connection detail between the GFRP top flange and the steel beam is investigated, along with strength and stiffness requirements for this type of structure under monotonic loading. The composite beam provides a ductile failure response through the yielding of the supporting steel girder, allowing for sufficient warning prior to failure of the GFRP flange. In addition, finite element (FE) modelling, analytical solutions and design recommendations are developed.

1.3.2 Column and Wall Members

Utilization of pultruded GFRP columns with closed sections and the proposed assembly of GFRP web-flange sandwich structures for load-bearing column and wall applications in building construction is further explored. In particular, GFRP composites with square hollow sections (SHS) are considered for column applications and their mechanical performance under concentric and eccentric compression needs to be understood. Furthermore, experimental investigations on the proposed web-flange sandwich units assembled using pultruded GFRP panels and SHS columns are also necessary to clarify their behaviour in compression. Meanwhile, analytical and FE modelling approaches can be employed after validation with the experimental results for understanding of load-bearing capacities, failure modes, load-displacement and load-strain responses. Such approaches are important for further parametric study on the effects of major design parameters for evaluation of the overall performance of such GFRP structural members.

In consideration of existing research results on GFRP columns reported in literature, Chap. 6 focuses on the effects of width-thickness ratio (b/t) as an important geometric parameter for the buckling behaviour of plates and therefore for GFRP SHS columns. The effects of b/t ratio on the failure modes and load-carrying capacities of pultruded GFRP SHS columns are investigated through experimental investigations and theoretical analysis. In this chapter, two SHSs with different b/t values are comparatively examined under concentric compression. Meanwhile, an analytical formulation of critical b/t values is established to determine the boundaries between different failure modes of GFRP SHS columns (such as local buckling and compressive failure) under compression, considering the different end constrain conditions of the SHS side plates. Design equations taking into account the effects of both non-dimensional slenderness and width-thickness ratio, are developed in this chapter to estimate the load-carrying capacities of GFRP columns with validation from experimental results.

In Chap. 7, web-flange sandwich wall units are assembled by two pultruded GFRP panels and two GFRP SHS columns in between through adhesive bonding or mechanical bolting. Experiments in concentric compression are conducted in order to understand the failure modes, load-bearing capacities, load-displacement and load-strain responses. The effects of different connection methods such as adhesive bonding and mechanical bolting and the spacing between SHS sections on the failure modes and load-bearing capacities are clarified in this chapter. Numerical analysis on the GFRP sandwich assemblies with bonded and bolted connections in compression is further carried out using FE approach. Geometric nonlinearity is considered and the modelling results are compared with experimental results. Furthermore, parametric studies based on the validated FE analysis are performed. In this way the effects of major design parameters, including the dimensions of GFRP SHSs, GFRP panel thickness, the spacing between two SHSs, initial imperfections, and material properties of GFRP SHSs and panels, on the overall performance of the sandwich assemblies are examined in this chapter.

Most of the published results in literature are about the performance of GFRP column members subjected to concentric compression. Chap. 8 therefore presents the mechanical performance of GFRP SHS columns under eccentric compression, with different eccentricities experimentally covered in this chapter. Furthermore, bolted sleeve joints are employed to connect the GFRP column specimens and loading end plates. Failure modes, load-displacements, load-strain responses are received. FE analysis is then developed and the modelling results are compared with the experimental results. The relationship between the load-bearing capacities of GFRP columns and the eccentricities is obtained and discussed. The interaction curve between compression load and bending moment is clarified based on the experiments and is further compared with design approaches. In addition, the effect of joint performance on the overall performance of FRP columns under compression and bending is highlighted.

1.3.3 Connections

Connecting individual pultruded GFRP members together with sufficient strength and stiffness is an important task for structural construction, where the anisotropic and brittle properties of the materials need to be well considered. Furthermore, the low elastic modulus of GFRP composites requires additional effort in order to provide sufficiently stiff joints and a closed section shape increases difficulties for bolting because of limited access space. These issues prompted researches into developing proper methods for connecting pultruded GFRP members and it may not be effective to simply mimic connection details in steel structures for GFRP composites. A sleeve connection configuration in cooperation with steel elements is introduced in this work to achieve both ductile failure behaviour and improved stiffness and strength for connections of GFRP composites. The proposed sleeve connection employs a sleeve connector formed by welding a steel tube to a steel endplate. The steel tube can be inserted into a GFRP member through adhesive bonding and/or mechanical bolting in between. The steel endplate can be then connected to a steel member or another GFRP member using bolts, as commonly used endplate connections.

Chapters 9 and 10 start with the investigation on the mechanical performance of bonded sleeve connections for pultruded GFRP to steel members through experimental and numerical studies. In the experimental studies, it can be demonstrated that GFRP to steel specimens with bonded sleeve connection may fail by the extensive yielding of the steel endplates, therefore providing a ductile failure manner prior to the damages observed on the GFRP member. In comparison of the mechanical performance of bonded sleeve connections and conventional steel angle connections, it can be found that when joining the GFRP beam and steel column, bonded sleeve connection may perform better than the conventional steel angle connection, in terms of the initial rotational stiffness and the elastic moment capacity. In Chap. 10, FE modelling approach is developed considering the bolt geometry, the contact and slip between different components, and the pretension of bolts, with satisfactory validation against experimental measured moment-rotation curves and failure modes of tested specimens. In this chapter, the parametric study indicates dominant effects of endplate thickness on both initial rotational stiffness and the elastic moment capacity of the bonded sleeve connections.

With the support from the mechanical performance of the proposed sleeve connections in pultruded GFRP to steel members, Chap. 11 applies the sleeve connections for GFRP beam to GFRP column scenarios. It can be found that the specimens with a longer bond length in sleeve connections may fail through yielding of the steel endplate while shorter bond length may induce cohesive failure. This highlights potential ductility and energy dissipation capacity of all GFRP structures assembled in this way of steel yielding. Therefore, in this chapter, the cyclic performance of bonded sleeve connections for GFRP beam to GFRP column are investigated through experimental and numerical studies. Specimens with different endplate thicknesses and number of bolts are examined. It can be found that specimens with proper endplate thickness may exhibit obvious ductility and energy dissipation capacity and this is contributed from the yielding of steel endplate. Further parametric study on the endplate thickness suggests that the reduction in endplate thickness may mitigate the cohesive failure initiation in company with the decrease in the initial rotation stiffness and moment capacity of the bonded sleeve connections.

With the experience gained from previous chapters, the proposed sleeve connection for beam column scenarios is extended to a splice configuration. Investigation of axial performance of the splice connection begins in Chap. 12 with a theoretical analysis of steel-FRP bonded sleeve joint components. A theoretical formulation is developed based on a bilinear elastic-softening bond-slip relation at the adhesive bond. Possible shear stress distributions at elastic limit and ultimate state are formulated, based on which the axial capacities of the bonded sleeve joints are derived. The theoretical analysis is validated by experimental results and FE analysis, covering a wide range of section geometries and bond lengths.

Chapters 13 and 14 introduce the proposed sleeve connection in a splice configuration for column to column applications. It consists of a steel bolted flange joint between two tubular section steel-FRP sleeve joints. To investigate the axial performance of such column-column connections, experimental studies of the connection components of bonded sleeve joints and bolted flange joints are individually presented in Chap. 13. It can be found that failure of the bonded sleeve joints is brittle within the adhesive layer and the joint capacity increases nonlinearly with the bond length, indicating the existence of an effective bond length. Failure of the bolted flange joints can be ductile by yielding of the steel flange-plates. Featuring a bilinear bond-slip relation, FE models are developed in this chapter and show capability in the characterization of the behaviour of these component joints. The FE modelling is then utilised to evaluate the axial behaviour of a complete splice connection integrating the component joints. Ductile failure can be realized through the yielding of the bolted flange joint prior to the brittle failure of the bonded sleeve joint.

Chapter 14 further investigates the flexural performance of the column-column splice connections subjected to cyclic loading. A cantilever setup is employed to subject the connections to combined bending and shear. Connection specimens are prepared in three configurations with difference in the bond length or in the bolt arrangement. The connection specimens experience different levels of yielding in the steel flange-plates, before the ultimate failure in mode of web-flange cracking of the GFRP member or fracture of the steel flange-plates near the weld toes. Energy dissipation performance is well demonstrated in a specimen where plastic deformation of the steel flange-plates is fully developed. The strain responses are analysed to identify damage in the adhesive bond and yielding in the flange-plates. FE modelling further incorporates damage accumulation in the adhesive bond and kinematic hardening (after yielding) of the steels, showing satisfactory agreement with the experimental results.

1.3.4 Fire Resistance

The GFRP web-flange sandwich assembles are developed as introduced in previous chapters to form modular multicellular web-flange sandwich structures. They have shown potential for applications as slabs and walls in buildings structures. However, the fire resistance performance of GFRP composites may be a concern for such applications. Indeed, at material level, GFRP composites experience glass transition around 90–120 °C; during and above which the material elastic modulus and strength degrade significantly. Further, the polymeric matrixes of GFRP materials thermally decompose when the decomposition temperature is approached at about 300 °C. This decomposition entails fibre degradation and delamination in GFRP composites, resulting in significant losses in stiffness and load-carrying capacity of GFRP structures as well as several fire reaction characteristics.

In consideration of the potential fire risks of GFRP structures, prefabricated fire-resistant panels may be used for their fire protections in reference to cold formed steel structures. In this way, fire resistant panels with low thermal conductivity are installed on the surface of GFRP web-flange sandwich structures to improve the fire resistance time of such structures. Chapter 15 introduces an experimental investigation on the fire resistance performance of large-scale GFRP multicellular web-flange sandwich structures with mechanical loading. These structures are formed using pultruded GFRP SHS sections in parallel incorporated between two GFRP flat panels by adhesive bonding, resulting the overall depth of 118 mm. Three types of single- or double-layers of fire-resistant panels can then be installed using screws at the lower surface of such GFRP multicellular sandwich assembles, including glass magnesium (GM) board, gypsum plaster (GP) board and lightweight calcium silicate (CS) board. The structures were subjected to mechanical loading in bending while their lower surfaces were exposed to ISO 834 fire simultaneously. In this way, the effects and mechanisms of the prefabricated fire-resistant panels for the improvement on fire resistance performance of such GFRP web-flange sandwich structures can be discussed and clarified.

It was found that the fire-resistant panels effectively improve the fire insulation performance and fire resistance performance for the GFRP web-flange sandwich structures under loading. With assistance from the GP fire-resistant panels of a thickness of 12 mm at the fire exposure side, the fire endurance time can be effectively extended to 83 min for the loaded GFRP structure, in comparison to the 54 min of fire endurance time for the one without any fire-resistance panels. When single layer of GM panels is used, the fire endurance time of 103 min can be achieved. The fire endurance time may be further extended to 113 min for such loaded GFRP structures when double layers of GP panels are used, or 158 min when double layers of GM panels are used, i.e. corresponding to a total thickness of 24 mm of fire-resistant panels. It can be seen that the assembly of GM fire-resistant panels at the fire exposed surfaces may provide better improvement on the fire resistance performance for such loaded GFRP structures. The mechanisms of the fire-resistant panels are mainly manifested by their heat insulation and fire shielding functions, depending on their thickness, moisture content, thermophysical properties, damage patterns in fire etc.

1.3.5 Large Scale Structural Applications

Based on the research experience and outcomes from Chaps. 2–5, a 9 m GFRP composite structure is presented in Chap. 16 as shown in Fig. 1.17 for pedestrian bridge applications. An experimental study is introduced for investigation of the mechanical performance of such an all-GFRP composite structure for positive and negative bending scenarios, with the latter specifically dedicated for the case of continuous spans. Again, the bridge deck is fabricated using pultruded GFRP SHS sections adhesively bonded between two GFRP flat plates. A bidirectional pultrusion orientation was present within the deck structure, where the pultrusion (i.e. longitudinal fibre) direction of the square tubes were placed perpendicular to the

Fig. 1.17 A GFRP pedestrian bridge using developed web-flange sandwich system as bridge deck

pultrusion direction of the flat plates. This orientation improves structural performance by enhancing the junction between the square hollow sections and face plates to prevent premature cracking. In this chapter, three different span lengths are examined, including two 4.5 m continuous spans, or one 6 m or 9 m single span. Composite action of the connections and the bidirectional deck is studied, along with shear lag under both positive and negative bending.

Compared with steel, pultruded GFRP composites exhibit lower elastic modulus, lower shear strength and lack of ductility. The low elastic modulus leads to low stiffness of structural members made of GFRP composites and is therefore more critical for serviceability. The lower shear strength implies that the susceptibility of such materials to shear failure may preclude full utilization of their potential. The lack of material ductility may prevent the incorporation of pre-failure warnings. It appears, however, that such concerns can be overcome through a space frame configuration, whereby the stiffness of such space frames is achieved at the structural level rather than only at the material level; the structural members in a space frame are subjected mainly to axial rather than shear loading; and dedicated compressive GFRP members in a robust space frame may indicate a warning of structural malfunctioning due to their potential instability and the resulting large second order deformation. Figure 1.18 shows an example where a 4.8 × 4.8 m space frame structure is assembled

Fig. 1.18 A GFRP space frame structure using developed sleeve connections to assemble pultruded GFRP members with space node joints

1 Introduction 21

by 72 GFRP members with the bonded sleeve connection at each end. The ends of such GFRP members are then bolted together at the space nodes using welded plate Octatube connectors made from steel or aluminium, forming the space frame with a height of 1.21 m.

Chapter 16 further presents an experimental study on a space frame structure assembled by pultruded circular hollow section GFRP members. This large scale structure is with the span length of 8 m, width of 1.6 m and depth of 1.13 m, but weighing only 773 kgf. Experimental results from the three-point bending experiment include the structural stiffness, load-carrying capacity, and failure modes. The structure showed satisfactory overall stiffness and load-carrying capacity for potential applications of supporting truss structures. It was further found that the second order bending of critical compressive members may induce large nonlinear deformation of the overall structure in a similar manner as structural ductility.

Supported by the research outcomes of the sleeve connections and the web-flange sandwich assemblies, a loft design of house structure is proposed for assembly using GFRP composites as shown in Fig. 1.19a. The house structure is about 4.8 m in height and covering an area of about 4.5×2.4 m, weighted approximately 1.2 t. Assembled from pultruded SHS GFRP members (see Fig. 1.19b), the primary structural frame consists of four equally spaced frames with steeply pitched rafters, two continuous base edge beams and upper beams in between the frames. The floor slabs can be supported onto the beams, and the roof and wall panels can be tied to the rafters and columns respectively. At each side of the frames, the columns are seated onto the base beam and bolted to footings through the use of bolted sleeve connectors welded with endplates. The upper beams are connected to the columns using sleeve connections in a likewise manner. Within each frame, the rafters and columns are joined together through sleeve connectors hidden inside the GFRP members. The GFRP web-flange sandwich panels, i.e. flat face sheets sandwiching hollow section cores, can be considered for the floor slabs and roof panels.

Because of their superior corrosion resistance to steel and steel reinforced concrete and also superior decay resistance to timber, GFRP composites have attracted strong interest for the construction in coastal regions. One of the applications is the retaining wall construction using such materials thanks to their lighter weightiness. It has

Fig. 1.19 A loft house design assembled using pultruded GFRP components with sleeve connections **a** in a conceptual design and **b** with structural frames

Fig. 1.20 Modular GFRP retaining wall assembled of web-flange plank and pile sections for **a** practice in coastal region and **b** with achieving extraordinary sea view

been reported [59] that the installation of a GFRP composites seawall system may not require heavy machinery, significantly saving construction time in comparison to traditional concrete seawalls. In practice, backfills create pressures on retaining walls, introducing bending stress and deformation to the wall members. Therefore, the seawall system must provide sufficient bending resistance and satisfy relevant strength and deformation requirements [60].

As GFRP composites are associated with relatively low material elastic modulus, appropriate design of sections is important to achieve adequate structural stiffness in bending, and individual sections are further required to be equipped with mechanism for convenient connection in order to form continuous retaining walls with possible changes in orientation. At last in Chap. 16, a modular GFRP retaining wall system is introduced with the use of web-flange plank and pile sections and assistance from a specific mechanical interlocking connection. This system has attracted a few applications in Australia with an example shown in Fig. 1.20, unlocking proper coastal land for building and receiving extraordinary sea view.

References

1. Irving PE, Soutis C (2019) Polymer composites in the aerospace industry. Woodhead Publishing
2. Graham-Jones J, Summerscales J (2015) Marine applications of advanced fibre-reinforced composites. Woodhead Publishing
3. Bank LC (2006) Materials and manufacturing. In: Composites for construction: structural design with FRP materials. John Wiley & Sons, pp 40–77
4. Hollaway LC, Teng JG (2008) Strengthening and rehabilitation of civil infrastructures using fibre-reinforced polymer (FRP) composites. Woodhead Publishing
5. Goldsworthy BW, Landgraf F (1959) Apparatus for producing elongated articles from fiber-reinforced plastic material. United States patent US2871911A
6. Exel Composites. Pultrusion (2018). https://www.exelcomposites.com/en-us/english/composites/manufacturingtechnologies/pultrusion.aspx. Accessed 22 May 2020
7. Strongwell. Fiberglass Building Solutions (2019) https://www.strongwell.com/wp-content/uploads/2013/04/Company-Portfolio.pdf. Accessed 14 June 2020

8. Hollaway L (2010) A review of the present and future utilisation of FRP composites in the civil infrastructure with reference to their important in-service properties. Constr Build Mater 24(12):2419–2445
9. CNR—Advisory committee on technical recommendations for construction (2008) Guide for the design and construction of structures made of FRP pultruded elements CNR-DT 205/2007. Rome, Italy: National Research Council of Italy
10. ASCE (American Society of Civil Engineers) (2010) Pre-standard for load & resistance factor design (LRFD) of pultruded fiber reinforced polymer (FRP) structures. ASCE, United States
11. Ascione L, Caron JF, Godonou P, van IJselmuijden K, Knippers J, Mottram T, Oppe M, Gantriis Sorensen M, Taby J, Tromp L (2016) Prospect for new guidance in the design of FRP EUR 27666 EN. Joint Research Centre of the European Commission, Ispra, Italy
12. Wikipedia. Monsanto House of the Future (2018) https://www.en.wikipedia.org/wiki/Monsanto_House_of_the_Future. Accessed 18 Oct 2019
13. Mahne KM (2015) Living in the Monsanto house of the future. Disney Avenue. http://www.dizavenue.com/2015/05/living-in-monsanto-house-of-future.html. Accessed 18 Oct 2019
14. Home M, Taanila M (eds) (2002) Futuro: Tomorrow's house from yesterday. Desura, Helsinki
15. Wikipedia. Futuro (2019). https://www.en.wikipedia.org/wiki/Futuro. Accessed 19 Oct 2019
16. Saba T (2003) IH169538. https://www.flickr.com/photos/97453745@N02/9689197391/in/photolist-9SKDtR-aWZNhD-89gViZ-9SjQX5-8igFCp-fLcEPr-9UEz3f-3s3Gs. Accessed 08 June 2020
17. Kärnä JP (2013) Futuro WeeGee Espoo. Wikimedia Commons. https://www.commons.wikimedia.org/wiki/File:Futuro_WeeGee_Espoo.jpg. Accessed 19 Oct 2019
18. Stewart R (2002) Pultrusion industry grows steadily in US. Reinf Plast 46(6):36–39
19. Rogers Stirk Harbour & Partners. The Millennium Dome (2010). https://www.rsh-p.com/projects/the-millennium-dome/. Accessed 23 Oct 2019
20. Kendall D (2007) Building the future with FRP composites. Reinf Plast 51(5):26–33
21. Butt M (2013) London MMB R9 millennium dome. Wikimedia Commons. https://www.commons.wikimedia.org/wiki/File:London_MMB_%C2%ABR9_Millennium_Dome.jpg. Accessed 08 June 2020
22. Evernden MC, Mottram JT (2012) A case for houses to be constructed of fibre reinforced polymer components. Proc Inst Civil Eng Constr Mater 165(1):3–13
23. Altair HyperWorks (2005) Using optimization to improve the design of a reproducible classroom environment. Altair Engineering. https://www.altairhyperworks.com/ResourceLibrary.aspx?category=Case+Studies&page=3. Accessed 10 Jan 2020
24. Holland Composites (2008) SpaceBox. https://www.hollandcomposites.nl/en/portfolio/temporary-livingspace-unit-spacebox/. Accessed 11 Nov 2019
25. Keller T, Haas C, Vallée T (2008) Structural concept, design, and experimental verification of a glass fiber-reinforced polymer sandwich roof structure. J Compos Constr 12(4):454–468
26. Foster & Partners. Steve Jobs Theatre (2018). https://www.fosterandpartners.com/projects/steve-jobs-theater/. Accessed 28 Jan 2020
27. Inside Composites (2017) Largest carbon-fibre roof for Apple HQ. https://www.insidecomposites.com/largest-carbonfibre-roof-for-apple-hq/. Accessed 28 Jan 2020
28. Strongwell (2019) Product brochure—COMPOSOLITE fibreglass building panel systems. https://www.strongwell.com/wp-content/uploads/2013/04/COMPOSOLITE-Brochure.pdf. Accessed 23 Oct 2019
29. Hollaway L, Head P (2001) Applications in advanced polymer composite constructions. In: Advanced polymer composites and polymers in the civil infrastructure. Elsevier Science Ltd, Oxford, pp 221–86
30. Hutchinson J, Singleton M (2007) Startlink composite housing. In: Proceedings of advanced composites in construction ACIC07 conference. University of Bath
31. Larkfleet Group (2014) PassiveHouse the future of building today. https://www.larkfleet.com/larkfleetgroup/assets/img/Downloads/Larkfleet%20Group/New/13-11-PassiveHouse.pdf. Accessed 23 Oct 2019

32. Zafari B (2012) Startlink building system and connections for fibre reinforced polymer structures. PhD thesis supervised by Prof. Mottram, University of Warwick
33. Bank LC (2006) Composites for construction: structural design with FRP materials. John Wiley & Sons
34. Strongwell application profile: Computer Testing Facility Constructed of DURASHIELD® Foam Core Panels & EXTREN®. https://www.strongwell.com/case-study-computer-testing-facility-constructed-of-durashield-foam-core-panels-extren/. Accessed 16 May 2020
35. Strongwell (2018) Case study: colossal cooling towers rebuilt using fiberglass components. https://www.strongwell.com/case-study-colossal-cooling-towers-rebuilt-using-fiberglass-components/. Accessed 25 Oct 2019
36. Strongwell. FRP Stairtower Designed to Meet the Challenge. 2019. Available: https://www.strongwell.com/case-study-frp-stairtower-designed-to-meet-the-challenge/ [accessed 17 November 2019].
37. Bank LC (2006) Pultruded connections. In: Composites for construction: structural design with FRP materials. John Wiley & Sons, pp 486–487
38. Fiberline. The Eyecatcher Building. 2019. Available: https://fiberline.com/cases/cases-doors-facades/glass-facades/the-eyecatcher-building/ [accessed 12 October 2019].
39. Keller T (1999) Towards structural forms for composite fibre materials. Struct Eng Int 9(4):297–300
40. Bai Y, Keller T, Wu C (2013) Pre-buckling and post-buckling failure at web-flange junction of pultruded GFRP beams. Mater Struct 46(7):1143–1154
41. Fang H, Bai Y, Liu WQ, Qi Y, Wang J (2019) Connections and structural applications of fibre reinforced polymer composites for civil infrastructure in aggressive environments. Compos B 164:129–143
42. Boscato G, Casalegno C, Russo S (2015) Design of FRP structures in seismic zone. Manual by Top Glass S.p.A. and IUAV University of Venice, Italy
43. CORE Group (2015) Applied projects—the clickhouse. http://www.coregroup.tecnico.ulisboa.pt/~coregroup.daemon/applied-projects/. Accessed 15 Dec 2019
44. Abdolpour H, Garzón-Roca J, Escusa G, Sena-Cruz JM, Barros JA, Valente IB (2016) Development of a composite prototype with GFRP profiles and sandwich panels used as a floor module of an emergency house. Compos Struct 153:81–95
45. Martins D, Sá MF, Gonilha JA, Correia JR, Silvestre N, Ferreira JG (2019) Experimental and numerical analysis of GFRP frame structures. Part 2: Monotonic and cyclic sway behaviour of plane frames. Compos Struct 220:194–208
46. Qi Y, Xie L, Bai Y, Liu W, Fang H (2019) Axial compression behaviors of pultruded GFRP-wood composite columns. Sensors 19(4):755
47. Zhu D, Shi H, Fang H, Liu W, Qi Y, Bai Y (2018) Fiber reinforced composites sandwich panels with web reinforced wood core for building floor applications. Compos B 150:196–211
48. Daniel RA (2010) A composite bridge is favoured by quantifying ecological impact. Struct Eng Int 20(4):385–391
49. Bai Y, Keller T (2009) Shear failure of pultruded FRP composites under axial compression. J Compos Constr 13(3):234–242
50. Bai Y, Vallée T, Keller T (2009) Delamination of pultruded glass fiber-reinforced polymer composites subjected to axial compression. Compos Struct 91(1):66–73
51. Bank L (2013) Progressive failure and ductility of FRP composites for construction: review. J Compos Constr 17(3):406–419
52. Keller T, Gürtler H (2005) Composite action and adhesive bond between fiber-reinforced polymer bridge decks and main girders. J Compos Constr 9(4):360–368
53. Bai Y, Zhang C (2012) Capacity of nonlinear large deformation for trusses assembled by brittle FRP composites. Compos Struct 94(11):3347–3353
54. Bai Y, Qiu C (2019) Load-dependent composite action for beam nonlinear and ductile behaviour. ASCE J Struct Eng
55. Smith SJ, Parsons ID, Hjelmstad KD (1998) An experimental study of the behavior of connections for pultruded GFRP I-beams and rectangular tubes. Compos Struct 42(3):281–290

56. Bank LC, Nadipelli M, Gentry TR (1994) Local buckling and failure of pultruded fiber-reinforced plastic beams. J Eng Mater Technol 116(2):233–237
57. Mottram JT (1992) Lateral-torsional buckling of a pultruded I-beam. Composites 23(2):81–92
58. Turvey GJ, Zhang Y (2006) Shear failure strength of web-flange junctions in pultruded GRP WF profiles. Constr Build Mater 20(1–2):81–89
59. Ashpiz ES, Egorov AO, Ushakov AE (2010) Application of composite materials for the protection of sea shores and engineering structures against the impact of waves. WIT Trans Ecol Envir 130:231–238
60. Bdeir Z (2001) Deflection-based design of fiber glass polymer (FRP) composite sheet pile wall in sandy soil. McGill University, Quebec

Chapter 2
Fibre Reinforced Polymer Built-Up Beams and One-Way Slabs

Sindu Satasivam, Yu Bai, and Xiao-Ling Zhao

Abstract An adhesively bonded modular GFRP web-flange sandwich system for use in building floor construction is described in this chapter. Sandwich units are developed by incorporating standard pultruded GFRP box (i.e. square hollow section) or I-profiles between two GFRP flat panels to form built-up modular sections with considerable improvement of bending stiffness. These modular sections may then be assembled in the transverse direction to form a one-way spanning slab system. Sandwich specimens with different span-to-depth ratios and core configurations were prepared via adhesively bonding the component profiles, and were then tested under four-point bending. It can be found that the span-to-depth ratio greatly influenced the failure mode, and that inserting foam into the core of the sandwich significantly improved the load-carrying capacity. Also, adhesive bonding was able to provide full composite action at both serviceability and ultimate loads, depending on the quality of the bond. Finally, structural theory was used to estimate the bending stiffness and load-carrying capacity of the sandwich specimens and good agreement with the experimental results was found.

2.1 Introduction

Glass fibre reinforced polymers (GFRPs) present favourable properties such as high strength, light weight and resistance to corrosion, and as a result are increasingly used in civil engineering applications [1, 2]. Furthermore, structures made of GFRP showed lower embodied energy compared to steel and concrete structures [3, 4]. However, the design of structures made of GFRP is usually governed by serviceability limit state (SLS) criteria rather than by ultimate strength design [2]. This is due to

---

Reprinted from Composite Structures, 111, Sindu Satasivam, Yu Bai, Xiao-Ling Zhao, Adhesively bonded modular GFRP web–flange sandwich for building floor construction, 381-392, Copyright 2014, with permission from Elsevier.

S. Satasivam · Y. Bai (✉) · X.-L. Zhao
Department of Civil Engineering, Monash University, Clayton, Australia
e-mail: yu.bai@monash.edu

the relatively low elastic modulus of GFRP composites, which is approximately 10–20% that of steel [1]. This low material stiffness can be compensated by improving the stiffness of GFRP at the structural level via sandwich construction. Sandwich construction improves the second moment of area, thereby increasing the bending stiffness of GFRP structures and compensating for its relatively low material stiffness.

Research into GFRP web-flange sandwich systems has predominantly been focused on the construction of bridge superstructures, whereby decks with built-up or cellular cross-sections are connected via adhesive bonding and/or shear studs to supporting steel girders [5–9]. In particular, the use of adhesive bonding as a means of connecting GFRP profiles has demonstrated its success [5, 6, 10–14]. For example, Liu et al. [10] tested built-up bridge decks fabricated from adhesively bonded GFRP components under service loads and to failure. No cracking of the adhesive joints was observed when the deck was subjected to service loads. Failure occurred in the pultruded components and not in the adhesive joints. The addition of foam as a core material has also been investigated, where it has been found that both strength and fatigue properties are enhanced [15–17].

The success of GFRP composites in bridge construction also stimulates their application in building floor construction. An early work was done by Sotiropolous et al. [18] to propose GFRP floor systems made of pultruded GFRP channels, cellular sections and wooden diaphragms connected together via either adhesive or steel bolts. A recent notable work was the Startlink modular system made by modular pultruded sections that can be assembled to form light-weight, low-cost housing [4, 19]. A CFRP floor panel was proposed by Gao et al. [20] for application of one-way spanning slabs with an open cross-section composed of a top plate and three webs. However, the experimental mechanical performance was examined through scaled test samples (7.5 mm depth and 25 mm width), adhesively bonded from four C-channels on a plate. Research on the use of GFRP composites for building floor construction is therefore still limited, and further investigations are necessary.

In terms of performance in fire (as an important concern for building elements), GFRP web-flange sandwiches with closed cross-sections are able to provide certain fire resistance because of the cellular configuration of the section and the endothermic decomposition of the polymer resin [21–23]. In addition, passive fire protection or active water-cooling systems can be used to further extend the fire resistance time to satisfy fire resistance of more than one or two hours [24, 25], and to significantly improve fire reaction properties such as rate of heat release, smoke production, CO and CO_2 release, and flame spread [26, 27]. GFRP sandwich structures may also satisfy other physical building requirements, such as those related to airborne and impact sound transmission. Adhesive bonding can provide airtight seals to prevent the transmission of sound vibrations, sound-absorbing materials such as polymeric foams can be inserted into the core of the structure, and additional floor coverings can be used as sound barriers [28, 29].

This chapter proposes a modular assembly of GFRP sandwich units for building slab construction. Compared to mechanical connections, adhesive bonding may be associated with several advantages in connecting GFRP components together to form built-up structures, without the cutting out of materials and the introduction of local

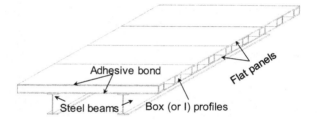

Fig. 2.1 Modular sandwich slab sections consisting of pultruded GFRP flat panels and box (or I-profiles) via adhesive bonding, connected to supporting steel beams and extended in the transverse direction

stress concentrations at boltholes [30, 31]. Therefore, this technique is employed to join individual pultruded GFRP shapes for the sandwich assembly in this application; while mechanical bolting is another way to form such built-up web-flange sandwich systems, with the convenience in on-site assembly and potential disassembly [32]. Figure 2.1 shows the modular sandwich configuration, whereby built-up sections consisting of standard pultruded GFRP box (i.e. square hollow section) or I-profiles are incorporated between two GFRP flat panels to form modular sandwich units, which are then assembled in the transverse direction to form a slab. These sandwich structures can then be connected to steel beams, as shown in Fig. 2.1. The sandwich section has an improved second moment of area compared to individual pultruded sections. Moreover, it permits adjustment of various geometric parameters such as the width and depth of the section, the spacing between the web-profiles and the thickness of the flat panels and webs, and thus offers greater design flexibility. This is a feature that is not seen in pre-existing pultruded decks which have fixed geometries. In addition, the modular nature of this system allows for easy integration of additional core materials such as prefabricated foam blocks or lightweight timber for the enhancement of structural performance.

In this study, five sandwich specimens consisting of flat panels and pultruded GFRP box or I-sections were prepared by connecting the components together using epoxy adhesive. The static properties (such as bending stiffness) of the sandwich specimens and the SLS, defined as a deflection limit, were quantified and compared with traditional one-way reinforced concrete (RC) slabs. The failure modes during the loading process and at the ultimate state were identified and the ultimate load-carrying capacity of the specimens were experimentally obtained and compared with calculated values estimated by existing structural theory. Various parameters, such as span length, bond quality, and the addition of prefabricated foam blocks within the core of the structure were also investigated.

2.2 Experimental Investigation

2.2.1 Overview

A series of single beam sections and sandwich specimens were fabricated and tested under four-point bending. Single beam sections were tested in order to characterise the failure modes of such single sections for a given geometry (either box or I-profile) and a given span-to-depth ratio. Sandwich specimens were then prepared and were designated according to the following naming convention: the first letter indicates the type of connection used in the sandwich (A = adhesive) and the second letter refers to the type of profile used within the core of the sandwich (B = box profile, I = I-profile). The third letter, if present, indicates information relevant to either the connection or the presence of additional core materials (D = discontinuous adhesive bonding, F = foam core). The number after the hyphen refers to the span length, given in metres.

In addition to testing the effects of profile shape and the inclusion of foam, the effect of span-to-depth ratio was also investigated. A number of span lengths were chosen, ranging from 1.25 to 2.7 m. Span lengths of 1.25 m were chosen to characterise shear failure modes, while span lengths of 2.7 m were chosen as these had span-to-depth ratios similar to structural members currently used in building applications. A span length of 1.65 m was also utilised to provide an intermediate span-to-depth ratio. The effect of bond quality on the structural performance of sandwich specimens was also investigated as a scenario of poor bond quality represented by discontinuous bonding. This may occur in practice for bonding of GFRP components on site, particularly along the lower faces of the specimen that are harder to access.

2.2.2 Materials

The modular built-up sandwich specimens were fabricated from pultruded GFRP materials. They consisted of E-glass fibres and polyester resin and the tensile properties of the GFRP materials were tested in accordance with ASTM D3039 [33], and the interlaminar shear strength was characterised via short-beam shear tests in accordance with ASTM 2344 [34]. The resulting properties of the GFRP materials are shown in Table 2.1, given as an average of five specimens for the tensile properties and an average of ten specimens for the short-beam shear properties. The tensile strength of Araldite 420 epoxy adhesive was 28.6 MPa, the tensile modulus was 1.9 GPa and the shear strength was 25 MPa according to the results given by Fawzia [35]. Divinycell P150 Foam was utilised as a core material, with nominal density of 150 kg/m^3, compressive modulus of 152 MPa, shear strength of 1.25 MPa and shear modulus of 40 MPa according to [36].

2 Fibre Reinforced Polymer Built-Up Beams and One-Way Slabs 31

Table 2.1 Dimensions and material properties of single-beam and sandwich specimens

Specimen ID	Shape of web section	Dimensions of web section (mm)	Span length (mm)	Depth (mm)	Span-to-depth ratio	Material properties of web section*			Material properties of flat panels**	
						Tensile strength (MPa)	Tensile Modulus (GPa)	Interlaminar shear strength (MPa)	Tensile strength (MPa)	Tensile Modulus (GPa)
SI-1.25	I-profile	152 × 152 × 9.5	1250	152	8.2	475.2 ± 30	29.1 ± 1.5	36.0 ± 1.0	–	–
SB1-1.25	Box	102 × 102 × 10	1250	102	12	306.5 ± 18	30.2 ± 1.4	26.7 ± 0.2	–	–
SB2-1.25	Box	102 × 102 × 10	1250	102	12	306.5 ± 18	30.2 ± 1.4	26.7 ± 0.2	–	–
SB3-2.7	Box	102 × 102 × 10	2700	102	26	306.5 ± 18	30.2 ± 1.4	26.7 ± 0.2	–	–
SB4-2.7	Box	102 × 102 × 10	2700	102	26	306.5 ± 18	30.2 ± 1.4	26.7 ± 0.2	–	–
AI-1.25	I-profile	152 × 152 × 6.5	1250	165	7.6	475.2 ± 30	29.1 ± 1.5	36.0 ± 1.0	305.7 ± 13 (L) 88.7 ± 1.6 (T)	20.9 ± 2.2 (L) 9.3 ± 0.4 (T)
AID-1.25	I-profile	152 × 152 × 9.5	1250	165	7.6	475.2 ± 30	29.1 ± 1.5	36.0 ± 1.0	305.7 ± 13 (L) 88.7 ± 1.6 (T)	20.9 ± 2.2 (L) 9.3 ± 0.4 (T)
AB-1.65	Box	102 × 102 × 5.3	1650	115	14	280.8 ± 11	18.3 ± 1.6	29.4 ± 1.1	305.7 ± 13 (L) 88.7 ± 1.6 (T)	20.9 ± 2.2 (L) 9.3 ± 0.4 (T)
AB-2.7	Box	102 × 102 × 10	2700	115	24	306.5 ± 18	30.2 ± 1.4	26.7 ± 0.2	393.1 ± 7.0 (L) 22.0 ± 2.1 (T)	31.7 ± 0.7 (L) 5.0 ± 0.2 (T)
ABF-2.7	Box	102 × 102 × 10	2700	115	24	306.5 ± 18	30.2 ± 1.4	26.7 ± 0.2	393.1 ± 7.0 (L) 22.0 ± 2.1 (T)	31.7 ± 0.7 (L) 5.0 ± 0.2 (T)

*Longitudinal fibre direction; **Longitudinal (L) or Transverse (T) fibre direction

2.2.3 Specimens

Five single-beam specimens were first tested up to failure. Four of these beams, designated SB1-1.25, SB2-1.25, SB3-2.7 and SB4-2.7, were box profiles with cross-section dimensions of 102 × 102 × 10 mm. SB1-1.25 and SB2-1.25 had an overall length of 1.5 m and a span length of 1.25 m; SB3-2.7 and SB4-2.7 had an overall length of 3 m and a span length of 2.7 m. The fifth beam, designated SI-1.25, was a 152 × 152 × 9.5 mm I-beam, with an overall length of 1.4 m and a span length of 1.25 m.

The details of the sandwich assemblies are summarised in Table 2.1. Specimens AI-1.25 and AID-1.25 were fabricated by incorporating two GFRP I-profiles between two GFRP flat panels. The I-profiles were 152 × 152 × 6.5 mm in size (for AI-1.25) and 152 × 152 × 9.5 mm (for AID-1.25), and for both specimens the flat panels were 6 mm thick and 345 mm wide. Figure 2.2a and b show the cross-section dimensions of AI-1.25 and AID-1.25 respectively, both of which had span lengths of 1.25 m. Specimen AID-1.25 had a discontinuous bond between the lower flat panel and the lower flange of the I-profiles to represent a situation of poor bonding quality. This poor bonding scenario was achieved by applying adhesive over a 250 mm-long region at each end to ensure that the lower panel did not move during testing. Between these bonding regions, along a length of 1 m, the lower panel was not adhered to the lower flanges of the I-profiles. However, the upper panel was fully adhered to the I-profiles.

Specimen AB-1.65 was fabricated by adhesively bonding two 102 × 102 × 5.3 mm box profiles between two GFRP flat panels, with cross-section as shown in Fig. 2.2c. This specimen had an overall length of 2 m and a span length of 1.65 m. Finally, specimens AB-2.7 and ABF-2.7 were both assembled by sandwiching two 102 × 102 × 10 mm box profiles between two flat panels of 345 mm width and 6 mm thickness. The cross-section dimensions are shown in Fig. 2.2d. These two specimens were both 3 m long, with a span length of 2.7 m. Specimen ABF-2.7 had

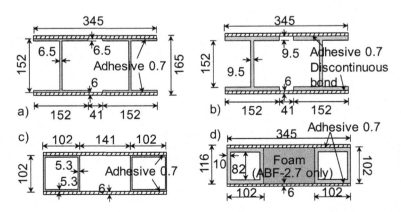

Fig. 2.2 Cross-section dimensions of specimens **a** AI-1.25 **b** AID-1.25 **c** AB-1.65 **d** AB-2.7 (without foam core) and ABF-2.7 (with foam core)

2 Fibre Reinforced Polymer Built-Up Beams and One-Way Slabs

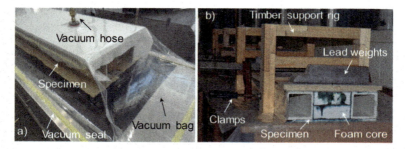

Fig. 2.3 **a** Fabrication of specimen AI-1.25 using vacuum bag. **b** Fabrication process of specimen ABF-2.7 using a timber support rig, clamps and lead weights

four prefabricated foam blocks inserted into its centre. The four 1.5 m-long blocks, with a width of 141 mm and a height of 50 mm, were adhesively bonded to the inner surfaces of the specimen.

A 0.7 mm thick layer of Araldite 420 A/B epoxy adhesive was used to bond the GFRP and foam components together. All adherend surfaces were sandblasted and degreased with a solvent prior to bonding, in accordance with the procedure outlined in the literature [37]. To control the thickness of the adhesive bond, spacer washers with a thickness of 0.7 mm were spaced in pairs at regular intervals along each bonding interface. AI-1.25 was fabricated via a vacuum bagging technique, as shown in Fig. 2.3a. This technique was used because it was deemed that, due to the shape of the I-profiles, the application of weights would not provide adequate clamping pressure during the curing process, particularly at the bond between the lower flat panel and the lower flanges of the I-profiles. A small wooden block was inserted between the two I-profiles to prevent them moving during the vacuum bag process, and suction was applied for 24 h, after which the vacuum bag was removed and the adhesive was allowed to continue curing at room temperature. The bond lines formed using this process were uniform, and no gaps were observed.

For specimens AB-1.65, AB-2.7 and ABF-2.7, weights were used to provide pressure while the adhesive cured. The amount of adhesive was deemed adequate, as there was generous squeeze-out in the joints during the bonding processes. Due to the longer length of specimens AB-2.7 and ABF-2.7, and the inclusion of foam into ABF-2.7, support rigs made of timber were used for those specimens and were clamped in place to ensure that the GFRP components remained aligned during the curing process, as shown in Fig. 2.3b. The adhesive was cured at room temperature, and the bond lines in all the joints were uniform and no gaps were present.

2.2.4 Experimental Setup and Instrumentation

The specimens with single sections SI-1.25, SB1-1.25 and SB2-1.25 and the sandwich specimens AI-1.25 and AID-1.25 were simply supported and loaded in four-point bending with a 5000kN Amsler/Shimadzu compression-testing machine. The point loads were transferred from the machine and applied to the specimens through two 50 mm-wide steel plates at a displacement control mode of 2 mm/min. The distance between the loads and supports was 400 mm for SI-1.25, SB1-1.25, SB2-1.25 and AI-1.25. For specimen AID-1.25, the loads were applied at a distance of 416 mm from the supports. Similarly, specimen AB-1.65 was loaded in four-point bending, but loading was applied under displacement control at a rate of 1 mm/min using an Instron 100kN actuator. The point loads were transferred from the actuator through a distribution steel beam and applied to the specimen via two 50 mm-wide steel plates at a distance of 550 mm from each support.

The strain gauge instrumentation for the sandwich specimens AI-1.25 and AID-1.25 is shown in Fig. 2.4a, and the instrumentation for AB-1.65 is shown in Fig. 2.4b. Strain gauges were placed in the longitudinal fibre direction along the upper and lower panels at midspan (S1–S5 and S21–S25 in Fig.2.4a and b) and at a position of span/6 from the support (S6–S10 and S26–S30) to determine the axial strain distribution along the width direction. To examine the degree of composite action between the upper and lower panels, strain gauges were also placed in the longitudinal direction at midspan and at span/6 (S11–S15 and S16–S20 respectively) along the depth of the specimen. For specimens SI-1.25, SB1-1.25 and SB2-1.25, strain gauges were placed in the longitudinal fibre direction along the upper flange and along the depth of the profile at midspan. Finally, the midspan deflection for all specimens was measured using a 100 mm linear potentiometer.

Single-beam specimens SB3-2.7 and SB4-2.7 and sandwich specimens AB-2.7 and ABF-2.7 were tested in a setup similar to that used for AI-1.25, but with the point loads applied at distance of 900 mm from each support. Furthermore, a displacement control mode of 3.5 mm/min was used during the loading process. The strain gauge configuration for single-beam specimens SB3-2.7 and SB4-2.7 was the same as that for SB1-1.25. Strain gauge positions for sandwich specimens AB-2.7 and ABF-2.7 are shown in Fig. 2.4c, where they were placed along the upper and lower panels (S1–S5 and S11–S15 respectively) to measure the axial strain distribution at midspan. They were also placed along the side of the specimen in the longitudinal fibre direction (S6–S10 in Fig. 2.4c) to evaluate composite action. Deformation shape was examined by placing five potentiometers along the length of the specimen at 450 mm intervals, the positions of which are also shown in Fig. 2.4c. At Def 1 and Def 5, the vertical deflections were measured using 100 mm linear potentiometers; and those for Def 2, Def 3 and Def 4 were measured by linear position transducers.

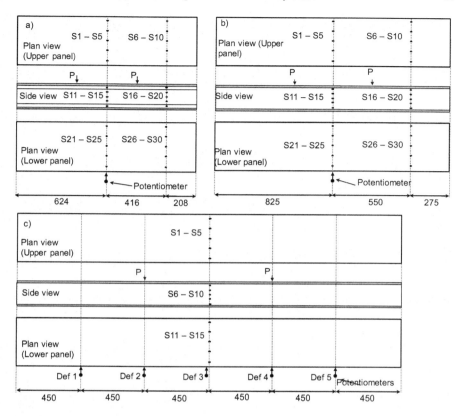

Fig. 2.4 Strain gauge and potentiometer instrumentation for **a** AI-1.25 and AID-1.25, **b** AB-1.65, **c** AB-2.7 and ABF-2.7

2.3 Experimental Results

2.3.1 Load–Deflection Response

The load–deflection responses for the 1.25 m-span specimens SB1-1.25, SB2-1.25, AI-1.25, and the 1.65 m-spanning sandwich specimen AB-1.65 are shown in Fig. 2.5a. Figure 2.5b shows the load-deflection responses for the 2.7 m-span specimens SB3-2.7, SB4-2.7, AB-2.7 and ABF-2.7. The SLS for sandwich specimen AI-1.25, which corresponded to a midspan deflection limit of span/300, was reached at a load of 36 kN (see Table 2.2). AI-1.25 also showed a slight nonlinear load-deflection response, as seen by the change in slope of the load-displacement curve.

Sandwich specimen AB-1.65 showed a linear load-deflection response until failure, and the serviceability load (based on a midspan deflection limit of span/300) was found to be 8.9 kN (see Table 2.2). AB-2.7 and ABF-2.7 also showed a linear

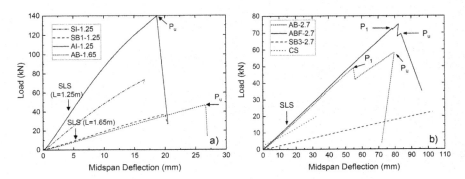

Fig. 2.5 **a** Load-deflection curves for SI-1.25, SB1-1.25 and AI-1.25, with a span length of 1.25 m, and AB-1.65 with a span length of 1.65 m. **b** load–deflection curves for AB-2.7, ABF-2.7, SB3-2.7 and a designed RC member (CS), with a span length of 2.7 m

load-deflection response up to material failure of the upper panel due to local buckling deformation (P_1, see Sect. 2.3.2), after which the stiffness reduced. Furthermore, the serviceability loads for AB-2.7 and ABF-2.7 were 7.4 kN and 8.2 kN respectively, based on a midspan deflection limit of 9 mm (span/300).

2.3.2 Failure Mode

The single-specimen beams SI-1.25, SB1-1.25 and SB2-1.25 failed via longitudinal shearing of the webs, as shown in Fig. 2.6a (SI-1.25) and 6b (SB1-1.25). This failure was sudden, with the shear crack propagating in the region of maximum shear (i.e. between a support and the nearest point load) and rapidly developing towards the midspan of the specimen. SI-1.25 failed at a load of 75 kN, SB1-1.25 failed at a load of 37 kN and SB2-1.25 failed at a load of 42 kN.

Similar failure modes were observed in the sandwich specimens AI-1.25, AID-1.25 and AB-1.65. The shear failure in AI-1.25 occurred at a load of 140 kN, with longitudinal shear cracks forming at the web–flange junction (see Fig. 2.6c). The same failure mode was also observed in AID-1.25 (with greater web thickness but a discontinuous bond, see Fig. 2.2b) at a load of 190 kN (see Fig. 2.6d). AB-1.65, which used box profiles of the same width and height as SB1-1.25 and SB2-1.25 but with a thinner web, also failed via shearing of its webs at a load of 47 kN. Again, the shear cracks were initiated in the region of maximum shear and then propagated within this region. Furthermore, this cracking caused the separation of the web and flange of one web-core box profile and then the separated web easily buckled outwards, as shown in Fig. 2.6e. For these sandwich specimens, neither failure nor local damage was observed in the upper or lower panels, nor at the adhesive bond.

Single-beam specimens SB3-2.7 and SB4-2.7 failed at loads of 23 kN and 26 kN respectively, and both failed via separation of the web–flange junction and buckling

Table 2.2 Results of four-point bending experiments on single-beam and sandwich specimens

Specimen ID	Weight (kg/m)	Second moment of area ($\times 10^6$ mm^4)	SLS Load* (kN)	Calculated stiffness EI at SLS ($\times 10^{11}$ Nmm^2)	Experimental stiffness EI at SLS ($\times 10^{11}$ Nmm^2)	Failure load P_u (kN)	Percentage shear deformability at SLS (%)	Strength (P_u) to weight ratio (kNm/kg)	Experimental stiffness EI to weight ratio at SLS ($\times 10^{10}$ Nmm^3/kg)
SI-1.25	8	16.5	21.0	4.8	7.1	75	53	9.4	8.9
SB1-1.25	7	5.25	7.5	1.6	1.4	37	15	5.3	2.0
SB2-1.25	7	5.25	8.0	1.6	1.5	42	15	6.0	2.1
SB3-2.7	7	5.25	2.0	1.6	1.6	23	4.1	3.3	2.3
SB4-2.7	7	5.25	2.0	1.6	1.6	26	4.1	3.7	2.3
AI-1.25	19	52.0	36.0	12.4	15.7	140	63	7.4	8.3
AB-1.65	16	18.7	8.9	3.7	3.0	47	15	2.9	1.9
AB-2.7	21	22.8	7.4	7.1	6.2	58	7.5	2.8	3.0
ABF-2.7	24	22.8**	8.2	7.1	6.9	70	8.3	2.9	2.9

*Based on a deflection limit of span/300 for a four-point bending test configuration; **Calculated without foam

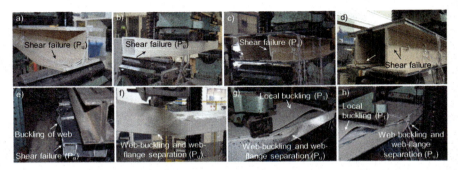

Fig. 2.6 **a** Shear failure of SI-1.25 at 75 kN (P_u), **b** shear failure of SB1-1.25 at 37 kN (P_u), **c** shear failure of AI-1.25 at 140 kN (P_u), **d** shear failure of AID-1.25 at 190kN (P_u), **e** shear failure and outward buckling of AB-1.65 at 47 kN (P_u), **f** web buckling and web–flange separation of SB3-2.7 at 23 kN (P_u), **g** material failure of buckled upper panel at 49 kN (P_1) and final failure at 58 kN (P_u) of AB-2.7, **h** material failure of buckled upper panel at 73 kN (P_1) and final failure at 70 kN (P_u) of ABF-2.7

of the web. It occurred near a loading point but was in the region of the beam under maximum bending as shown in Fig. 2.6f. This failure mode was also observed in both AB-2.7 and ABF-2.7 at the ultimate load; before that, these sandwich specimens also experienced local buckling in the upper panel. For specimen AB-2.7, the development of out-of-plane deformation due to local buckling caused material failure of the upper panel at a load of 49 kN (P_1), as shown in Fig. 2.6g. After this, the load dropped suddenly to 42 kN, but then increased again until failure at 58 kN (P_u). Final failure was via separation of the web–flange junction on the compression side of the specimen and an associated outward buckling of the web as shown in Fig. 2.6g. It was observed that the local buckling region of the upper panel elongated at P_u.

In a manner similar to AB-2.7, specimen ABF-2.7 experienced material failure of the upper panel due to local buckling deformation at a load of 73 kN (P_1), and this buckling region increased before reaching a maximum load of 75 kN. However, while this buckling occurred near a load point, it arose outside the region of maximum bending. At this point, the load suddenly dropped to 68 kN, but then increased again until failure of the web at 70 kN (P_u), when the load-carrying capacity of the entire specimen was lost. As in AB-2.7, the final failure modes included the separation of the web–flange junction and outward buckling of the web, as shown in Fig. 2.6h.

For both AB-2.7 and ABF-2.7, although some damage was seen in the adhesive layer, this damage arose only at the final failure, i.e. 58 kN (P_u) for AB-2.7 and 70 kN (P_u) for ABF-2.7, because of the elongation of the buckling region of the upper panel in addition to the web–flange separation. At P_1, where the material failure of the upper panel occurred due to local buckling effects, damage was seen only in the upper panel where it split within the fibre layer, and no damage was seen in the adhesive layer.

2.3.3 Axial Strain Along Specimen Depth

The axial strain distributions at midspan along the depth of the sandwich specimens AI-1.25 and AID-1.25 are shown in Fig. 2.7a, and those for AB-1.65, AB-2.7 and ABF-2.7 are shown respectively in Figs. 2.7b–d. At SLS (see Table 2.2 for the corresponding load levels) specimens AI-1.25, AB-1.65, AB-2.7 and ABF-2.7 showed a linear axial strain distribution from the upper panel through the depth of the specimen to the lower panel at midspan, indicating that full composite action was achieved with adhesive connections. The full composite action at midspan was maintained for specimens AI-1.25, AB-1.65 and ABF-2.7 until the ultimate load P_u was reached, as evidenced by the linear strain distributions along the depth of the specimens in Figs. 2.7a, b and d respectively. Specimen AB-2.7 demonstrated full composite action until the load level P_1 was reached (see the linear distribution of axial strain in Fig. 2.7c). After that, partial composite action occurred because of material failure of the upper panel due to local buckling (see Fig. 2.6g), corresponding to the nonlinear distribution of axial strain as shown in Fig. 2.7c. The results indicate that, by virtue of adhesive bonding, full composite action could be achieved before failure if bond quality provided.

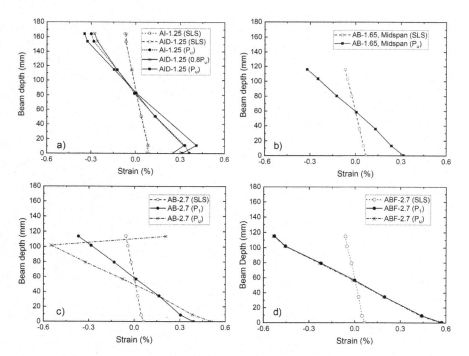

Fig. 2.7 Axial strain distribution along depth of a cross-section at serviceability and failure loads (including $0.8P_u$ for AID-1.25) for **a** AI-1.25 and AID-1.25, **b** AB-1.65, **c** AB-2.7, **d** ABF-2.7 at midspan

Partial composite action between the lower panel and the I-profile at midspan was identified for AID-1.25 with a discontinuous bond in this region (representing poor bond quality). The partial composite action was not obvious at a low load level (SLS, see Fig. 2.7a). However, it became significant at higher load levels, as seen for example in the strain distribution at $0.8P_u$ in Fig. 2.7a, suggesting that the non-uniform strain distribution was a result of discontinuous bonding rather than failure. This non-uniform strain distribution was maintained until the failure load P_u. As shown in Fig. 2.7a, specimen AID-1.25 still demonstrated full composite action up to failure P_u between the upper panel and the upper flanges of the I-profiles where no discontinuous bond was present.

2.3.4 Axial Strain Along Specimen Width

Figures 2.8a and b show the midspan axial strain distribution at SLS and $0.8P_u$ along the upper panel and lower panel respectively for all sandwich specimens tested. The axial strain distributions along the upper and lower panels were uniform at SLS and at ultimate failure for most of the specimens. The only exception was AB-2.7. As evidenced in Fig. 2.8a at a load of $0.8P_u$ (46 kN), the value of compressive strain at the centre of the upper panel ($y/b = 0$) was much lower than those at the sides ($y/b = \pm 1$) because the buckling there introduced strain in tension. Detailed inspection of the strain results (measured by S3 at the region of local buckling of the upper panel of AB-2.7) indicated that the initiation of local buckling occurred at a load of 42 kN, when the compressive strain started to decrease. For ABF-2.7, the axial strains at the midspan were uniformly distributed even after local material failure of the upper panel at 73 kN (P_1) because this failure occurred outside the region of maximum bending.

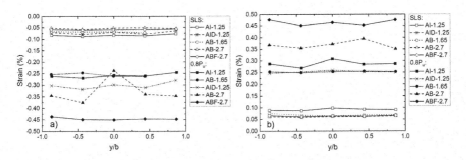

Fig. 2.8 Axial strain distributions at SLS and $0.8P_u$ for sandwich specimens AI-1.25, AID-1.25, AB-1.65, AB-2.7 and ABF-2.7 at midspan along **a** the upper panel (negative values in compression), **b** the lower panel (positive values in tension), with the centre of the specimen width corresponding to $y/b = 0$

2.4 Discussion and Comparison

2.4.1 Bending Stiffness

The bending stiffness of the sandwich specimens can be calculated using Timoshenko beam theory [38]. The central deflection δ of a simply supported beam under four-point bending is given in Eq. (2.1):

$$\delta = \frac{PL^3}{24EI}\left(\frac{3a}{L} - \frac{4a^3}{L^3}\right) + \frac{Pa}{GA} \tag{2.1}$$

where P is the applied load, L is the span length, a is the distance from the support and a point load, E is the elastic modulus, G is the shear modulus and A is the cross-sectional area of the webs. According to the material properties given in Table 2.1 (3 GPa for G based on the results of similar pultruded profiles [39] and the manufacturer data) and the experimental setup (see Sect. 2.4), the experimental bending stiffness EI of the sandwich specimens were found at SLS and are presented in Table 2.2. Table 2.2 also gives the calculated stiffness values that were based on the elastic modulus and the geometry of the specimen. There is an agreement between the calculated stiffness and experimental stiffness for the specimens with span-to-depth ratios of 12 or greater (see Table 2.1). However, a noticeable difference between the calculated and experimental stiffness values was found for specimens SI-1.25 and AI-1.25 with smaller span-to-depth ratios. This difference is therefore most likely due to the fact that shear effects are more significant for short beams (the deformability caused by shear is given as a percentage of total deflection in Table 2.2).

The stiffness values at serviceability for the single-beam specimens, the second moment of area and the weight of all sections are summarised in Table 2.2, along with the strength-to-weight ratios and stiffness-to-weight ratios of all tested specimens. These single specimens have a slightly larger strength-to-weight ratio compared to the sandwich specimens. However, stiffness is of more importance because the design of GFRP structures is usually governed by SLS criteria rather than strength. In terms of stiffness for AB-2.7 and ABF-2.7, the stiffness-to-weight ratio is larger than the single specimens SB-3 and SB-4. This was because the sandwich configuration enhanced the bending stiffness of the section by increasing the second moment of area. The stiffness-to-weight ratio for SI-1.25 is only slightly larger than AI-1.25. This suggests that the sandwich configuration may be more efficient in the case of larger span-to-depth ratios where bending deformation is more dominant.

2.4.2 Structural Load-Carrying Capacity

Single beam sections SI-1.25, SB1-1.25 and SB2-1.25, which all spanned 1.25 m, all failed via shearing of the web. Similarly, sandwich beams AI-1.25, AID-1.25 and AB-1.65 also failed in the same manner, whereby shearing of the webs occurred in the region of maximum shear (see Fig. 2.6). Regardless of the profile shape, these beams, which had relatively low span-to-depth ratios (see Table 2.1) failed in the same manner. Hence, it is the span-to-depth ratio, rather than the profile shape, that is significant in determining the failure mode.

The load-carrying capacity is plotted against the web area for specimens SB1-1.25, SB2-1.25 and sandwich specimen AB-1.65 and is shown in Fig. 2.9 (specimens SI-1.25, AI-12.5 and AID-1.25 are not included because of a considerably higher in-plane shear strength). Results from previous research [40–43] with similar failure modes and shear strengths are also plotted in Fig. 2.9. A linear relationship can be seen between shear force at failure and web-area, suggesting that the failure mechanism was the same and that the ultimate load can be governed accordingly. The load-carrying capacity of such web-flange sandwich sections with small span-to-depth ratios can therefore be estimated according to the maximum shear stress at the web (or in some cases the web-flange junction [40]), or approximately as the multiple of the web area and the corresponding in-plane or interlaminar shear strength [44].

Different failure modes were seen in the longer specimens SB3-2.7, SB4-2.7, AB-2.7 and ABF-2.7, as introduced in Sect. 2.3.2. The span-to-depth ratio of specimens SB3-2.7 and SB4-2.7 was 26, and the span-to-depth ratio of AB-2.7 and ABF-2.7 was 24. These specimens finally failed because of separation of the web–flange junction and buckling of the web (see Fig. 2.6f for SB3-2.7, Fig. 2.6g for AB-2.7 and Fig. 2.6h for ABF-2.7). These failure modes arose in the region where compressive stresses were predominant. Thus span-to-depth ratio may be considered an important parameter in the governing conditions of failure modes and the consequent design equations of such a failure mode can also be referred from [44].

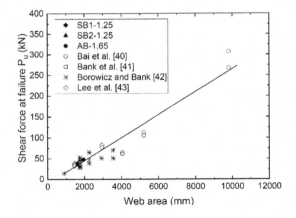

Fig. 2.9 Shear force at failure P_u against web area for specimens SB1-1.25, SB2-1.25, AB-1.65 and results from previous research work

Fig. 2.10 Upper panel of AB-2.7 as a uniaxially loaded plate for buckling load estimation, with simple supports on the loaded edges and two clamped edges

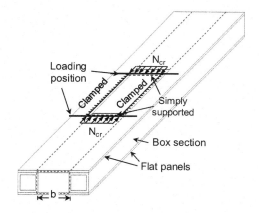

However, the buckling of the upper panel of AB-2.7 and ABF-2.7 before final failure is of more importance in design of such sandwich units, since such a failure caused the loss of structural integrity. The buckling load of AB-2.7 can be calculated by assuming that the upper panel within the maximum bending region (between the two loading points) acts as a thin orthotropic plate subjected to a uniformly distributed compressive load, as shown in Fig. 2.10. The two loaded edges (corresponding to the position of the point loads) are simply supported and the two unloaded edges are clamped. The buckling load is given in Eq. 2.2 [45],

$$N_{cr} = \frac{24}{b^2}\left[1.871\sqrt{(D_{11})(D_{22})} + (D_{12}) + 2(D_{66})\right] \quad (2.2)$$

where b is the distance between the two innermost webs as shown in Fig. 2.10, here taken as 151 mm, and D_{ij} ($i, j = 1,2,6$) are the bending stiffness coefficients of the upper panel and can be calculated based on material properties given in Table 2.1. The Poisson ratio v_{12} was assumed to be 0.3 [46]. From Eq. (2.2), the calculated buckling load N_{cr} is 600 N/mm, corresponding to a load level of 44 kN under four-point bending. The experimental load at which initiation of buckling along the upper panel occurred was 42 kN (see Sect. 2.3.4), and compared well with the calculated value (44 kN).

2.4.3 Effect of Foam Core

The addition of foam was shown to provide a slight increase in stiffness. At SLS, the stiffness of ABF-2.7 (with foam core) was about 10% greater than that of AB-2.7 (without foam core), as demonstrated in Table 2.2 and also evidenced by the load-deflection graphs of Fig. 2.5b. The foam core was associated with a small elastic modulus of 152 MPa, corresponding to only 0.5% of that of the GFRP box profiles

(30.2 GPa, see Table 2.1). However, the presence of foam significantly increased the failure load, as demonstrated in Fig. 2.5b. The load P_1 at which failure of the upper panel occurred due to local buckling was 73 kN for specimen ABF-2.7, representing an increase of about 50% in comparison to AB-2.7 (at 49 kN). This increase in load was likely due to the fact that the whole upper panel of ABF-2.7 was restrained in the out-of-plane direction from its inner side, as it was adhesively bonded to both the box profiles and the foam core. In contrast, AB-2.7 was bonded only to the box profiles, giving a smaller restrained flange width. This comparison demonstrated that the addition of a lightweight core material (even with a small elastic modulus) may be a means of significantly improving the load-carrying capacity of GFRP sandwich structures with a small increase in weight (about 15% increase of weight in this study).

2.4.4 Effect of Discontinuous Bonding

Bonding quality has an effect on the degree of composite action. The strain results shown in Fig. 2.7 for specimens AI-1.25, AB-1.65, AB-2.7 and ABF-2.7 demonstrated that full composite action was provided by uniform bonding between the GFRP panels and the web-core profiles. However, sandwich specimen AID-1.25 was associated with partial composite action between the lower panel and the I-profiles because of the discontinuous bond there. It appeared, however, that the poor bond quality in the middle region of the tension side was not detrimental to the ultimate capacity of the specimen. AI-1.25 failed at a load of 140 kN. This value was about 75% of the failure load of AID-1.25, which failed at 190 kN. This difference was due to AI-1.25 having a smaller web area than AID-1.25. The web area of AI-1.25 was 1807 mm^2, about 70% of the web area of AID-1.25 (at 2527mm^2). Therefore, although the poor bond quality in the middle region of the tension side decreased the composite action, it did not affect the ultimate capacity in this study.

2.4.5 Comparison to RC One-Way Spanning Slab

Considering that a span-to-depth ratio of 24 is commonly used in RC slabs, the sandwich specimens AB-2.7 and ABF-2.7, with this same span-to-depth ratio, provides appropriate examples for comparison with RC one-way spanning slabs. A RC member, designated CS, was designed according to AS3600 [47] with the same cross-section dimensions (345 × 115 mm) and same span length (2.7 m) as those of AB-2.7 and ABF-2.7. Due to a 20 mm-thick concrete cover, the effective depth was 90 mm; the compressive strength f'_c of the concrete was assumed to be 32 MPa. Six N10 ($f_{sy} = 500$ MPa) bars were used as tensile reinforcement and four N10 bars as compressive reinforcement.

Under four-point bending, the unfactored failure load of CS was calculated to be 20 kN per point load, which is equivalent to an ultimate moment capacity of 18 kNm. The cracking moment of the concrete slab was calculated as 2.8 kNm, corresponding to a load of 3kN under four-point bending. After the cracking moment, deflections were calculated based on the effective second moment of area of the cracked section [47]. The load-deflection curve of CS is provided in Fig. 2.5b, with a resulting ultimate load of 20 kN.

It can be seen that GFRP sandwich specimens showed favourable properties compared to the one-way spanning RC slab of the same sectional size. The sandwich specimens AB-2.7 and ABF-2.7 were about 80% lighter than CS (which weighed 100 kg/m), but their ultimate loads were about 2–2.5 times greater than that of CS. Furthermore, prior to cracking, the bending stiffness of CS was 13.9×10^{11} Nmm^2, which was larger than that of sandwich specimens AB-2.7 and ABF-2.7 (at 6.2×10^{11} Nmm^2 and 6.9×10^{11} Nmm^2 respectively, see Table 2.2). However, once cracking occurred, its effective section and therefore the stiffness was greatly reduced as shown in Fig. 2.5b. The SLS is reached in this post-cracking stage, where the effective bending stiffness of the cross-section is 5.2×10^{11} Nmm^2, up to 25% less than the stiffness of the sandwich specimens. This results in a 30% larger SLS load for AB-2.7 and ABF-2.7 (as 8.1 kN/m and 9.0 kN/m respectively) compared to that of the CS member (6.7 kN/m) if designed under a uniformly distributed load.

Taking into account long-term deflection behaviour and the design procedure outlined in [47], the SLS load for the RC one-way slab is reached at a load 3 kN/m. For the GFRP sandwich system, if a partial safety factor is applied based on a time-dependent coefficient of viscosity of 0.71 given in [48], the serviceability limit state is reached at a load of 4.7 kN/m and 5.3 kN/m for specimens AB-2.7 and ABF-2.7 respectively. The comparison of GFRP sandwich structures to RC one-way slabs in this regard seems favourable, although further experimental validation on the creep behaviour of adhesively-bonded FRP sandwich structures is important.

2.5 Conclusions

This chapter has presented an adhesively bonded modular system of GFRP web-flange sandwiches for use in building floor construction. Modular units consisting of pultruded box or I-profiles sandwiched between two flat panels and connected via adhesive bonding were fabricated and tested under four-point bending. Further, the effects of span-to-depth ratio, foam core material and quality of the bond were examined. From this work, the following conclusions can be drawn:

1. The experiments demonstrated a typical shear failure for the sandwich specimens AI-1.25, AID-1.25 and AB-1.65. These three specimens were associated with a relatively lower span-to-depth ratio of 7.6 (for AI-1.25 and AID-1.25) and 14 (for AB-1.65), and therefore shear stresses became dominant. Such shear failure was also evidenced by the single box and I-sections (the web sections

used in the sandwich specimens AI-1.25, AID-1.25 and AB-1.65) with span-to-depth ratios ranging from 8.2 to 12. In contrast, sandwich specimens AB-2.7 and ABF-2.7, with span-to-depth ratios of 24, showed final failure of web and web–flange separation in the region of dominant compressive stress due to bending. This failure mode was similar to that seen in single box sections, which had span-to-depth ratios of 26.

2. The shear failure mechanism can be estimated based on existing beam theory. For the specimens with larger span-to-depth ratios, the local buckling of the upper panel is of more relevance in designing such sandwich units as this failure caused the loss of structural integrity. The corresponding buckling load can be predicted by assuming the upper panel within the maximum bending region (between the two loading points) acts as a thin orthotropic plate subjected to a uniformly distributed compressive load. The predicted value showed good agreement with the experimental buckling load.

3. The addition of foam greatly enhanced the load-carrying capacity of the sandwich unit with a larger span-to-depth ratio, since the presence of foam prevented the upper panel in compression from premature out-of-plane buckling by restraining the whole panel in this direction. Material failure of the buckled upper panel occurred at 73 kN for specimen ABF-2.7 with a foam core. This was about 50% greater than that for AB-2.7, in which local failure of the upper buckled panel occurred at 49 kN. The addition of foam increased the weight by only about 15% because of its own light weight, and increased the stiffness by about 10% because of its low elastic modulus.

4. Uniform adhesive bonding provided full composite action at both serviceability and ultimate limit states. For the cases of uniform bonding applied between the flat panels and the web box or I-profiles, a linear strain distribution in the depth direction was observed before buckling of the upper flat panel for specimens AB-2.7 and ABF-2.7. Discontinuous bonding introduced partial composite action. However, it appears that a discontinuous bond in the middle region of the tension side did not affect the load-carrying capacity of AID-1.25.

In comparison to a designed one-way RC member, the proposed all-composite adhesively bonded GFRP sandwich configuration in this chapter demonstrated considerable reduction in weight (by about 80%), with a significant improvement in load-carrying capacity (up to 250%). The GFRP sandwich assemblies further showed higher serviceability stiffness and load because the SLS of such concrete structures is reached in the post-cracking stage and the effective inertial moment of area of the concrete section is therefore greatly reduced. The work presented in this study investigates the potential use further in steel-FRP composite systems, as a design often seen in FRP bridge superstructures. However, this system can be developed further by replacing the steel girders with FRP girders, and this may be investigated in the future.

References

1. Hollaway LC (2010) A review of the present and future utilisation of FRP composites in the civil infrastructure with reference to their important in-service properties. Constr Build Mater 24:2419–2445
2. Bakis CE, Bank LC, Brown VL, Cosenza E, Davalos JF, Lesko JJ et al (2002) Fiber-reinforced polymer composites for construction—state-of-the-art review. J Compos Constr 6:73–87
3. Halliwell S (2010) Technical papers: FRPs—the environmental agenda. Adv Struct Eng 13:783–791
4. Hutchinson JA, Singleton MJ (2007) Startlink composite housing. Advanced composites for construction (ACIC). University of Bath
5. Keller T, Gürtler H (2005) Composite action and adhesive bond between fiber-reinforced polymer bridge decks and main girders. J Compos Constr 9:360–368
6. Keller T, Gürtler H (2005) Quasi-static and fatigue performance of a cellular FRP bridge deck adhesively bonded to steel girders. Compos Struct 70:484–496
7. Keller T (2003) Use of fibre reinforced polymers in bridge construction (SED 7): IABSE
8. Jeong J, Lee Y-H, Park K-T, Hwang Y-K (2007) Field and laboratory performance of a rectangular shaped glass fiber reinforced polymer deck. Compos Struct 81:622–628
9. Keelor DC, Luo Y, Earls CJ, Yulismana W (2004) Service load effective compression flange width in fiber reinforced polymer deck systems acting compositely with steel stringers. J Compos Constr 8:289–297
10. Liu Z, Majumdar PK, Cousins TE, Lesko JJ (2008) Development and evaluation of an adhesively bonded panel-to-panel joint for a FRP bridge deck system. J Compos Constr 12:224–233
11. Hayes MD, Ohanehi D, Lesko JJ, Cousins TE, Witcher D (2000) Performance of tube and plate fiberglass composite bridge deck. J Compos Constr 4:48–55
12. Zhou A, Coleman JT, Temeles AB, Lesko JJ, Cousins TE (2005) Laboratory and field performance of cellular fiber-reinforced polymer composite bridge deck systems. J Compos Constr 9:458–467
13. Zhou A, Lesko JJ, Coleman JT, Cousins TE (2002) Failure modes and failure mechanisms of fiber reinforced polymer composite bridge decks. In: 3rd International conference on composites in infrastructure. San Francisco
14. Majumdar PK, Liu Z, Lesko JJ, Cousins TE (2009) Performance evaluation of FRP composite deck considering for local deformation effects. J Compos Constr 13:332–338
15. Zi G, Kim BM, Hwang YK, Lee YH (2008) An experimental study on static behavior of a GFRP bridge deck filled with a polyurethane foam. Compos Struct 82:257–268
16. Zi G, Kim BM, Hwang YK, Lee YH (2008) The static behavior of a modular foam-filled GFRP bridge deck with a strong web-flange joint. Compos Struct 85:155–163
17. Moon DY, Zi G, Lee DH, Kim BM, Hwang YK (2009) Fatigue behavior of the foam-filled GFRP bridge deck. Compos B Eng 40:141–148
18. Sotiropoulos SN, GangaRao HVS, Mongi ANK (1994) Theoretical and experimental evaluation of FRP components and systems. J Struct Eng 120:464–485
19. Evernden MC, Mottram JT (2012) A case for houses to be constructed of fibre reinforced polymer components. Proc ICE Constr Mater 165:3–13
20. Gao Y, Chen J, Zhang Z, Fox D (2013) An advanced FRP floor panel system in buildings. Compos Struct 96:683–690
21. Bai Y, Hugi E, Ludwig C, Keller T (2011) Fire performance of water-cooled GFRP columns. I: Fire endurance investigation. J Compos Constr 15:404–412
22. Bai Y, Vallée T, Keller T (2008) Modeling of thermal responses for FRP composites under elevated and high temperatures. Compos Sci Technol 68:47–56
23. Bai Y, Vallée T, Keller T (2007) Modeling of thermo-physical properties for FRP composites under elevated and high temperature. Compos Sci Technol 67:3098–3109

24. Correia JR, Branco FA, Ferreira JG, Bai Y, Keller T (2010) Fire protection systems for building floors made of pultruded GFRP profiles: Part 1: Experimental investigations. Compos B Eng 41:617–629
25. Keller T, Tracy C, Hugi E (2006) Fire endurance of loaded and liquid-cooled GFRP slabs for construction. Compos A Appl Sci Manuf 37:1055–1067
26. Correia JR, Branco FA, Ferreira JG (2010) The effect of different passive fire protection systems on the fire reaction properties of GFRP pultruded profiles for civil construction. Compos A Appl Sci Manuf 41:441–452
27. Ohlemiller TJ, Shields JR (1999) The effect of surface coatings on fire growth over composite materials in a corner configuration. Fire Saf J 32:173–193
28. Harris D (1991) Noise control manual, Springer, US
29. Sagartzazu X, Hervella-nieto L, Pagalday JM (2008) Review in sound absorbing materials. Arch Comput Meth Eng 15:311–342
30. Baldan A (2004) Adhesively-bonded joints and repairs in metallic alloys, polymers and composite materials: adhesives, adhesion theories and surface pretreatment. J Mater Sci 39:1–49
31. Banea MD, da Silva LFM (2009) Adhesively bonded joints in composite materials: an overview. Proc Inst Mech Eng Part L J Mater Des Appl 223:1–18
32. Satasivam S, Bai Y (2014) Mechanical performance of bolted modular GFRP composite sandwich structures using standard and blind bolts. Compos Struct 117:59–70
33. ASTM D3039 (2000) Standard test method for tensile properties of polymer matrix composite materials. West Conshohocken, United States
34. ASTM D2344 (2000) Standard test method for short-beam strength of polymer matrix composite materials and their laminates. West Conshohocken, United States
35. Fawzia S, Zhao X-L, Al-Mahaidi R (2010) Bond-slip models for double strap joints strengthened by CFRP. Compos Struct 92:2137–2145
36. DIAB (2011) Divinycell P technical data. http://www.diabgroup.com/aao/a_literature/a_lit_ds.html
37. Clarke JL (1996) Structural design of polymer composites: EUROCOMP design code and handbook. E & FN Spon
38. Timoshenko S, Goodier JN (1969) Theory of elasticity, 3rd ed. McGraw-Hill
39. Mottram JT (2004) Shear modulus of standard pultruded fiber reinforced plastic material. J Compos Constr 8:141–147
40. Bai Y, Keller T, Wu C (2012) Pre-buckling and post-buckling failure at web-flange junction of pultruded GFRP beams. Mater Struct 1–12
41. Bank LC, Nadipelli M, Gentry TR (1994) Local buckling and failure of pultruded fiber-reinforced plastic beams. J Eng Mater Technol 116:233–237
42. Borowicz D, Bank LC (2011) Behaviour of pultruded fiber-reinforced polymer beams subjected to concentrated loads in the plane of the web. J Compos Constr 229–238
43. Lee J, Kim Y, Jung J, Kosmatka J (2007) Experimental characterization of a pultruded GFRP bridge deck for light-weight vehicles. Compos Struct 80:141–151
44. Bank LC (2007) Composites for construction: structural design with FRP materials. John Wiley & Sons, Inc
45. Qiao P, Shan L (2005) Explicit local buckling analysis and design of fiber-reinforced plastic composite structural shapes. Compos Struct 70:468–483
46. Pecce M, Cosenza E (2000) Local buckling curves for the design of FRP profiles. Thin-Walled Struct 37:207–222
47. Australian Standards (2009) AS3600: Concrete structures. SAI Global
48. Sá MF, Gomes AM, Correia JR, Silvestre N (2011) Creep behavior of pultruded GFRP elements – Part 2: Analytical study. Compos Struct 93:2409–2418

Chapter 3
Fibre Reinforced Polymer Composites Two-Way Slabs

Sindu Satasivam, Yu Bai, Yue Yang, Lei Zhu, and Xiao-Ling Zhao

Abstract This chapter presents a modular assembly of sandwich structures using glass fibre reinforced polymer (GFRP) components for two-way slab applications. The sandwich assemblies were built-up sections made from pultruded web-core box (i.e. square hollow section) profiles incorporated between two flat panels, connected via adhesive bonding or novel blind bolts. Two different pultrusion orientations were achieved and examined i.e. flat panels with pultrusion directions either parallel (unidirectional orientation) or perpendicular (bidirectional orientation) to the web-core box profiles. The effects of pultrusion orientation, shear connection and the presence of foam core materials on strength and stiffness were investigated by experimental testing of two-way slab specimens until failure. Sandwich slabs with unidirectional orientation showed premature cracking of the upper panel between fibres, whereas those with bidirectional orientation showed local out-of-plane buckling of the upper flat panel. The unidirectional slab also showed greater bending stiffness than bidirectional slabs due to the weak in-plane shear stiffness provided by the web-core profiles, which introduced partial composite action between the upper and lower flat panels of the bidirectional slab. Finite element models and analytical techniques were developed to estimate deformation, failure loads and the degree of composite action, and these showed reasonable agreement with the experimental results.

Reprinted from Composite Structures, 184, Sindu Satasivam, Yu Bai, Yue Yang, Lei Zhu, Xiao-Ling Zhao, Mechanical performance of two-way modular FRP sandwich slabs, 904-916, Copyright 2018, with permission from Elsevier.

S. Satasivam · X.-L. Zhao
Department of Civil Engineering, Monash University, Clayton, Australia

Y. Bai (✉)
Department of Civil Engineering, Monash University, Clayton, Australia
e-mail: Yu.Bai@monash.edu

Y. Yang
Department of Civil Engineering, Tsinghua Univeristy, Beijing, China

L. Zhu
School of Civil and Transportation Engineering, Beijing University of Civil Engineering and Architecture, Beijing, China

3.1 Introduction

Over the past several decades the use of glass fibre reinforced polymers (GFRP) has become alternatives as primary load-carrying structural members in civil engineering applications. GFRP has high strength and offers advantages over traditional construction materials, such as corrosion resistance, low thermal conductivity and light weight [1, 2]. Furthermore, the embodied energy of GFRP structures is lower than that of equivalent structures made of steel or concrete [3, 4]. However, due to its low elastic modulus [2], the material stiffness of GFRP is 10–20% that of steel. Hence the design of GFRP is usually governed by serviceability limits rather than strength. Moreover, the fibre architecture of pultruded GFRP results in materials in which strength and stiffness in the pultrusion direction (also known as the longitudinal fibre direction, where the majority of the fibres run) are much greater than in the direction perpendicular (transverse) to the pultrusion direction. Web-flange sandwich systems may address both these challenges at the structural level by improving the second moment of area (and hence increasing bending stiffness) and by incorporating multiple pultrusion directions into the one sandwich assembly [5–7].

Extensive studies into GFRP web-flange sandwich systems have focused on bridge deck construction, where it has been shown that strength and stiffness criteria can be met [8–12]. These decks are usually composed of built-up or cellular cross-sections with webs and flanges, supported on two sides by steel girders (i.e. as one-way systems), and connected via adhesive and/or shear studs. The decks are assembled so that their pultrusion directions, which exert the majority of the strength, lie in the direction perpendicular to traffic. Due to the orthotropic material properties and the cellular configuration of pultruded FRP decks, differences in mechanical properties in the pultrusion direction and transverse pultrusion direction (i.e. the direction of traffic) are apparent. For example, rectangular modules loaded in four-point bending in the pultrusion direction show linear load–deflection behaviour, but when loaded in four-point bending in the transverse direction these rectangular modules exhibit non-linear load–deflection responses [13–15] with much less load capacities. Partial composite action is also present between the upper and lower flat panels of rectangular or trapezoidal decks when loaded in the transverse direction because transverse webs provide reduced in-plane shear stiffness [16, 17]. Case studies have also shown cracking along the fibres in the pultrusion direction [18]. Debonding of joints connecting adjacent GFRP modules have also been observed in the longitudinal pultrusion direction in road bridges, caused by tensile stresses developed within joints when traffic travels across the deck [19, 20]. This results in cracking of wearing surfaces in the direction transverse to traffic, i.e. along the pultrusion direction of the GFRP components. Such consequences are primarily a result of the relatively low strength in the transverse pultrusion direction compared to that in the longitudinal direction.

One way to prevent premature cracking is to provide additional reinforcement in the perpendicular pultrusion directions within the one sandwich structure. This is achieved in the modular GFRP web-core sandwich system introduced in this chapter. Furthermore, existing results about the bending behaviour of GFRP decks has focused predominantly on one-way systems, such as for bridge decks. These cases then are different to two-way floor slabs which can be supported on three or four sides. It is important, therefore, that the structural performance in both directions is characterised for two-way FRP slabs. While studies have been performed on one-way FRP floors [21, 22] and on FRP building systems such as the Startlink modular system for housing [23, 24], studies of two-way slabs with FRP for building applications have primarily focused on hybrid FRP systems and strengthening [25, 26]. The results on pultruded GFRP in two-way slab systems are therefore limited.

This chapter presents modular two-way web-core sandwich slabs as built-up sections consisting of GFRP box (i.e. square hollow section) profiles incorporated between two flat panels, connected via adhesive or novel blind bolts. The pultrusion direction of the flat panels is perpendicular to the pultrusion direction of the box profiles, creating a slab with bidirectional pultrusion orientation. This feature is not often seen in existing pultruded GFRP members, where the majority of the fibres, and hence the member's strength, lie in the same direction, since only one pultrusion direction is achieved. With bidirectional pultrusion orientation, it is expected that the structural performance of the deck can be improved in the transverse slab direction by enhancing the junction between the GFRP face panels and box profiles, thereby preventing cracking between fibres along the pultrusion direction of GFRP components under transverse bending. An experimental investigation was undertaken into such modular GFRP sandwich structures under two-way bending when subjected to a static load up to failure. The effects of pultrusion orientation, connection type (adhesive or mechanical blind bolts) and the addition of lightweight foam core materials on strength and stiffness were examined. Following this, finite element (FE) models were used to predict structural stiffness, and analytical techniques were developed to evaluate structural deformation and estimate load-carrying capacity.

3.2 Experimental Investigation

Four two-way modular GFRP sandwich slab specimens were fabricated and tested under static loads. Parameters that were tested were the pultrusion orientation of the GFRP components, the connection between GFRP components and the effect of foam as an additional core material. Box profiles were used as the web-core, with specimens designated according to the following naming convention: the first letter refers to the pultrusion orientation (U = unidirectional, B = bidirectional), the second letter refers to the connection type (A = adhesive, B = bolts) and the third letter, if present, refers to the presence of additional core materials (F = foam). Finally, a fixed span-to-depth ratio of 24 was adopted to illustrate bending behaviour.

3.2.1 Materials

All sandwich slabs were fabricated with 50 × 50 × 6 mm box profiles and 6 mm-thick flat panels, and consisted of E-glass fibres and polyester resin. ASTM D3171, Procedure G [27], was used to determine the resin volume fraction (52%) and the fibre volume fraction (48%) of the flat panels. The fibre architectures of both the box profiles and flat panels were similar, consisting of a roving layer embedded between two mat layers. The resin in the box profiles and flat panels were also the same. The tensile properties of the box profiles and flat panels are given in Table 3.1 as an average of five specimens. Tensile properties were characterised in accordance with ASTM D3039 [28], taken as an average of five specimens. The interlaminar shear strength of the box profile was 28 MPa, found as an average of 10 specimens via short-beam shear tests in accordance with ASTM D2344 [29].

The in-plane shear modulus of the flat panel was found via off-axis coupon tests [30]. Five coupons with an off-axis fibre direction of 10° of 250 mm length and 25 mm width were tested in tension using an Instron 100 kN testing machine at a loading rate of 2 mm/min. The in-plane shear modulus was found to be 3.5 GPa. Appropriately sized off-axis coupons could not be cut from the box profiles, and hence the in-plane shear modulus of the box profiles was assumed to be the same as that of the flat panels due to the similar fibre architecture. Furthermore, the in-plane shear strength of the box profile was characterised by testing two single box-profiles under four-point bending. These beams were simply supported and loaded with a Baldwin 500 kN testing machine. The overall length of each profile was 500 mm, and the span length was 450 mm. The point loads were transferred from the machine and applied directly onto the single profiles at a distance of 155 mm from the support, and loading was applied at a displacement control of 1 mm/min. The corresponding in-plane shear strength was 28 MPa.

Araldite 420 epoxy adhesive was used as the adhesive connection or M10 (class 10.9) blind bolts were used as the mechanical connection. The Araldite 420 epoxy had a tensile strength of 28.6 MPa, a tensile modulus of 1.9 GPa and a shear strength of 25 MPa [31]. The blind bolts had a bolt diameter of 10 mm, an overall length of 140 mm and a thread length of 35 mm. In addition, the nominal tensile strength was 1000 MPa and the proof load stress was 900 MPa according to the manufacturer. Divinycell P120 foam was used as a core material. The nominal density was 120 kg/m^3, the compressive modulus was 115 MPa, the shear modulus was 32 MPa and the shear strength was 0.91 MPa according to manufacturer data [32].

Table 3.1 Measured material properties of 50 × 50 × 6 mm box profiles and flat panels

Material	Tensile strength (MPa)[a]	Tensile modulus (GPa)[a]
Box profiles	362.5 ± 21 (L)	32.2 ± 2.1 (L)
Flat panels	393.1 ± 7.0 (L)	31.7 ± 0.7 (L)
	22.0 ± 2.1 (T)	5.0 ± 0.2 (T)

[a] Longitudinal (L) or transverse (T) fibre (pultrusion) direction

3.2.2 Specimens

Four GFRP sandwich slab specimens were fabricated and tested. All specimens had the same geometry, where seven 50 × 50 × 6 mm GFRP box profiles were sandwiched between 6 mm-thick flat GFRP panels as shown in Fig. 3.1. Each sandwich slab had a total depth of 62 mm, an overall length and width of 1.5 × 1.5 m, a span of 1.45 × 1.45 m, and was supported on all four sides by steel rollers with a diameter of 30 mm. The steel rollers allowed both longitudinal and transverse sliding. Figure 3.2 shows the pultrusion orientations of the slab specimens. In all cases, the box profiles were placed in the longitudinal slab direction (along the x-axis in Fig. 3.2). Three sandwich slabs (UA, BA and BAF) had an adhesively bonded connection. Sandwich slab UA had a unidirectional pultrusion orientation with adhesive bonding, as shown Fig. 3.2a where the pultrusion direction of the upper and lower flat panels lay parallel to the pultrusion direction of the box profiles (i.e. along the x-axis).

Sandwich slabs BA, BAF and BB all had bidirectional pultrusion orientations, where the pultrusion direction of the upper and lower flat panels lay in the transverse slab direction (along the y-axis), perpendicular to that of the box profiles as shown in Fig. 3.2b. Sandwich slabs BA and BAF both had an adhesive connection, but BAF

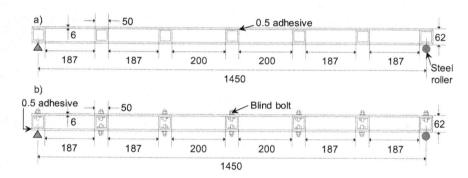

Fig. 3.1 Cross-sections of sandwich slabs **a** UA, BA and BAF, **b** BB

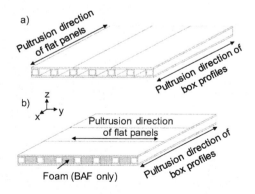

Fig. 3.2 Pultrusion directions of **a** unidirectional sandwich slab UA, **b** bidirectional sandwich slabs BA, BAF and BB

also had six prefabricated foam blocks inserted into the core of the structure between each box profile. These foam blocks were adhesively bonded to the inner surfaces of the specimen. All GFRP adherend surfaces were sandblasted and degreased with isopropanol in accordance with procedures outlined in [33]. The thickness of the adhesive layers was controlled by placing spacer wire 0.5 mm thick and 5 mm long at regular intervals along the adherend surfaces. Finally, sandwich slab BB was fabricated by connecting the GFRP components with blind bolts at a practical spacing of 150 mm along the longitudinal slab direction to achieve partial shear interaction. The bolt hole diameter was 10.5 mm, and a torque of 30 N·m was used to tighten each bolt. An adhesive connection was used between the lower flat panels and the lower flanges of the outermost box profiles at the ends of the slab to allow the slab to be supported onto the testing rig, as shown in Fig. 3.1b.

3.2.3 Setup and Instrumentation

A localised load was applied onto the slab using an Instron 250 kN load cell. The load was applied at the centre of the slab specimens onto a 35 mm-thick steel plate with dimensions of 550 × 550 mm at a displacement control mode of 1.5 mm/min. The dimensions of the loading area are shown in Fig. 3.3. These steel plate dimensions were used so that the load could be transferred onto the third, fourth and fifth web-core box profiles (W3 to W5, see Fig. 3.3), thereby preventing any punching failure through the flat panels. The steel plate also had holes of diameter 35 mm drilled in to it, which corresponded to the position of the bolts on the upper surface of specimen BB. This was so that the load could be applied directly onto the slab itself, rather than onto the bolts. The midspan deflection was measured using a 100 mm linear potentiometer.

To measure the degree of composite action provided by the adhesive and bolted shear connections, strain gauges were placed in the longitudinal slab direction (i.e.

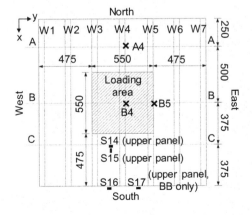

Fig. 3.3 Plan view of slab showing dimensions of loading area and strain gauge instrumentation

the longitudinal direction of the box profiles, see the x-axis in Fig. 3.2) along the depth of the slab at the eastern face of the web-core profile $W4$ at cross-section A-A, which was 250 mm from the northern edge of the slab (location A4 in Fig. 3.3). These strain gauges were placed in the longitudinal slab direction on the upper panel (S1) through the depth of the box profile (S2-S4) to the lower panel (S5). Longitudinal strain gauges were also placed along the depth at cross-section B-B, 750 mm from the northern edge at location B4 (eastern face of web-core profile $W4$), and at location B5 (eastern face of web-core profile $W5$). The distribution of strains along the slab surface was also measured by placing strain gauges S14 and S15 at a position 375 mm from the southern edge, between box profiles $W3$ and $W4$ on the upper panel in both the longitudinal and transverse slab directions. Finally, for slab BB, strain gauges S16 and S17 were placed on the southern edge of the upper panel along the transverse slab direction to measure the load at which out-of-plane local buckling occurred, as shown in Fig. 3.3.

3.3 Simplified FE Analysis

FE models were established for all the sandwich slabs to evaluate the bending stiffness considering effects of pultrusion orientation, the presence of a foam core and connection type. All GFRP components of the sandwich slabs (i.e. the flat panels and box profiles) were modelled with ANSYS finite element software using Shell181 elements: a four-node element, with each node having six degrees of freedom (translations and rotations in the x, y and z directions). Element size was 10 mm, giving a total of 66,300 shell elements. The material properties were taken based on experimentally measured values (as described in Sect. 3.2.1). Poisson's ratio was assumed to be 0.3 [34]. Figure 3.4a shows the FE models for sandwich slabs UA, BA and BB. As shown in Fig. 3.4b for BAF, the 0.5 mm-thick adhesive was modelled using Solid185 elements, an eight-node element with three degrees of freedom on each node (translations in the x, y and z directions). The number of adhesive elements within UA and BA was 10,500, whereas that for BAF was 45,300. The foam core of BAF was also modelled using Solid185 elements, using an element size of 10 mm to give 87,000 elements within the foam core.

The bolted connections in BB were modelled by coupling the coincident nodes at the interface between the flat panels and profiles at the positions of the bolts (i.e. at a bolt spacing of 150 mm), as shown in Fig. 3.4c. At each bolt position, eight nodes were coupled for all degrees of freedom. Four of these nodes were situated on the flat panel and four were on the box profile, which corresponded to a 10 mm bolt diameter. The boundary conditions were modelled by constraining displacements in the z direction on all four sides where the roller supports were situated. Additionally, displacements in the x-direction and y-direction were constrained at the eastern and southern supports respectively to prevent lateral movement of the slab. Rotations were allowed at all supports. Loading was applied in ten incremental steps, and the FE analyses were stopped when the maximum shear stress of the box profiles

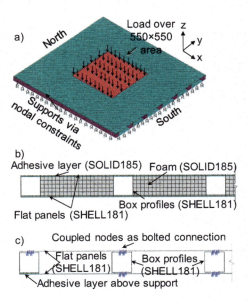

Fig. 3.4 **a** FE models for sandwich slabs UA, BA and BB, **b** FE model of BAF with adhesive bonding layer and foam core modelled with Solid185 elements, **c** node coupling as bolted connections for bolted slab BB

reached 28 MPa, which corresponded to the experimentally determined in-plane shear strength of the box profiles (see Sect. 3.2.1). The load–deflection curves from the FE analysis are provided and compared with the experimental results in Fig. 3.5.

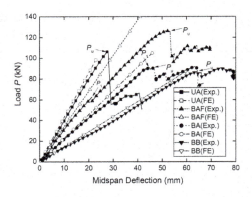

Fig. 3.5 Load–deflection curves for GFRP sandwich slabs

3.4 Experimental Results and Comparison with FE Modelling

3.4.1 Load–Deflection Responses and Bending Stiffness

The sandwich slab with the largest stiffness was UA (with unidirectional pultrusion orientation and adhesive bonding), as shown by the slope of the load deflection curves in Fig. 3.5. When the pultrusion direction of the flat panel was aligned so that it lay perpendicular to that of the box profiles (as in BA), the bending stiffness was reduced by 80%. This significant reduction in bending stiffness occurred because the webs of the box profiles provided low in-plane shear stiffness in the transverse slab direction, reducing the level of composite action between the upper and lower panels. Therefore, the flat panels in the transverse slab direction (which for the bidirectional slabs provided the majority of the bending stiffness) partially contributed to the overall bending stiffness of the structure. The load–deflection responses in Fig. 3.5 also show the ultimate load corresponding to a loss of load-carrying capacity (P_u) and local damage loads (P_1) prior to ultimate failure. The stiffness at failure loads of both P_u (for UA) and P_1 (BA) was slightly less than that measured at 20 kN, but only by 5% or less. Furthermore, the bending rigidities from the FE models for all specimens had a percentage difference of less than 10% compared to the initial experimental bending rigidities measured at 20 kN (prior to any localised failure), the values of which are summarised in Table 3.2.

The distribution of the centrally applied load P in the longitudinal slab direction (P_x) and transverse slab direction (P_y) was found from the FE analysis by determining the reaction forces at each support. The distribution of loading for all slabs is provided in Table 3.2. As can be seen, for UA the majority of the load (93% of P) was distributed

Table 3.2 Summary of experimental and FE results for sandwich slab parameters

Results	UA	BA	BAF	BB
Pultrusion orientation	Unidirectional	Bidirectional	Bidirectional	Bidirectional
Connection type	Adhesive	Adhesive	Adhesive	Blind bolts
Additional core material	–	–	Foam	–
Weight (kg)	70	71	86	74
Bending rigidity at 20 kN, Exp. (kN/mm)	3.92	2.20	3.19	1.37
Ultimate failure load P_u (kN)	107	90	126	87
Ultimate failure mode	Shear	Shear	Shear	Bending
Bending rigidity, FE (kN/mm)	4.30	2.28	3.37	1.46
Proportion of load P in x-axis, FE (P_x)	0.93P	0.78P	0.64P	0.87P
Proportion of load P in y-axis, FE (P_y)	0.07P	0.22P	0.36P	0.13P

along the longitudinal slab direction (P_x). Of this, 90% of the load P_x was carried by the three central box profiles $W3$, $W4$ and $W5$, which were situated directly under the loading plate (see Fig. 3.3). Only 7% of the load P was carried in the transverse slab direction (P_y) since the bending stiffness in this direction contributed very little to the overall bending stiffness of the slab. In the bidirectional slab BA, the majority of the load (78%) was still distributed in the longitudinal slab direction (P_x), where over 80% of the load was carried by the central box profiles $W3$ to $W5$. However, the transverse slab direction in the bidirectional slab BA contributed to overall bending by a greater amount than that in UA. This was also the case for BAF and BB.

The stiffness of the slab BAF, with the foam core and adhesive bonding, was 45% greater than that of BA and 20% less than that of UA. The addition of foam provided a significant increase in stiffness. This is demonstrated from slopes of the load–deflection responses in Fig. 3.5 prior to 20 kN, even though the foam core was associated with a low elastic modulus of 115 MPa (corresponding to only 0.4% of that of the GFRP box profiles, 32 GPa, see Sect. 3.2.1). The effect of the foam core on bending stiffness in the two-way spanning slab was different from that seen in similar GFRP modular sandwich beams and one-way spanning slabs presented in Chap. 2, where the addition of a similar foam core material provided an increase in bending stiffness of only 10%. In the present study, foam greatly affected bending stiffness in the transverse slab direction because it improved the in-plane shear stiffness of the web, which in turn improved the degree of composite action in the transverse slab direction between the flat panels. This effect has also been observed in foam-filled rectangular FRP bridge deck modules under transverse bending [13, 15]. However, unlike UA or BA, BAF showed a reduction in stiffness of 13% after 63 kN (P_1) because of localised damages arising inside the slab that were likely caused by damages of the adhesive due to the presence of voids (see Sect. 3.4.2 for details). These premature damages were not modelled in the FE analysis, hence resulting in the large difference (about 20%) between the FE and experimental bending stiffness at load levels greater than P_1 in Fig. 3.5.

The slab with the lowest stiffness was BB, in which the stiffness was 40% less than that of BA. Adhesive bonding provided full composite action, whereas bolted connections provided partial composite action based on nonlinear strain distributions along the specimen depth, as further discussed in Sect. 3.4.4. Partial composite action in the longitudinal slab direction was shown at all load levels up to failure for the mechanically bolted sandwich slab BB, resulting in lower bending stiffness than that in BA. Moreover, no significant changes in load–deflection responses were caused by local buckling of the upper panel (see Sect. 3.4.2 for details). Finally, the bending stiffness from the FE analysis was in good agreement with the experimental bending stiffness (with a percentage difference less than 10%), indicating that modelling bolted connections via node coupling was an effective means of describing the bending stiffness of a bolted two-way sandwich slab. It should be noted that out-of-plane buckling of the upper panels was not taken into account in the FE modelling, because this did not introduce much change in the overall bending stiffness as evidenced by the experimental results. Therefore, this modelling approach is a simplified linear analysis to describe bending stiffness and the load distribution of the centrally applied load in both slab directions. It further allows a theoretical formulation and design approach based on the grillage model as presented in Sect. 3.5. Non-linear FE analysis considering various initial imperfections for the bending performance of pultruded FRP profiles before buckling and during post-buckling process was developed and discussed in [35].

3.4.2 Failure Modes and Load-Carrying Capacity

A large crack became visible along the pultrusion direction on the upper panel of the unidirectional sandwich slab UA at 65 kN. This crack arose above the edge of box profile W2 as shown in Fig. 3.6a because the pultrusion direction (and hence fibres) of the upper panel was placed in the longitudinal slab direction. Bending in the transverse slab direction caused cracking between the fibres within the flat panel. This longitudinal crack continued to propagate as loading increased. Slab UA failed at the ultimate load P_u of 107 kN, with failure occurring via in-plane shearing of box profiles W4 and W5 on the southern side of the slab. There was no failure in the adhesive at this point. Shear failure of the box profiles resulted in a significant drop in load-carrying capacity.

No premature local cracking was observed in the bidirectional slab BA. Instead, local out-of-plane buckling arose at both the northern and southern edges of the slab due to differences in the longitudinal and transverse elastic moduli of the flat panels (as discussed in Sect. 3.4.3). This local buckling, which is shown at the southern edge

Fig. 3.6 **a** Longitudinal cracking at 65 kN and shear failure of webs at 107 kN (P_u) in slab specimen UA, **b** shear failure of web-core profile and delamination of upper panel at 90 kN (P_u) in specimen BA, **c** local out-of-plane buckling and shear failure at 126 kN (P_u) in specimen BAF, **d** local buckling, failure of web-core profiles and pull-out of bolts at 87 kN (P_u) in specimen BB

of the slab in Fig. 3.6b, became visible at approximately 60 kN, and the associated deformation continued to increase until 90 kN (P_1, see Fig. 3.5), when delamination occurred between the upper panel and the central box profile ($W4$). Progressive failures arose in this delaminated region until in-plane shear failure of box profile $W3$ occurred at the ultimate load of 90 kN (P_u, see Fig. 3.5), after which the load dropped to 80 kN.

The failure mode of the bidirectional sandwich slab with foam core BAF is shown in Fig. 3.6c. The sandwich slab BAF showed premature damages at 63 kN (P_1, see Fig. 3.5), characterised by a loud cracking noise and a localised permanent deformation. It arose inside the slab between box profiles $W1$ and $W3$ (see Fig. 3.3 for profile positions), and was most likely caused by failure of the adhesive due to the presence of voids in this region. There was a slight drop in load to 59 kN, but the load then increased. At approximately 70 kN visible local out-of-plane buckling of the upper panel arose on the southern side of the structure between box profiles $W3$ and $W5$. This was a result of imperfect bond quality in this region, where it was observed that the upper panel had not fully adhered to the box profiles and foam. As shown in Fig. 3.6c, no buckling occurred on the northern side of the slab, due to superior bond quality which restrained the upper panel from out-of-plane deformation. This was the case even after the ultimate load P_u was reached at 126 kN as a result of in-plane shear failure of the central box profile $W4$ on the southern edge of the slab specimen (near the buckled region). The corresponding ultimate load was 40% greater than that of BA ($P_u = 90$ kN), which had no foam core, even though both showed the same failure mechanism of in-plane shearing of the webs. The ultimate load of BAF was also 20% greater than that of UA (107 kN). This arose because the shear stresses were distributed across both the webs of the GFRP box profiles and the foam core, unlike in UA or BA where the shear stresses were distributed in the webs of the box profiles only. This resulted in a higher failure load in BAF. These findings suggest that the inclusion of foam can greatly increase both strength and stiffness of a two-way modular GFRP sandwich slab, with a small increase in weight. In this study, the inclusion of foam increased the weight of the structure by 20% (86 kg compared to 71 kg, see Table 3.2).

Sandwich slab BB showed local out-of-plane buckling of the upper panel at the northern and southern edges of the slab, similar to that seen in BA. However, this local buckling arose at 45 kN. This load was found by analysing strain gauges S16 and S17 situated at the southern edge of the upper panel, which is presented in Sect. 3.4.3. Failure of the web-core profiles arose at 87 kN at the compression fibre of web-flange junctions of box profiles $W3$, $W4$ and $W5$ at positions directly beneath the edges of the loading plate, as shown in Fig. 3.6d. Propagation of the local buckling region also resulted in pull-out of the blind bolts. This ultimate failure mode was different from that seen in the adhesively bonded slab specimens UA, BA and BAF, due the lower bending stiffness of BB (caused by the partial composite action provided by the bolted connections, see Sect. 3.4.1). The low bending stiffness in BB caused greater bending strains and hence resulted in bending failure being the dominant failure mode.

3.4.3 Stresses on Slab Surface

Strains were measured at S14 and S15 at a position 375 mm from the southern edge (see Fig. 3.3). The bending stresses in the longitudinal slab direction (σ_x) and in the transverse slab direction (σ_y) were then obtained from these strain results by taking into account the corresponding elastic moduli values of the upper panel in each direction. At 60 kN in UA, the compressive stresses on the upper panel in the longitudinal slab direction ($\sigma_x = -28$ MPa) were much greater than those in the transverse slab direction ($\sigma_y = -1$ MPa). This was due to the small transverse elastic modulus (5 GPa) compared to the longitudinal (pultrusion direction) elastic modulus (32 GPa) of the upper panel. In the bidirectional slabs BA and BAF, however, the stresses in both the longitudinal and transverse slab directions were approximately equal (i.e. $\sigma_x \approx \sigma_y$), and hence buckling loads were reached at a much lower load than in UA. In BA, the stresses in both directions were -9 MPa when the load was 60 kN. However, after 60 kN this was not the case, as the stresses in the longitudinal slab direction did not increase further, whereas stresses in the transverse slab direction increased (i.e., at 90 kN, $\sigma_x = -9$ MPa and $\sigma_y = -13$ MPa) as a result of the initiation of local buckling of the upper panel in this region (see Sect. 3.4.2). In BAF, $\sigma_x = -6$ MPa and $\sigma_y = -5$ MPa at a load level of 60 kN, prior to the onset of local failures.

Strain gauges S16 and S17 (see Fig. 3.3 for positions) were used to measure the onset of local buckling of the upper panel at the southern edge of slab BB. The stress distributions derived from these strain measurements are given in Fig. 3.7. Prior to 45 kN, compressive stresses were present at both S16 and S17. At 45 kN, however, compressive stresses began to decrease at S17 because local out-of-plane buckling in this region introduced tensile stresses. The onset of out-of-plane local buckling in the bolted slab BB (at 45 kN) occurred at a load 25% lower than that in the adhesively bonded BA (at 60 kN, see Sect. 3.4.2). This was due to the differences in the buckling plate width. As was the case for the sandwich beams in [5, 6], the upper panel can be assumed as a thin orthotropic plate. Based on the buckling plate widths of the sandwich beams, the sandwich slab BA had a plate width of 206 mm, taken as the

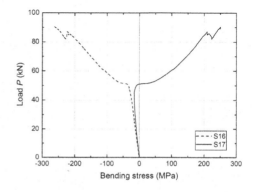

Fig. 3.7 Bending stresses σ_y on upper panel of slab BB measured from strain gauges S16 and S17

distance between the centres of adjacent webs in the buckling region, whereas the plate width of BB was 230 mm, taken as the distance between two washers on the bolts. In thin orthotropic plates, the buckling stress is inversely proportional to the buckling plate width squared [36, 37]. Assuming that the material properties and the support conditions of the buckled upper panel are the same, then a buckling plate width of 230 mm (as in BB) will result in a buckling stress that is 25% lower than that with the width of 206 mm (as in BA).

3.4.4 Axial Strains Along Depth of Section

The axial strain distributions measured in the longitudinal slab direction at location A4 (see Fig. 3.3 for locations) along the depth of the sandwich slabs are shown in Fig. 3.8a, b for load levels 20 kN and ultimate failure P_u respectively. Slabs UA, BA and BAF, all with adhesive bonding, showed linear strain distributions and therefore full composite action at all load levels up to P_u. Full composite action was also seen at all load levels up to P_u for the adhesively bonded slabs at locations B4 and B5. As full composite action was maintained in the adhesively bonded sandwich slabs UA and BA, the load–deflection responses in Fig. 3.5 remained linear (see Sect. 3.4.1). Sandwich slab BAF showed nonlinear load–deflection responses, but this was most likely the result of local failures (as discussed in Sect. 3.4.2), rather than composite action of the adhesive. This effect of full composite action providing linear load deflection responses in adhesively bonded web-flange sandwich structures has also been observed in one-way modular web-core sandwich slabs [5].

The axial strain distribution for specimen BB was nonlinear along the depth of the slab at all load levels, as shown for location A4 in Fig. 3.8, thus suggesting that partial composite action was provided by the blind bolts. As there was no significant change in the degree of composite action, the load–deflection response for BB was linear until the failure load P_u (refer to Fig. 3.5). Blind-bolted connections provided these

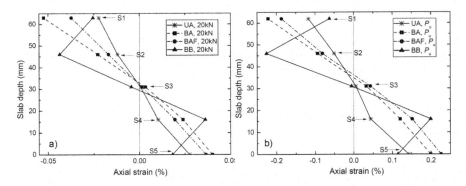

Fig. 3.8 a Strain distributions along depth in the longitudinal slab direction at 20 kN, b ultimate failure P_u at location A4

different structural responses also depending on the number of blind bolts and the longitudinal shear force at the flat panel and web-profile interface.

3.5 Analytical Approach and Discussion

3.5.1 Grillage Analysis

A symmetrical grillage model can be used to simplify the sandwich slabs as four beam members of length 725 mm (i.e. half the span length, measured between a support and the point load). The width of each member was taken as 550 mm (i.e. the width of the loading plate), as it was shown in Sect. 3.4.1 that the majority (more than 80%) of the load was carried by the box profiles in this region. $EI_{F,x}$ is the bending stiffness in the longitudinal slab direction and $EI_{F,y}$ is the bending stiffness in the transverse slab direction, as shown in Fig. 3.9. Both $EI_{F,x}$ and $EI_{F,y}$ were calculated by assuming full composite action between the GFRP components, and these values are summarised in Table 3.3. Additional factors α_x (see Sect. 3.5.2) and α_y (see Sect. 3.5.3) take into account the level of composite action in the longitudinal and transverse slab directions respectively.

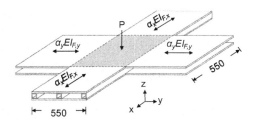

Fig. 3.9 Simplified slab model for grillage analysis

Table 3.3 Calculated bending stiffness in grillage analysis

Parameter	UA	BA	BAF	BB
Bending stiffness $EI_{F,x}$ (kN·m^2)	198	59.5	60.0	59.5
Bending stiffness $EI_{F,y}$ (kN·m^2)	26.0	165	165	165
Experimental degree of composite action α_x	1	1	1	0.65
Calculated degree of composite action α_x	1	1	1	0.62

3.5.2 Degree of Composite Action in Longitudinal Slab Direction

In slabs UA, BA and BAF, full composite action was provided by the adhesive connection in the longitudinal slab direction (i.e. along the x-axis, see Fig. 3.2), as shown in the axial strain distributions in Fig. 3.8. Therefore, $\alpha_x = 1$ for these slabs. Partial composite action was observed in the bolted slab BB due to non-linear strain distribution in Fig. 3.8, and hence $\alpha_x < 1$ to account for the reduction in bending stiffness. α_x is given as:

$$\alpha_x = \frac{EI_{P,x}}{EI_{F,x}} \quad (3.1)$$

where $EI_{P,x}$ is the bending stiffness of the partial composite section in the longitudinal slab direction. Due to the two-way action present in the system, determination of the experimental α_x was based on examination of the experimental strain data according to a three-layered beam analysis by Goodman and Popov [38], with the following assumptions: i) friction at the interfaces is ignored, ii) each layer has the same curvature, and iii) the strain distributions are linear throughout the depth of each layer.

The sandwich slab is similar to the three-layered system analysed in [38], but the central layer (layer 2, i.e. the box profiles) has greater thickness than the outer layers (layers 1 and 3, i.e. the flat panels). This is shown in Fig. 3.10, along with the theoretical strain distribution of a partial composite section. In the case of partial composite action, slip occurs between the flat panels and the box profiles. The slip strain $\varepsilon_{slip,1}$ is defined as the difference in strains at the interface between the upper flat panel (ε_1) and the upper flanges of the box profiles (ε_2, T), and slip strain $\varepsilon_{slip,2}$ is defined as the difference in strains at the interface between the lower flat panel (ε_3) and the lower flanges of the box profiles (ε_2, B), as depicted in Fig. 3.10.

From [38], the sum of the moments is given in Eq. 3.2, assuming that the curvatures of each layer are the same.

$$M = \kappa[(EI)_1 + (EI)_2 + (EI)_3] - C \cdot D_{N,1} + T \cdot D_{N,3} \quad (3.2)$$

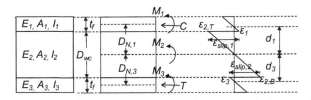

Fig. 3.10 Theoretical strain distribution for sandwich cross-section in the longitudinal slab direction with partial composite action between panel and box profile interfaces

where $D_{N,1}$ and $D_{N,3}$ are the distances from the geometric centroid of the box profiles to the geometric centroids of the upper or lower flat panels respectively, κ is the curvature, C is the resultant compressive force in the upper panel and T is the resultant tensile force in the lower panel. The distance from the neutral axis of the upper flat panel to the neutral axis of the box profiles is d_1, and the distance from the neutral axis of the box profiles to the neutral axis of the lower flat panel is d_3, as shown in Fig. 3.10. This results in:

$$d_1 = d_3 = \frac{\varepsilon_{\text{slip},1}}{\kappa} = \frac{\varepsilon_{\text{slip},2}}{\kappa} \quad (3.3)$$

Taking into account the strains ε_1, $\varepsilon_{2,T}$, $\varepsilon_{2,B}$, ε_3 and slip strains $\varepsilon_{\text{slip},1}$ and $\varepsilon_{\text{slip},2}$ from [38], Eqs. 3.4 and 3.5 can be derived:

$$C = \kappa (EA)_1 \cdot [D_{N,1} - d_1] \quad (3.4)$$

$$T = \kappa (EA)_3 [D_{N,3} - d_3] \quad (3.5)$$

$(EA)_1 = (EA)_3$ because the flat panels have the same geometric and material properties, and $C = T$ due to force equilibrium, thus giving that $d_1 = d_3$. Substituting Eqs. 3.4 and 3.5 into Eq. 3.2 gives:

$$EI_{P,x} = EI_{N,x} + 2 \cdot D_{N,1} \cdot (EA)_1 (D_{N,1} - d_1) \quad (3.6)$$

where $EI_{N,x}$ is the bending stiffness assuming no composite action. Analysis of the experimental strain curves given in Fig. 3.8 for BB provides that $d_1 = 23$ mm, which gives $EI_{P,x}$ as 38.7 kNm^2 and an α_x of 0.65. This is summarised in Table 3.3.

The bending stiffness presented above was determined based on experimental data. $EI_{P,x}$ can also be obtained via theoretical analysis based on Annex B of Eurocode 5 [39]. This approach is used to analyse simply supported beams under distributed loads that provide either sinusoidal or parabolic bending moments, but has been shown to provide good approximations when used for members with linearly distributed bending moments [40]. This was therefore used to calculate α_x for sandwich slab BB. From [39], the bending stiffness of the beam $EI_{P,x}$ is given as:

$$EI_{P,x} = \sum_{i=1}^{3} (E_i I_i + \gamma_i E_i A_i D_{N,i}^2) \quad (3.7)$$

where

$$\gamma_i = \left[1 + \frac{\pi^2 E_i A_i s}{K_i L^2} \right], i = 1, 3 \quad (3.8)$$

$$\gamma = 1 \quad (3.9)$$

L is the span length, s is the spacing of the bolts in the longitudinal slab direction and K is the slip modulus of the mechanical fastener (shear stiffness of the connection per shear plane [41]). $D_{N,1} = D_{N,3}$ and $D_{N,2} = 0$ due to symmetry about the axis of bending.

For timber structures, the slip modulus can be determined from the slope of the load–displacement curve of bolted joints [42]. Wu et al. [43] performed bolt tests on the same type and diameter of blind bolt and with the same bolt hole clearance as that used in the bolted slab BB. Furthermore, the GFRP plates used in [43] were composed of the same materials as used in the upper and lower panels of BB. The joints that were tested were double-lap GFRP joints in tension, with blind bolts spaced in pairs (i.e. two columns) of either one, two or three rows. The stiffness of the shear connector was found as the slope of the linear elastic region of the load–displacement curve for the joints with the single row of blind bolts, giving an average stiffness of 7600 kN/m. To use this value in the calculation of the bending stiffness of BB, the stiffness found in [43] was divided by four to obtain the stiffness of one bolt per shear plane. The shear connector stiffness K was found to be 1800 kN/m. The calculated $EI_{P,x}$ was therefore 37.1 kNm^2, with an α_x of 0.62. This calculated method gave an underestimation of the bending stiffness by 5%, providing a good estimation of the bending stiffness in the longitudinal slab direction.

3.5.3 Degree of Composite Action in Transverse Slab Direction

The composite action between the upper and lower flat panels in the transverse slab direction (i.e. along the y-axis, see Fig. 3.2) was influenced both by the shear connections between the flat panels and the box profiles and by the in-plane shear stiffness provided by the box profiles. Figure 3.11a shows the bending rigidity of the adhesively bonded slabs UA, BA and BAF for varying values of α_y when α_x

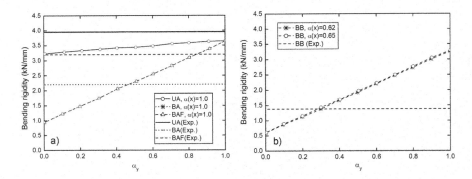

Fig. 3.11 **a** Bending rigidities of adhesively bonded sandwich slabs UA, BA and BAF ($\alpha_x = 1.0$), **b** bolted sandwich slab BB ($\alpha_x = 0.62$ and $\alpha_x = 0.65$) for various α_y values

= 1 (full composite action in the longitudinal slab direction, see Sect. 3.5.2). The bending rigidity of the bolted slab BB is shown in Fig. 3.11b for varying values of α_y, calculated at $\alpha_x = 0.65$ (experimental value, see Sect. 3.5.2) or $\alpha_x = 0.62$ (calculated value). The experimental bending rigidities are also plotted in Fig. 3.11 (see Table 3.2 for experimental values).

As shown in Fig. 3.11a, the underestimation of bending rigidity with various α_y values in comparison to the experimental bending rigidity for specimen UA very likely suggests full composite action between the upper and lower panels. Therefore, in the grillage analysis $\alpha_y = 1$ was used for specimen UA.

From Fig. 3.11a, the value of α_y for slab BA was identified as 0.48 based on the experimental bending rigidity and the grillage analysis procedure described in Sect. 3.5.1. This was lower than that for UA, even though both utilised adhesive bonded connections and had the same arrangement of web-core box profiles. The difference in composite action was therefore related to the differences in the elastic modulus of the upper and lower panels, in that the upper and lower panels of BA were associated with an elastic modulus of 32 GPa in the transverse slab direction whereas those of UA were associated with an elastic modulus of 5 GPa in the transverse slab direction.

Given a specified in-plane shear stiffness of the web sections, the difference in composite action due to different moduli of upper and lower panels can be illustrated using a method from Keller and Gürtler [40], where the Eurocode 5 method described in Sect. 3.5.2 was adapted to take into account the partial composite action between the upper and lower panels of an FRP bridge deck. Equation 3.8 was adapted to give Eq. 3.10 [40]:

$$\gamma_1 = [1 + \frac{\pi^2 E_1 A_1}{k_d L^2}] \quad (3.10)$$

$$\gamma_3 = 1 \quad (3.11)$$

where L is the span length in the transverse slab direction. In [40], a normalised shear deck stiffness $\hat{k}$ was used, the equivalent of a shear stress G_{yz} that causes a unit displacement ($u = 1$ mm) between the upper and lower flat panels, as shown in Fig. 3.12. The normalised stiffness of the box profiles $\hat{k}$ is given in Eq. 3.12 [40]:

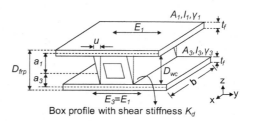

Fig. 3.12 In-plane shear stiffness of box profiles in transverse slab direction

$$\hat{k} = G_{yz} \cdot \frac{u}{D_{wc}} = G_{yz} \cdot \frac{1}{D_{wc}} \qquad (3.12)$$

where G_{yz} is the shear modulus of the box profiles in the y–z plane. This normalised $\hat{k}$ is then multiplied by the width of the transverse section b to give the in-plane shear stiffness of the deck K_d:

$$K_d = \hat{k}\, b \qquad (3.13)$$

This analysis can be used for slabs with adhesive bonding (such as UA and BA) between the GFRP components due to full composite action of the joint. However, it does not take into account any partial composite action provided by the connections between the GFRP components (such as bolted connections like those in BB). The bending stiffness in the transverse slab direction $EI_{P,y}$ is calculated from Eq. 3.14, where a_1 and a_3 are the distances from the centroids of the flat panels to the neutral axis, and are calculated using Eqs. 3.15 and 3.16 respectively (adapted from [39, 40] to take into account the depth of the box profile D_{wc}). $EI_{P,y}$ is then substituted into Eq. 3.17 to obtain the degree of composite action α_y factor in the transverse slab direction, assuming that the material and geometric properties of the upper and lower flat panels are the same (i.e. $E_1 = E_3$, $A_1 = A_3$ and $I_1 = I_3$).

$$EI_{P,y} = E_1 I_1 + E_3 I_3 + \gamma_1 E_1 A_1 \alpha_1^2 + \gamma_3 E_3 A_3 \alpha_3^2 \qquad (3.14)$$

$$\alpha_1 = D_{N,1} + \frac{t_f}{2} - \frac{\gamma_1 E_1 A_1 t_f}{2(\gamma_1 E_1 A_1 + \gamma_3 E_3 A_3)} \qquad (3.15)$$

$$\alpha_3 = D_{N,3} + \frac{t_f}{2} - \frac{\gamma_1 E_1 A_1 t_f}{2(\gamma_1 E_1 A_1 + \gamma_3 E_3 A_3)} \qquad (3.16)$$

$$\alpha_y = \frac{EI_{P,y}}{EI_{F,y}} \qquad (3.17)$$

Figure 3.13a shows γ_1, calculated using Eq. 3.10, and Fig. 3.13b shows α_y, both of which are plotted against various levels of E_1 (the elastic modulus in the transverse slab direction of both the upper and lower panels, where $E_1 = E_3$). In these calculations, b was assumed to be 550 mm, L to be 1450 mm and A_1 to be 3300 mm as the case for UA and BA. A range of specified K_d values were analysed, from 50 MPa (i.e. a flexible connection) to a fully rigid connection. It can be seen that for a given K_d, as the elastic modulus of the upper panel increases the degree of composite action between the upper and lower flat panels, and therefore the bending stiffness, decreases. $E_1 = 5$ GPa in slab UA, whereas $E_1 = 32$ GPa in slab BA. Hence, the bidirectional slab BA showed lower transverse bending stiffness than that in the unidirectional slab UA, even though the box profiles in both slabs provided the same shear stiffness.

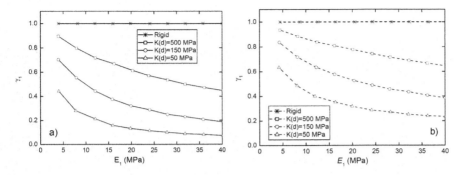

Fig. 3.13 a Factor γ_1 between upper and lower flat panels, b degree of composite action factor α_y in transverse slab direction for various levels of elastic modulus of upper and lower flat panels with constant length and cross-sectional geometry

From Fig. 3.11a, the value of α_y for slab BAF was identified as 0.86 based on the experimental bending rigidity and the grillage analysis procedure described in Sect. 3.5.1. This is higher than that for BA due to the presence of the foam core, which, as discussed in Sect. 3.4.1, increased the in-plane shear stiffness and hence the degree of composite action between the upper and lower panels in the transverse slab direction. It should be noted that the foam core did not directly increase bending stiffness in the longitudinal slab direction due to its low elastic modulus. This is shown in Fig. 3.11a, where the bending rigidities of BA (without foam) and BAF (with foam) are almost identical at a certain α_y value when $\alpha_x = 1$. The α_y factor for BB was 0.3 from Fig. 3.11b, which was lower than that for BA because the bolted connection provided partial composite action between the flat panels and box profiles in the transverse slab direction, in addition to the partial composite action provided by the web-core profiles.

3.5.4 Load-Carrying Capacity

Shear failure was observed in sandwich slabs UA, BA and BAF. The shear failure could be predicted by analysing only a 550 mm wide portion (as shown in Fig. 3.9) because the loading plate was applied directly over this width and was where the first shear failure modes arose. As presented in Sect. 3.4.2, the upper panel of UA did not experience any local out-of-plane buckling. Therefore, at failure the entire upper panel is included in the calculation of failure load. However, both BA and BAF experienced local out-of-plane buckling and subsequent delamination of the upper panel at loads lower than ultimate failure P_u, as discussed in Sect. 3.4.2, and hence the upper panel is not taken into account in the calculation of shear failure load.

The ultimate failure load in the longitudinal slab direction $P_{u,x}$ according to the shear failure criterion can be calculated from Eq. 3.18:

Table 3.4 Failure loads of sandwich slabs and comparisons to the calculated values based on shear failure

Results	UA	BA	BAF	BB
Neutral axis z_n (mm)[a]	31	17	27[b]	30
$\int E_x(z) \cdot z \cdot dA (\times 10^3$ kN·m)	3.78	1.69	1.05[b]	0.88
Bending stiffness EI_x (kN·m^2)	198	57.5	45.2[b]	35.3
b_w (mm)	36	36	36	36
Calculated ultimate load $P_{u,x}$ (kN)	106	68	87	81
Calculated ultimate load P_u (kN)	114	87	137	93
Experimental ultimate load P_u (kN)	107	90	126	87
Calc. P_u/Exp. P_u	1.08	0.97	1.09	1.07

[a]Calculated from base of structure, [b]Calculated with foam

$$P_{u,x} = \frac{2\tau_f \cdot EI_x \cdot b_w}{\int E_x(z) \cdot z \cdot dA} \quad (3.18)$$

where $E_x(z)$ is the elastic modulus of the flat panels and box profiles in the longitudinal slab direction measured at position z, which is the distance along the depth of the beam. EI_x is the bending stiffness in the longitudinal slab direction and b_w is the width of the box-profile webs where maximum shear stress occurs, taken as 36 mm for all the slab specimens. The in-plane shear strength τ_f of each box profile was 28 MPa, which was derived via experimental testing (see Sect. 3.2.1). The failure load P_u was then calculated by taking into account the relevant load distributions given in Table 3.2. Predicted shear failure values are summarised in Table 3.4. The calculated values show good agreement with the experimental results. In all sandwich beams that failed in shear, the percentage error of the calculated failure loads was 10% or less, showing reasonable agreement with the experimental results. Discrepancies between the experimental and calculated loads may be due to the fact that the load distributions used from Table 3.2 were based on cross-sections without buckling or other premature failure.

The failure load of BB was also calculated based on the shear failure criterion and is shown in Table 3.4. As in BA and BAF, the upper panel of BB buckled, and hence was not included in the analysis. The bending stiffness EI_x was calculated from Eq. 3.7. The calculated failure load was 93 kN assuming in-plane shear failure of the webs, which was 7% greater than the experimental failure load (87 kN). However, the maximum bending stress at the compression fibre of the box profiles at the edge of the loading plate was 406 MPa, calculated for a partial composite section as described in Eurocode 5 [39]. This exceeded the compressive strength $f_{c,f}$ of the box profiles (assumed to be equal to the tensile strength 363 MPa, see Table 3.1), verifying that BB failed via shear/bending interaction rather than by shear alone. Therefore, when calculating failure load, an interactive failure criterion considering both shear and bending stresses at the web-flange junctions under the edge of the loading plate (where failure occurred, see Sect. 3.4.2) may be more reasonable. Such a failure criterion is given in Eq. 3.19 [1].

$$\left(\frac{\sigma_x}{f_{c,f}}\right)^2 + \left(\frac{\tau_{xz}}{\tau_f}\right) = 1 \qquad (3.19)$$

where τ_{xz} is the in-plane shear stress of the web. Using this failure criterion, the calculated failure load of BB was 80 kN. This was 8% less than the experimental failure load, thus showing a reasonable comparison. This failure criterion was not exceeded in UA, BA or BAF at the corresponding ultimate loads calculated based on in-plane shear failure criterion (see Table 3.4).

3.6 Conclusions

This chapter investigated two-way spanning GFRP sandwich slabs for building structures. The effects on mechanical performance in terms of strength and stiffness were investigated based on the parameters of i) pultrusion orientation, ii) adhesive or blind bolts as the shear connection and iii) the presence of foam core materials. FE models and analytical techniques were developed to evaluate or predict deformation, failure loads and the degree of composite action. The following conclusions can be drawn:

1. The unidirectional slab UA had the highest stiffness because the major pultrusion directions of both the box profiles and flat panels lay in the longitudinal slab direction. In the bidirectional slab BA, the majority of the stiffness was provided by the flat panels in the transverse slab direction. However, the weak in-plane shear properties of the box profiles in the transverse slab direction introduced partial composite action between the upper and lower flat panels, thereby reducing the transverse bending stiffness. Composite action in the transverse slab direction was further reduced in the slab with the bolted connections BB due to the additional partial composite action provided by the bolts.
2. The unidirectional slab UA experienced longitudinal cracking in the pultrusion direction of the upper panel via local failure in the resin between fibres; and this was not observed in the bidirectional slabs BA, BB and BAF. Instead, the bidirectional slabs showed development of out-of-plane local buckling of the upper panel in compression as a result of the large differences in elastic modulus of the flat panel in each direction. Furthermore, the load at which local out-of-plane buckling of the upper panel commenced was lower in the bolted slab BB than in the adhesively bonded slab BA by 25%, due to the greater buckling plate width in BB.
3. The addition of a foam core significantly increased both strength and stiffness of the BAF slab. This occurred because the foam improved the in-plane shear stiffness of the web-core, which improved the composite action in the transverse slab direction between the upper and lower panels. The ultimate load of BAF was

greater than that of BA since the foam core distributed in-plane shear stresses over a larger area. Furthermore, local out-of-plane buckling of the upper panel was restrained by the foam core, provided that good bond quality was present.

4. Adhesively bonded slabs UA, BA and BAF all failed via in-plane shearing of the web-core box profiles. The bolted slab BB failed via bending in the box profiles directly beneath the edge of the loading plate. This was attributed to the lower bending stiffness of BB due to the partial composite action provided by the bolts, producing greater bending strains than those found in the adhesively bonded slabs.

5. FE analysis was performed on the sandwich slabs to predict load–deflection responses. Although local premature failure modes and out-of-plane buckling of the upper panels were not take into account in the FE modelling, satisfactory estimations of bending stiffness were found for all the tested sandwich slabs. A theoretic grillage analysis was further formulated to evaluate the bending stiffness of the sandwich slabs. The degree of composite action in both the longitudinal and transverse slab directions was quantified using theory developed from Eurocode 5 for timber structures. The degree of composite action in the longitudinal slab direction was dependent on connection type, whereas that in the transverse slab direction was dependent on the in-plane shear stiffness of the box profiles, the elastic modulus of the upper and lower panels and the type of connection. Ultimate shear failure was predicted with consideration of shear failure criterion or a shear and normal bending stress interactive failure criterion. Reasonable agreement with the experimental results was found.

Further work may be useful to describe other possible damage initiations or failure modes, such as the onset of local buckling as this may cause a loss of structural integrity. Experimental and modelling work has been conducted in literature, providing insights into the load redistribution and the effect on the ultimate failure modes that arise after local damages.

References

1. Bakis CE, Bank LC, Brown VL, Cosenza E, Davalos JF, Lesko JJ, Rizkalla SH, Triantafillou TC (2002) Fiber-reinforced polymer composites for construction - State-of-the-art review. J Compos Constr 6:73–87
2. Hollaway LC (2010) A review of the present and future utilisation of FRP composites in the civil infrastructure with reference to their important in-service properties. Constr Build Mater 24:2419–2445
3. Halliwell S (2010) Technical papers: FRPs—the environmental agenda. Adv Struct Eng 13:783–791
4. Mara V, Haghani R, Harryson P (2014) Bridge decks of fibre reinforced polymer (FRP): a sustainable solution. Constr Build Mater 50:190–199
5. Satasivam S, Bai Y, Zhao X-L (2014) Adhesively bonded modular GFRP web–flange sandwich for building floor construction. Compos Struct 111:381–392
6. Satasivam S, Bai Y (2014) Mechanical performance of bolted modular GFRP composite sandwich structures using standard and blind bolts. Compos Struct 117:59–70

7. Satasivam S, Bai Y (2016) Mechanical performance of FRP-steel composite beams for building construction. Mater Struct 49(10):4113–4129
8. Alagusundaramoorthy P, Harik IE, Choo CC (2006) Structural behavior of FRP composite bridge deck panels. J Bridg Eng 11:384–393
9. Johnson AF, Sims GD, Ajibade F (1990) Performance analysis of web-core composite sandwich panels. Composites 21:319–324
10. Jeong J, Lee Y-H, Park K-T, Hwang Y-K (2007) Field and laboratory performance of a rectangular shaped glass fiber reinforced polymer deck. Compos Struct 81:622–628
11. Foster DC, Richards D, Bogner BR (2000) Design and installation of fiber-reinforced polymer composite bridge. J Compos Constr 4:33–37
12. Turner MK, Harries KA, Petrou MF, Rizos D (2004) In situ structural evaluation of a GFRP bridge deck system. Compos Struct 65:157–165
13. Zi G, Kim BM, Hwang YK, Lee YH (2008) An experimental study on static behavior of a GFRP bridge deck filled with a polyurethane foam. Compos Struct 82:257–268
14. Park K-T, Kim S-H, Lee Y-H, Hwang Y-K (2005) Pilot test on a developed GFRP bridge deck. Compos Struct 70:48–59
15. Zi G, Kim BM, Hwang YK, Lee YH (2008) The static behavior of a modular foam-filled GFRP bridge deck with a strong web-flange joint. Compos Struct 85:155–163
16. Keller T, Gürtler H (2006) In-plane compression and shear performance of FRP bridge decks acting as top chord of bridge girders. Compos Struct 72:151–162
17. Keller T, Gürtler H (2005) Composite action and adhesive bond between fiber-reinforced polymer bridge decks and main girders. J Compos Constr 9:360–368
18. Keller T, Bai Y, Vallée T (2007) Long-term performance of a glass fiber-reinforced polymer truss bridge. J Compos Constr 11:99–108
19. Berman J, Brown D (2010) Field monitoring and repair of a glass fiber-reinforced polymer bridge deck. J Perform Constr Facil 24:215–222
20. Hong T, Hastak M (2006) Construction, inspection, and maintenance of FRP deck panels. J Compos Constr 10:561–572
21. Sotiropoulos SN, GangaRao HVS, Mongi ANK (1994) Theoretical and experimental evaluation of FRP components and systems. J Struct Eng 120:464–485
22. Gao Y, Chen J, Zhang Z, Fox D (2013) An advanced FRP floor panel system in buildings. Compos Struct 96:683–690
23. Evernden M, Mottram J (2012) A case for houses to be constructed of fibre reinforced polymer components. Proc ICE - Constr Mater 165(1):3–13
24. Hutchinson J, Hartley J (2011) STARTLINK lightweight building systems—wholly polymeric structures. In: Whysell C, Halliwell S, Mottram J (eds) Advanced composites in construction (ACIC). University of Warwick
25. Dawood M, Taylor E, Rizkalla S (2010) Two-way bending behavior of 3-D GFRP sandwich panels with through-thickness fiber insertions. Compos Struct 92(4):950–963
26. Keller T, Kenel A, Koppitz R (2013) Carbon fiber-reinforced polymer punching reinforcement and strengthening of concrete flat slabs. ACI Struct J 110(6):919–927
27. ASTM. D3171 (2011) Standard test methods for constituent content of composite materials. West Conshohocken, United States
28. ASTM. D3039 (2000) Standard test method for tensile properties of polymer matrix composite materials. West Conshohocken, United States
29. ASTM. D2344 (2000) Standard test method for short-beam strength of polymer matrix composite materials and their laminates. West Conshohocken, United States
30. Chamis CC, Sinclair JH (1977) Ten-deg off-axis test for shear properties in fiber composites. Exp Mech 17:339–346
31. Fawzia S, Zhao XL, Al-Mahaidi R (2010) Bond-slip models for double strap joints strengthened by CFRP. Compos Struct 92:2137–2145
32. DIAB (2011) Divinycell P technical data. http://www.diabgroup.com/aao/a_literature/a_lit_ds.html

33. Clarke JL (1996) Structural design of polymer composites: Eurocomp design code and handbook. Spon, New York
34. Pecce M, Cosenza E (2000) Local buckling curves for the design of FRP profiles. Thin-Walled Struct 37:207–222
35. Bai Y, Keller T, Wu C (2013) Pre-buckling and post-buckling failure at web-flange junction of pultruded GFRP beams. Mater Struct 46:1143–1154
36. Qiao P, Shan L (2005) Explicit local buckling analysis and design of fiber-reinforced plastic composite structural shapes. Compos Struct 70:468–483
37. Kollár L (2003) Local buckling of fiber reinforced plastic composite structural members with open and closed cross sections. J Struct Eng 129:1503–1513
38. Goodman JR, Popov EP (1968) Layered beam systems with interlayer slip. J Struct Div 94:2535–2548
39. CEN. Eurocode 5 (2004) Design of timber structures. Brussels: European Committee for Standardisation (CEN). London, British
40. Keller T, Gürtler H (2006) Design of hybrid bridge girders with adhesively bonded and compositely acting FRP deck. Compos Struct 74:202–212
41. Porteous J, Kermani A (2013) Structural timber design to Eurocode 5, 2nd ed. Wiley-Blackwell, Sussex 28(5)
42. Clouston P, Bathon L, Schreyer A (2005) Shear and bending performance of a novel wood–concrete composite system. J Struct Eng 131:1404–1412
43. Wu C, Feng P, Bai Y (2015) Comparative study on static and fatigue performances of pultruded GFRP joints using ordinary and blind bolts. J Compos Constr 19(4):04014065

Chapter 4
Steel- Fibre Reinforced Polymer Composite Beams

Sindu Satasivam and Yu Bai

Abstract This chapter presents an experimental and modelling investigation into modular composite beam structures using web-flange fibre reinforced polymer (FRP) and steel for building floor construction. The modular FRP slabs are formed from adhesively bonding pultruded box profiles (i.e. square hollow sections) sandwiched between two flat panels. They are then connected via adhesive or one-sided bolted connections to steel beams to form a composite system. Two different fibre (pultrusion) configurations are investigated in this chapter: flat panel pultrusion with direction either parallel or perpendicular to the box profiles. Composite beams were tested under four-point bending and evaluated for bending stiffness, load-carrying capacity, and the degree of composite action within the FRP web-flange sandwich slab and that provided by the shear connections. All the composite beams showed ductile load–deflection responses, with yielding of the composite beam commencing prior to failure of the FRP slabs. Furthermore, adhesive bonding provided full composite action, but the novel bolted connections with a certain spacing provided either full or partial composite action, dependent on the pultrusion configuration of the FRP slab. An analytical procedure is also developed to evaluate the bending stiffness and load-carrying capacity of the composite beams. Finite element analysis was further employed in this chapter, showing good comparisons to the experimental results.

Reprinted by permission from Springer Nature: Springer Nature, Materials and Structures, Mechanical performance of modular FRP-steel composite beams for building construction, Sindu Satasivam and Yu Bai, Copyright 2016.

S. Satasivam
Department of Civil Engineering, Monash University, Clayton, Australia

Y. Bai (✉)
Department of Civil Engineering, Monash University, Clayton, Australia
e-mail: yu.bai@monash.edu

© The Author(s), under exclusive license to Springer Nature Singapore Pte Ltd. 2023
Y. Bai (ed.), *Composites for Building Assembly*, Springer Tracts in Civil Engineering,
https://doi.org/10.1007/978-981-19-4278-5_4

4.1 Introduction

Fibre-reinforced polymer (FRP) composites have been highlighted due to their favourable properties such as high strength and corrosion resistance [1, 2] and low thermal conductivity and lower life cycle costs [2, 3]. In addition, due to the lightness in weight of FRP materials, structures made of FRP have a lower embodied energy than equivalent structures made of steel or concrete [3, 4]. Construction times for FRP structures are usually quicker than those of steel or concrete structures, leading to further cost savings [5]. In addition, due to the automated and continuous manufacturing process known as pultrusion, the cost of manufacturing FRP sections has been significantly reduced [1, 6].

A number of studies have been conducted of FRP-steel composite systems (i.e. FRP web-flange sandwich decks supported by steel girders through shear connectors), mainly for bridge superstructure construction. These systems consist of pultruded FRP decks of built-up or cellular cross-section connected to supporting steel girders via adhesive bonding or mechanical connections [7–11]. The modular cross-sections are assembled so that their pultrusion directions (which exert the majority of the strength) lie in the direction transverse to traffic, and have been shown to provide sufficient strength and stiffness [5, 12–15]. However, some case studies involving bridge decks made from pultruded FRP sections have shown challenges in joints connecting these decks, resulting in cracking of wearing surfaces in the direction transverse to traffic, i.e. along the pultrusion direction of the FRP components [16–20]. Longitudinal cracking has also been observed in FRP components along the pultrusion direction in an all-composite pedestrian bridge [21].

Appropriate shear connections are an essential requirement for FRP-steel composite systems. To ensure sufficient load transfer between these components, development of composite action between FRP and steel components needs to be provided. FRP-steel composite beam systems with adhesive shear connections were developed and investigated by Keller and Gürtler [7, 8], where it was found that adhesive bonding provided full composite action. Mechanical connections are also used in FRP-steel composite construction, the most common of which are shear studs pre-welded onto steel beams [11, 14, 22]. Shear stud connections have been shown to provide full composite action at a low load level (such as service loads) [9], but installation of grouted shear studs is time consuming, requiring additional cut-outs in the FRP deck to facilitate the placement of grout and the use of foam blocks or cardboard to prevent the grout from leaking out during curing [23].

FRP web-flange sandwich structures are also promising for building construction [24, 25]. A notable development is the Startlink modular system, through the assembly of pultruded web-flange sandwich sections for low-cost housing [26, 27]. It seems, however, that FRP-steel composite beams have not yet been well introduced in building applications, even though their counterpart steel–concrete composite beams are used extensively in building floor construction. The fire performance of the former may be a concern. Recent work has shown that web-flange sandwich configurations with closed cross-sections can provide certain fire resistance [28], and that fire resistance time can be extended for more than one to two hours with the use of passive or active fire protection systems [29]. Experimental studies on FRP-steel composite beams for building floor applications are very limited. Successful experience from FRP-steel composite systems in bridge construction may provide impetus for the use of such systems in building construction. Further developments of modular slab assemblies with different pultrusion directions as well as the use of appropriate adhesive or novel bolted shear connections may further improve the performance of FRP-steel composite beams in building applications.

Work on modular FRP web-flange sandwich structures was performed in [30, 31] and introduced in previous chapters, where modular FRP sandwich beams were fabricated by adhesively bonding or bolting pultruded GFRP box or I-profiles between GFRP plates to form a web-flange sandwich structure. It was shown that such modular web-flange sandwich structures met sufficient strength and stiffness requirements. A further development would be to assemble them onto supporting steel beams to form FRP-steel composite systems for building floor applications. In this chapter, a modular FRP-steel system is proposed that allows for two pultrusion directions within the one slab or deck. The fibre direction of the flat panels lies perpendicular to the fibre direction of the box profiles, creating a slab or deck with bidirectional pultrusion (fibre) orientation. This feature is not seen in existing pultruded FRP modular units, where only one fibre direction (or pultrusion direction) can be achieved. With a bidirectional fibre orientation, it is expected that the structural performance of the deck can be improved in the transverse direction by enhancing the junction between the GFRP face panels and box profiles, thereby preventing longitudinal cracking along the pultrusion direction of the FRP components in the transverse beam direction. Moreover, due to the flexibility in design of this system, the geometry and number of face panels or web-profiles spanning the transverse direction can be altered in order to satisfy design requirements. Another development in this chapter is the application of novel one-sided bolts as shear connections. One-sided bolts may be more convenient than shear studs since they allow easy installation and disassembly, as no grout is needed and access to only one face of the structure is required [32, 33].

FRP-steel composite beams were fabricated and were then tested under four-point bending. The FRP sandwich slabs were fabricated by adhesively bonding FRP box profiles between two FRP flat panels. The effect of adhesive or one-sided bolts as the shear connection between the FRP sandwich slab and the supporting steel beam was investigated. Two different pultrusion configurations within the FRP sandwich slabs were also investigated (i.e. flat panel fibre directions either parallel or perpendicular to the box profiles). These configurations may be observed in a building floor framing system where FRP slabs are connected to primary or secondary steel beams. Finally, theoretical and finite element (FE) analyses were performed to evaluate the mechanical performance of the FRP-steel composite beams.

Table 4.1 Material properties of GFRP materials

Material	Fibre direction	Tensile strength/MPa	Tensile modulus/GPa
Box profile	Longitudinal	362.5 ± 21	32.2 ± 2.1
Flat panel	Longitudinal	393.1 ± 7.0	31.7 ± 0.7
	Transverse	22.0 ± 2.1	5.0 ± 0.2

4.2 Experimental Investigation

4.2.1 Materials

The modular sandwich slabs were fabricated from pultruded GFRP materials. These materials were consisted of E-glass fibres and polyester resin. The volume fraction for the flat panels was 52% polyester resin, 34% unidirectional roving and 14% chopped strand mat (CSM), as determined using ASTM D3171, Procedure G [34]. The flat panels and the box profiles had similar fibre architecture, with a roving layer embedded between two CSM layers. The tensile properties of the GFRP materials were tested in accordance with ASTM D3039 [35]. The resulting tensile properties of the GFRP materials are shown in Table 4.1, given as an average of five samples.

Universal 150UB18.0 steel beams were used, with an elastic modulus of 200 GPa and yield strength of 320 MPa. The tensile strength of the Araldite 420 epoxy adhesive used was 28.6 MPa, the tensile modulus was 1.9 GPa and the shear strength was 25 MPa, according to the results given by Fawzia [36]. M10 one-sided bolts, supplied by the Blind Bolt Company UK, were used, and had the diameter of 10 mm, nominal tensile strength of 1000 MPa, proof load stress of 900 MPa and shear capacity of 23.2 kN (over thread) and 15.9 kN (over slot), as provided by the manufacturer.

4.2.2 Specimens

Four FRP-steel composite beams were fabricated. The following naming convention was used for the composite beams. The first letter refers to the pultrusion configuration (B = bidirectional, U = unidirectional) and the second letter refers to the connection type between the FRP sandwich slab and the steel beam (A = adhesive, B = bolted). A reference steel beam was also tested.

The overall length of each specimen was 3 m, and the span length was 2730 mm. The depth of each sandwich slab was 62 mm, and they were connected to steel beams with a depth of 155 mm, giving an overall composite beam depth of 217 mm. The width of each FRP slab was 500 mm, as shown for the bidirectional beams BA in Fig. 4.1a and BB in Fig. 4.1b, and for the unidirectional beams UA and UB in Fig. 4.1c, d respectively. The bidirectional composite beams BA and BB were fabricated with twenty-nine 50 × 50 × 6 mm box profiles, each with a length of

4 Steel- Fibre Reinforced Polymer Composite Beams 79

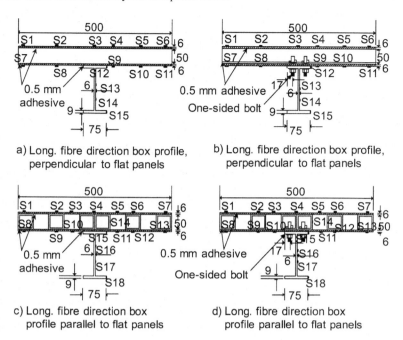

Fig. 4.1 Cross-sections of composite beams **a** BA, **b** BB, **c** UA, **d** UB with strain gauge instrumentation

500 mm, positioned in the transverse beam direction at a spacing of 55 mm (see Fig. 4.2a). This spacing was chosen because it was similar to existing FRP deck

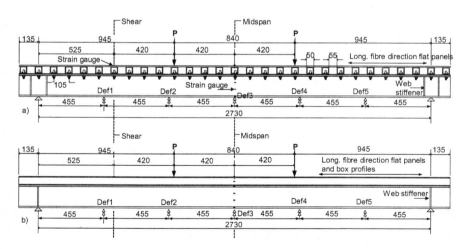

Fig. 4.2 Instrumentation and loading set up for **a** composite beam BB (same instrumentation as used for BA), **b** composite beam UA (same instrumentation as used for UB)

geometries and also because it constrained local out-of-plane buckling of the upper panel (which might arise if the distance between two adjacent box profiles was too large). The upper and lower flat panels were placed in the longitudinal beam direction so that their fibres were positioned perpendicular to the fibres of the box profiles. The unidirectional composite beams UA and UB were fabricated with six 50 × 50 × 6 mm box profiles positioned in the longitudinal beam direction, parallel to the fibre direction of the upper and lower panels, which were also placed in the longitudinal beam direction, as shown in Fig. 4.2b.

The FRP box profiles and flat panels of all the sandwich slabs were adhesively bonded together by first sandblasting all adherend surfaces and then degreasing with isopropanol, according to surface preparation procedures outlined in [37]. The thickness of the adhesive was 0.5 mm, and was controlled by placing 5 mm-long pieces of wire 0.5 mm in thickness at regular intervals along the adherend surfaces. Weights were placed to provide clamping pressure while the adhesive cured at room temperature. The sandwich slabs were then adhesively bonded to the steel beam in the composite beams BA and UA. The steel beam was sandblasted and cleaned with acetone prior to adhesion of the FRP slab.

The composite beams BB and UB used one-sided bolts, which have been shown to be a suitable connection technique in modular FRP web-flange sandwich sections [31]. The bolt holes for composite beams BB and UB were 10.5 mm in diameter, and were placed in pairs with gauge spacing of 41 mm (i.e. the distance between bolt hole centres in the transverse beam direction). The main design constraint of the bolt spacing was the geometry of the FRP sandwich slab. The bolts were placed so as to clamp the flanges of the box profiles and the lower flat panel to the steel beam. Therefore, a spacing of 105 mm was used because this corresponded to the position of the box profiles in composite beam BB, as shown in Fig. 4.2a. The same bolt configuration was therefore used for UB. A torque of 30 Nm was applied to tighten all the bolts.

Web stiffeners (8 mm in thickness) were welded to the web of the steel beam at the supports to prevent localised failure of the steel beam during the loading process. For composite beams BB and UB, two web stiffeners were welded onto each side of the supports, on either side of the bolts at the support positions, to facilitate application of the bolts. This is shown in Fig. 4.2a for composite beam BB. One web stiffener was welded onto each side of the supports for the adhesively bonded composite beams BA and UA, and for the reference steel beam, as shown in Fig. 4.2b (for UA).

4.2.3 Experimental Setup and Instrumentation

All specimens were simply supported and loaded in four-point bending with a 5000 kN Amsler/Shimadzu testing machine. The distance between the supports and the point load was 945 mm, as shown in Fig. 4.2. The load was applied as a displacement control at a rate of 3.5 mm/min. The point loads were transferred from the machine to the specimens through steel plates (20 mm in thickness with a width of 50 mm) for the composite beams BA, UA, UB, and the reference steel beam, while a width of 200 mm was used for the composite beam BB (to prevent local crushing of the box profile at the loading point).

The strain gauge instrumentation and deflection measurements along the sides of the composite beams are shown in Fig. 4.2a for the bidirectional BA and BB and in Fig. 4.2b for the unidirectional UA and UB. The deformed shapes were measured by placing five 100 mm linear potentiometers along the length of all specimens at 455 mm intervals (Def 1 to Def 5). Strain gauges were placed along the depth of the composite beam at midspan to investigate the degree of composite action, as shown in Fig. 4.2. Strain gauges were also placed along the width of the upper and lower flat panels of the FRP slabs (see Fig. 4.1) to measure the axial strain distribution and to evaluate the effective width. These were placed at the midspan region and the shear region, the latter of which was positioned 525 mm from the support (as shown in Fig. 4.2).

4.3 Experimental Results and Discussion

4.3.1 Load–Deflection Response

The load–deflection responses for the composite beams and reference steel beam are shown in Fig. 4.3. By comparing the slopes of the load–deflection curves within the linear-elastic region, it can be seen that composite beam UA (unidirectional with adhesive bonding) has the highest bending stiffness. The stiffness of composite beam UA was 130% larger than the reference steel beam. The stiffness of composite beams BA and UB were 60% larger than the reference steel beam. The composite beam with the lowest stiffness was BB, in which the slope of its load–deflection curve was 55% greater than that of the reference steel beam. Furthermore, as shown in Fig. 4.3, all

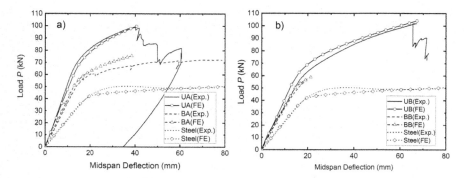

Fig. 4.3 Experimental and FE load–deflection responses for **a** adhesively bonded composite beams BA and UA, **b** bolted composite beams BB and UB under four-point bending, also including the reference steel beam

specimens showed linear load–deflection behaviour up to a certain stage, followed by non-linear responses due to yielding of the steel beam (see Sect. 4.3.2).

The differences in bending stiffness were due to both the pultrusion orientation of the slab and the type of shear connection between the FRP slab and the steel beam. In the composite beams with the same shear connection, the bending stiffness of UA was 40% greater than that of BA (see Fig. 4.3a), and that of UB was 5% greater than that of BB (see Fig. 4.3b). This is because the box profiles in a unidirectional deck (as in UA and UB, see Fig. 4.1c, d) contributed to the bending stiffness. However the contribution of box profiles in a bidirectional deck (as in BA and BB, see Fig. 4.1a, b) was negligible due to their transverse placement and the low modulus in their transverse direction. These transversely-placed box profiles also introduced partial composite action between the upper and lower flat panels for composite beams BA and BB due to their weak shear stiffness. This is discussed further in Sect. 4.3.5.

In the beams with the same FRP slab configuration, the bending stiffness of BA was only 6% greater than that of BB, whereas that of UA was 40% greater than that of UB. This was due to the different degrees of composite action offered by the shear connections. The degree of composite action was also a function of the longitudinal shear force at the FRP-steel interface (see more detailed evidence and discussion in Sect. 4.3.4).

4.3.2 Yielding of Composite Beams

Table 4.2 shows the loads at which yielding (P_y) of the steel beam commenced. This load is also considered as the load-carrying capacity of the beams since the material began to fail. Yielding occurred at midspan at the extreme tension fibre of the steel beam, where bending stresses were the highest. The yield load was specified as the load at which the stresses in this region reached 320 MPa (the yield stress of steel), as determined from analysis of strain gauge data. The composite beam BA had a yield load of 49 kN and the composite beam BB had a yield load of 46 kN. Both yield loads were greater than that of the reference steel beam (P_y at 37 kN), by 30% for BA and 20% for BB. Composite beams UA and UB had higher yield loads, at 51kN for UA and 54kN for UB. These values were greater than the yield load of the reference steel beam by 40% and 50% respectively.

No failure occurred in any FRP materials or the connections prior to yielding of the composite beams. At yield, the maximum compressive bending stresses that developed within the midspan region of the FRP slabs of BA and BB were 30 MPa and 26 MPa respectively, as shown in Fig. 4.4. Similarly, for composite beams UA and UB the maximum compressive bending stresses that arose in the upper panel of the FRP slab when the steel commenced yielding were 22 MPa and 35 MPa respectively. Composite beam UB also had tensile stresses within the FRP slab. At yield, the maximum tensile stress that developed in the lower panel of the FRP slab in beam UB was 5 MPa. The bending stresses that developed in the FRP slab were very low, therefore indicating that such FRP-steel beams can be designed so that the

4 Steel-Fibre Reinforced Polymer Composite Beams

Table 4.2 Summary of experimental, theoretical and FE results of composite beams

Parameter		Region	BA	BB	UA	UB
Experimental yield load (kN)	P_y	–	49	46	51	54
Experimental maximum load (kN)	P_{max}	–	72	79	99	102
Shear lag factor	β_m	Midspan	0.85	0.91	0.91	0.86
	β_s	Shear	0.80	0.89	0.73	0.75
Effective width FRP deck (mm)	b_e	Midspan	425	455	455	430
		Shear	400	445	365	375
Effective width box profiles (mm)	$b_{e,box}$	Midspan	–	–	273	258
		Shear	–	–	219	225
Second moment of area ($\times 10^7$ mm^4)	I_m	Midspan	1.61	1.65	2.11	2.06
	I_s	Shear	1.57	1.63	1.94	1.96
Modulus ratio	n	–	6.3	6.3	6.3	6.3
Degree of composite action	λ	–	1.0	1.0	1.0	0.4
Bending stiffness ($\times 10^{12}$ Nmm^2)	EI_m	Midspan	3.21	3.30	4.21	2.78
	EI_s	Shear	3.14	3.26	3.89	2.69
Theoretical yield load (kN)	P_y	–	51	51	56	49
FE yield load (kN)	P_y	–	51	48	53	53

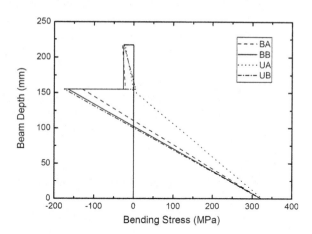

Fig. 4.4 Bending stress distribution along composite beam depth at yield P_y for all composite beams

steel yields first, providing ductile failure responses prior to failure of the brittle FRP materials.

4.3.3 Failure of Composite Beams

Composite beam BA experienced premature local crushing of a box profile situated directly beneath a loading point, as shown in Fig. 4.5a. This was due to the width of the steel plate (50 mm), which concentrated the load onto one box profile. The local crushing occurred at a load of 53 kN, after which there was a slight decrease in load to 52 kN. This was also associated with failure of the upper panel in the transverse direction along the edge of the steel plate and of the lower panel in the longitudinal direction directly underneath the loading point, as shown in Fig. 4.5a. At that point, no visible failure was observed in the adhesive bonding between the flat panels and the box profiles, or between the FRP slab and the steel beam. Further increase in load was associated with a large increase in deformation. At 72 kN, there was no observable increase in load and loading was stopped manually because of the large deformation. This load (72kN) was recorded as its maximum load (P_{max}), and was 40% higher than the maximum load of the reference steel beam (52 kN).

For composite beam BB, the data acquisition system that recorded deflection and strain measurements stopped functioning when the load reached 55 kN. As a result, the load–deflection response in Fig. 4.3b is shown only up to this load. However, the testing machine continued to measure the load beyond this point, recording a maximum load of 79 kN when the experiment was manually stopped because of excessive large bending deformations. Failure of BB was via in-plane shearing of the FRP slab, as shown in Fig. 4.5b. Due to this deformation, shear cracks formed at the web-flange junctions of the box profiles. The premature local crushing of box profiles beneath the point load (that was observed in BA) was avoided in BB by the use of a greater steel plate width of 200 mm, which distributed the load over a larger area.

The maximum failure load of each unidirectional composite beam UA and UB was reached when the beams failed via longitudinal shear of the upper and lower panels. This failure is shown in Fig. 4.6a for beam UA and Fig. 4.6b for beam UB. The maximum loads P_{max} were 99 kN and 102 kN for UA and UB respectively, as summarised in Table 4.2. In both composite beams (UA and UB), the onset of longitudinal shear failure at P_{max} led to a drop in load-carrying capacity. These

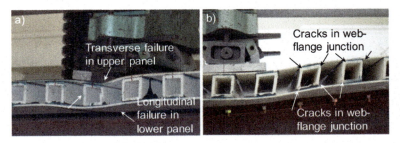

Fig. 4.5 a Local crushing of box profile directly beneath point load at 53 kN for composite beam BA, b failure in web-flange junctions of box profiles within shear region in composite beam BB

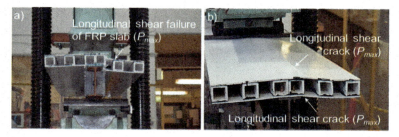

Fig. 4.6 a Longitudinal shear failure and local failure of upper panel of UA at P_{max} (99 kN), **b** longitudinal shear failure of UB at P_{max} (102 kN)

maximum loads were almost identical (3% difference), even though different shear connection techniques were used. Furthermore, these maximum loads were 90% greater than the maximum load of the reference steel beam.

4.3.4 Composite Action at FRP-Steel Interface

The axial strain distributions along the composite beam depth at yield P_y in the midspan region (see Fig. 4.2), are given in Fig. 4.7. Strains are at gauge positions S4, S9 and S12 to S15 for BA and BB (see Fig. 4.1a, b), and at gauge positions S5, S11 and S14-S18 for UA and UB (see Fig. 4.1c, d). Full composite action is identified between the steel beam and the lower panel of the FRP slab for the adhesively bonded composite beams BA and UA. This is evidenced by the linear strain distributions, shown in Fig. 4.7a, b respectively, from the bottom fibre of the steel beam (S15 in BA, or S18 in UA) to the lower panel of the FRP slab (at strain gauge S9 in BA, or S11 in UA).

The bolted composite beam BB presented full composite action in the midspan region (see Fig. 4.7c), indicated by the linear strain distribution between the steel beam and the lower flat panel. However, the bolted composite beam UB showed partial composite action, as shown by the difference in strain between the lower panel of the FRP slab and the top of the steel beam in Fig. 4.7d. This difference in composite action may be due to a different longitudinal shear force per unit length generated in the FRP-steel interface. The bidirectional composite beam BB was associated with a lower shear force at the FRP-steel interface compared to UB, therefore requiring fewer shear connectors (i.e. one-sided bolts) to achieve a certain degree of composite action.

The level of composite action at the FRP-steel interface explains the differences in bending stiffness, as discussed in Sect. 4.3.1. The full composite action provided by the adhesive shear connection in UA resulted in a greater bending stiffness than that in UB, where the bolt provided partial composite action. The bolted connection in BB also provided full composite action, and therefore there was a small difference in bending stiffness between BA and BB.

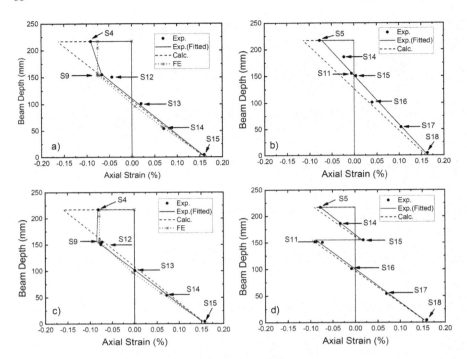

Fig. 4.7 Axial strain distributions along the depth of a cross-section at midspan region and shear region, and theoretical strain distributions at midspan, for composite beams **a** BA at P_y (49 kN), **b** UA at P_y (51 kN), **c** BB at P_y (46 kN), **d** UB at P_y (54 kN)

4.3.5 Composite Action of FRP Web-Flange Sandwich Slabs

As shown from the axial strain distributions in Fig. 4.7b, d, the strain values between the lower panels (S11, see Fig. 4.1c, d) and the upper panels (S5) is linear in the unidirectional FRP slabs UA and UB. In this case, the upper panel of the slab fully participated in bending for all load levels up to P_y. However, this was not the case for the bidirectional slabs BA and BB, as shown in Fig. 4.7a and c respectively. At the midspan, the axial strains distribution along the entire depth of the beam is not linear. The upper panels of composite beams BA and BB contribute partially to the overall bending stiffness of the structure. This is because the webs of the box profiles were placed in the transverse direction, and therefore their shear stiffness was much smaller than that of the webs of the box profiles when they were placed in the longitudinal direction in composite beams UA and UB. It has also been reported in the literature [7, 38, 39] that partial composite action such as that seen in BA and BB is common in pultruded cellular FRP slabs with rectangular or trapezoidal configurations where the webs are transverse to the span direction.

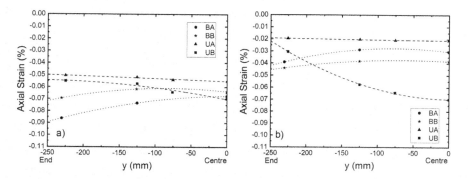

Fig. 4.8 Experimental strain distributions and fitted parabolic curves for the composite beams along the upper panel at $0.8P_y$ measured in **a** the midspan region and **b** the shear region (half the width is shown due to structural symmetry)

4.3.6 Axial Strain Along Specimen Width

Strains were measured along the beam width at strain gauge positions S1–S6 for beams BA and BB (see Fig. 4.1a, b), and at S1–S7 for beams UA and UB (see Fig. 4.1c, d). These were measured in both the midspan region and the shear region (see Fig. 4.2 for positions). The measured axial strain distribution along the upper flat panel from the centre of the beam width ($y = 0$) to the end of the beam width ($y = \pm 250$ mm) of the composite beam at midspan is shown in Fig. 4.8a, along with fitted parabolic curves. For all composite beams, the strain distributions were not uniform across the width of the beam. This was also the case for the strain distribution in the shear region, as shown in Fig. 4.8b. This non-uniform normal strain distribution is caused by the shear lag mechanism within a composite beam, resulting in an effective width of the slab contributing to the overall bending stiffness. Moreover, the extent of non-uniform distribution is more pronounced in the shear region than in the midspan region due to these shear lag effects, as evidenced in Fig. 4.8.

4.4 Analytical Evaluation

4.4.1 Effective Width and Shear Lag

A non-uniform strain distribution as a result of shear lag is seen in Fig. 4.8. Hence an understanding of the effective width is required. The width at which a constant strain is equal to the maximum value ε_{max} of the actual strain distribution ε_x is known as the effective width b_e [40]. The effective width can be calculated by employing a shear lag factor β, as shown in Eq. 4.1.

$$b_e = \beta b \tag{4.1}$$

where

$$\beta = \frac{2\int_0^{b/2} \varepsilon_x dy}{b\varepsilon_{max}} \tag{4.2}$$

The shear lag factor was derived from Eq. 4.2 by fitting a second-order polynomial to the experimental axial strains measured on the upper panel at midspan and in the shear region. Shear lag factors β were determined at loads of $0.4P_y$, $0.6P_y$, $0.8P_y$, and P_y. There was very little variation between the shear lag factors at the different load levels, so the average of these shear lag values is used in this analytical evaluation. The shear lag factors used in subsequent theoretical calculations for the midspan region (β_m) and the shear region (β_s) are given in Table 4.2. Again, it can be seen that the shear lag factors obtained in the shear region are lower than those obtained at midspan.

4.4.2 Bending Stiffness

Section transformation was used to evaluate the bending stiffness of the composite beams. Figure 4.9a, b show the transformation of the bidirectional composite beams BA and BB and of the unidirectional beams UA and UB, respectively. The box profiles can be ignored in the analysis of BA and BB as they were placed in the transverse direction and provided little bending resistance. However, the box profiles in UA and UB were placed in the longitudinal beam direction and therefore are taken into account. Using the shear lag factors determined in Sect. 4.4.1, the effective width of the FRP slab was calculated from Eq. 4.1, and then transformed into equivalent steel

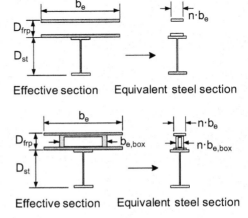

Fig. 4.9 Transformation of effective composite beam sections to equivalent steel sections for **a** bidirectional composite beams BA and BB, **b** unidirectional composite beams UA and UB

section as shown in Fig. 4.9. The effective width of each FRP beam is given in Table 4.2.

A few assumptions were made in the analytical evaluation [41]: i) plane sections remained plane; ii) the FRP slab and the steel beam had the same curvature; iii) shear deformations were ignored; and iv) the elastic case was considered (i.e. up to the yielding of the steel beam). The slip strain $\varepsilon_{\text{slip}}$ was defined as the difference between the strains at the top fibre of the steel beam ε_{st} and the bottom fibre of the FRP slab ε_{frp}. This was calculated using Eq. 4.3 [42]:

$$\varepsilon_{\text{slip}} = d \cdot \kappa \tag{4.3}$$

where d is the distance between the neutral axes of the steel beam and FRP slab and κ is the curvature [41].

The bending stiffness EI of a composite beam is given in Eq. 4.4, adapted from [41]:

$$EI = EI_F - \overline{EI} \cdot D_N^2 (1 - \lambda) \tag{4.4}$$

where λ is a factor that describes the degree of composite action of the shear connection as defined in Eq. 4.5.

$$\lambda = (1 - \frac{d}{D_N}) \tag{4.5}$$

and

$$\frac{1}{\overline{EA}} = \frac{1}{E_{\text{frp}} A_{\text{frp}}} + \frac{1}{E_{\text{st}} A_{\text{st}}} \tag{4.6}$$

EI_F is the bending stiffness of the composite beam assuming full composite action, D_N is the distance between the geometric centres of the FRP slab and the steel beam, and A_{st} and A_{frp} are the cross-sectional areas of the steel and FRP components respectively.

4.4.3 Degree of Composite Action Factor Λ

The composite action factor λ ranges from 0 for a non-composite case to 1 for a full composite case, and $0 < \lambda < 1$ for the partial composite case. This factor was determined from Eq. (4.5) for the section based on the distance d determined from the strain distributions. Full composite action existed between the steel beam and the FRP slab in composite beams BA, BB and UA, and therefore $\lambda = 1$. The composite beam UB showed partial composite action, where $\lambda = 0.4$ (as for $d = 69$ mm). These values are summarised in Table 4.2. The degree of composite action (40%) of the bolted connections for UB is comparable to a degree of composite action of 46% in Park et al. [43].

4.4.4 Evaluation of Deflections

Using the moment area method, the deflection δ of the composite beam in the region between the support and the point loads ($x \leq a$) can be calculated from Eq. 4.7, and the deflection between the point load and midspan ($a < x \leq L/2$) can be calculated from Eq. 4.8, both of which take into account the difference in bending stiffness values in the midspan (EI_m) and shear regions (EI_s):

$$\delta = \frac{Px}{6EI_s}(3a^2 - x^2) - \frac{Pa}{8EI_m}(8ax - 4Lx), x \leq a \tag{4.7}$$

$$\delta = \frac{Pa^3}{3EI_s} + \frac{Pa}{8EI_m}[(L^2 - 4a^2) - (L - 2x)^2], a \leq x \leq L/2 \tag{4.8}$$

where P is the applied load, L is the span length (2730 mm), x is the point along the beam length where deflection is to be calculated, and a is the distance between the point load and its nearest support (945 mm in this study). EI_m and EI_s are the bending stiffness of the midspan and shear regions respectively, and are calculated from Eq. 4.4. The calculated values are summarised in Table 4.2.

A comparison between the experimental and theoretical deformed shapes is shown in Fig. 4.10. The theoretical deflections for BA and BB are up to 25% greater than the experimental deflections, as shown in Fig. 4.10a. This is because the partial composition action between the upper and lower panels in the bidirectional FRP slabs of

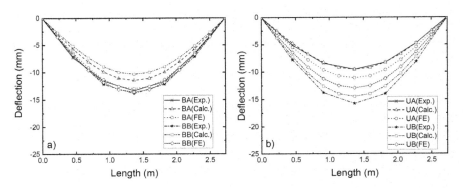

Fig. 4.10 Experimental, theoretical and FE deformed shapes at P_y of **a** bidirectional composite beams BA and BB, **b** unidirectional composite beams UA and UB

Fig. 4.11 Theoretical strain distributions for composite beams with partial composite action between the steel and FRP slab

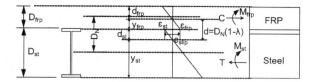

BA and BB (see Sect. 4.3.5) was not considered in the theoretical evaluation, invalidating the assumption that the FRP slab and the steel beam had the same curvature. However, the theoretical values of both UA and UB (where full composite action was provided within the unidirectional FRP slabs) showed very good agreement with the experimental results in Fig. 4.10b, with a difference of less than 10%.

4.4.5 Yielding Load P_y

Steel yielding first arises due to the maximum bending moment in the midspan region. Beams BA, BB and UA showed full composite action in the FRP-steel interface in the midspan region, therefore the yield failure load P_y could be calculated using beam theory, as given in Eq. 4.9:

$$P_y = \frac{EI_m \varepsilon_y}{y_{st} a} \quad (4.9)$$

where ε_y is the yield strain of steel and y_{st} is the distance between the extreme tension fibre of the steel beam and the neutral axis of the beam, calculated based on an equivalent steel composite beam section at midspan.

The composite beam UB showed partial composite action in the FRP-steel interface in the midspan region, and hence the yield load had to be calculated by finding y_{st} based on a partial composite section. As shown in Fig. 4.11 for the partial composite action case, the resultant compressive force C in the FRP slab and the tensile force T in the steel beam are given in Eqs. 4.10 and 4.11 respectively [41]:

$$C = \kappa \left(d_{frp} - \frac{D_{frp}}{2} \right) E_{frp} A_{frp} \quad (4.10)$$

$$T = \kappa \left(\frac{D_{st}}{2} - d_{st} \right) E_{st} A_{st} \quad (4.11)$$

where d_{frp} is the distance between the top of the composite beam and the neutral axis of the FRP slab, d_{st} is the distance between the neutral axis of the steel beam and the FRP-steel interface, D_{frp} is the depth of the FRP slab and D_{st} is the depth of the steel beam, as shown in Fig. 4.11. By equating these forces ($C = T$), and combining with

Eq. (4.3), the distance between the neutral axis of the steel beam to the FRP-steel interface d_{st} can be obtained from Eq. 4.12:

$$d_{st} = \left(\frac{1}{E_{frp}A_{frp} + E_{st}A_{st}}\right)\left[E_{st}A_{st}\frac{D_{st}}{2} - E_{frp}A_{frp}\frac{D_{frp}}{2} + E_{frp}A_{frp}D_N(1-\lambda)\right] \quad (4.12)$$

The distance d_{frp} is calculated in Eq. 4.13:

$$d_{frp} = D_{frp} + d_{st} - D_N(1-\lambda) \quad (4.13)$$

The position of the neutral axis of the steel beam y_{st} for the partial composite case can then be found in Eq. 4.14, and P_y can then be calculated using Eq. 4.9.

$$y_{st} = D_{frp} - d_{st} \quad (4.14)$$

The calculated yield loads for each composite beam are given in Table 4.2, with 51 kN for both BA and BB, 56 kN for UA, and 49 kN for UB. These values compared well with the experimental yield loads (see Table 4.2 for values), with differences between theoretical and experimental values 10% or less.

The theoretical strain distributions calculated at P_y using this method are shown in Fig. 4.7, with very good agreement with the experimental results for the unidirectional beams UA and UB. This method was also able to describe the strain distribution from the extreme tension fibre of the steel beam to the lower panel of the FRP slab of composite beams BA and BB, as shown in Fig. 4.7a, c respectively. However, this method was not able to model the axial strain in the upper panels of beams BA and BB due to the partial composite action present within the FRP slab of bidirectional beams BA and BB. At P_y in the midspan region, the experimental strain values on the upper panel for both BA and BB were 50% lower than the calculated values.

4.5 Numerical Comparison

The composite beams and reference steel beam were modelled with FE software ANSYS to predict load–deflection responses and yield loads. Furthermore, these models were used to consider the partial composite action within the FRP slab of the bidirectional beams BA and BB. Shell181 elements were adopted as a four-node element, with each node having six degrees of freedom (translations and rotations in the x, y and z directions). The material properties were taken based on the experimental results given in Table 4.1, and the element size was 10 mm. Figure 4.12a shows the FE models for composite beams BA and BB, and Fig. 4.12b shows the FE models for composite beams UA and UB. The adhesive connections were modelled by coupling interface coincident nodes between the sandwich slab and the steel beam.

Fig. 4.12 FE models for **a** bidirectional composite beams BA and BB, **b** bidirectional composite beams UA and UB, **c** Node coupling as bolted connections for a bolted beam BB

For the bolted connections, nodes were coupled at the bolt locations only, as shown in Fig. 4.12c. The load–deflection curves comparing the experimental values with the FE results are shown in Fig. 4.3, in which the FE results show good agreement with the experimental results from all composite beams and the steel reference beam.

The FE models were used to predict the stiffness of the composite beams. Comparing the slopes of the load–deflection curves within the elastic region, the bending stiffness from FE analysis differed from the experimental results by less than 10% for beams BA, BB and UA. The FE model predicted a bending stiffness 12% greater than the experimental value for composite beam UB, providing a slight overestimation due to inaccurate modelling of the partial composite action delivered by the bolted connection in this specimen. Further approaches to model the bolted connections to allow for interlayer-slip at the FRP-steel interface may be performed for more accurate bending stiffness predictions. However, these approaches are usually more computationally expensive and require additional shear-stiffness properties of the connection [44].

The axial strain distributions along the beam depth of bidirectional beams BA and BB were obtained from the FE models and are given in Fig. 4.7a, c respectively. The FE analysis was able to model the partial composite action present within the FRP slab as a result of the weak in-plane shear stiffness of the webs of the box profiles placed in the transverse beam direction (as discussed in Sect. 4.3.5). The FE models were further used to predict the yield loads. These values are summarised in Table 4.2. For all beams, the yield loads determined from FE analysis differed from the experimental yield loads by 5% or less. This indicates that the FE modelling approach could be used effectively to predict stiffness and load-carrying capacity (i.e. yield load) of steel-FRP composite beams with unidirectional and bidirectional pultrusion directions, and with adhesive or one-sided bolted shear connections.

4.6 Conclusions

This chapter presented a modular FRP-steel composite beam system in which the FRP slab was assembled by incorporating FRP box profiles between pultruded FRP plates with parallel or perpendicular arrangement of pultrusion directions. The effects on mechanical performance of i) adhesive or novel bolted connections as the shear connection and ii) the arrangement of pultrusion directions within the FRP slab were examined. These included effective width, bending stiffness, degree of composite action and yielding load. The following conclusions can be drawn from this work:

1. The unidirectional composite beams UA and UB were stiffer than the bidirectional composite beams BA and BB. This was because the major fibre (pultrusion) direction of the box profiles within BA and BB lay in the direction transverse to bending, and hence provided negligible contribution to the bending stiffness. The box profiles within UA and UB were in the longitudinal span of the beam, and therefore provided considerable contribution to bending stiffness.

2. The composite beams BA and BB had bidirectional fibre orientations, where the pultrusion direction of the box profiles was perpendicular to the flat panels. Composite beam BA experienced premature local crushing of box profiles situated directly underneath the loading point due to the small width of the loading steel plate. Composite beam BB failed via in-plane shearing of the slab as a more representative mode. The maximum loads of these two composite beams were 40–50% greater than the maximum load of the reference steel beam. In comparison, composite beams UA and UB (in which the pultrusion direction of the box profiles was parallel to the flat panels) failed via longitudinal shear of the upper and lower panels, at a position above the end of the flange of the steel beam. The maximum load of beams UA and UB was 90% greater than that of the reference steel beam. Ductile load–deflection responses were achieved in all beams, with yielding of the steel beam occurring prior to any failure in the FRP slab or the connections.

3. Adhesive bonding between the FRP slabs and steel beams provided full composite action. The one-sided bolts could provide full or partial composite action, depending on the number of bolts associated with the longitudinal shear force per unit length at the FRP-steel interface. Given the same number of bolts as shear connectors in this study, full composite action was provided in composite beam BB, which had a bidirectional pultrusion orientation associated with a lower longitudinal shear force at the FRP-steel interface. However, partial composite action was observed in composite beam UB, which had the same number and spacing of bolted connections but a larger longitudinal shear force at the FRP-steel interface caused by the unidirectional pultrusion orientation of flat panels and box profiles.

4. Partial composite action was observed between the lower and upper panels of the FRP slab for the bidirectional BA and BB composite beams, but full composite action was observed in the slab for beams UA and UB. This was because the webs of the box profiles in BA and BB were positioned in the transverse beam direction, providing weaker shear stiffness in comparison to beams UA and UB where the webs of the box profiles were placed in the longitudinal direction.

5. The analytical procedure based on beam theory showed good agreement with the experimental results for the composite beams UA and UB with unidirectional fibre orientation. The degree of composite action between the FRP slab

and steel beam could be quantified with a composite action factor λ, which took into account the distance between the neutral axes of the FRP slab and steel beam. This could be used to calculate bending stiffness and load-carrying capacity through steel yielding. However, ignoring the partial composite action of the bidirectional FRP slab of composite beams BA and BB would lead to overestimation of the bending stiffness by up to 25% in this study.

6. FE analysis was performed on the composite beams to predict load–deflection responses and yield loads. The shear connections were modelled using a simplified approach via node coupling. This approach was effective in modelling beams with both adhesive and bolted shear connections, and with unidirectional and bidirectional orientations. FE analysis further allowed modelling of the partial composite action within the bidirectional FRP slab in composite beams BA and BB. Good estimations of load–deflection responses, composite action within the FRP slab and yield loads were found for all the tested FRP-steel composite beams.

Incorporating two different pultrusion directions within the one slab structure was shown to provide sufficient mechanical performance under static loading and to prevent the formation of longitudinal cracking commonly observed in existing FRP structures. Ductile load–deflection responses were also achieved as a result of the steel beam yielding, which commenced prior to failure of the FRP slabs. In addition, the flexibility in design of this system means that the geometry and number of face panels or web-core profiles can be altered to satisfy design requirements. Finally, the one-sided bolts with retractable anchors are a suitable form of shear connection, allowing for an appropriate level of composite action while still providing easy on-site installation compared to adhesive bonding. As such, the use of such one-sided bolts in FRP-steel composite beam construction may be further explored.

References

1. Bakis CE, Bank LC, Brown VL, Cosenza E, Davalos JF, Lesko JJ et al (2002) Fiber-reinforced polymer composites for construction—State-of-the-art review. J Compos Constr 6:73–87
2. Hollaway LC (2010) A review of the present and future utilisation of FRP composites in the civil infrastructure with reference to their important in-service properties. Constr Build Mater 24:2419–2445
3. Mara V, Haghani R, Harryson P (2014) Bridge decks of fibre reinforced polymer (FRP): a sustainable solution. Constr Build Mater 50:190–199
4. Halliwell S (2000) Technical papers: FRPs—the environmental agenda. Adv Struct Eng 13:783–791
5. Foster DC, Richards D, Bogner BR (2000) Design and installation of fiber-reinforced polymer composite bridge. J Compos Constr 4(1):33–37
6. Bank LC (2007) Composites for construction: structural design with FRP materials: John Wiley & Sons, Inc.
7. Keller T, Gürtler H (2005) Composite action and adhesive bond between fiber-reinforced polymer bridge decks and main girders. J Compos Constr 9:360–368
8. Keller T, Gürtler H (2005) Quasi-static and fatigue performance of a cellular FRP bridge deck adhesively bonded to steel girders. Compos Struct 70:484–496

9. Keelor DC, Luo Y, Earls CJ, Yulismana W (2004) Service load effective compression flange width in fiber reinforced polymer deck systems acting compositely with steel stringers. J Compos Constr 8:289–297
10. Jeong J, Lee Y-H, Park K-T, Hwang Y-K (2007) Field and laboratory performance of a rectangular shaped glass fiber reinforced polymer deck. Compos Struct 81:622–628
11. Zhou A, Keller T (2005) Joining techniques for fiber reinforced polymer composite bridge deck systems. Compos Struct 69:336–345
12. Alagusundaramoorthy P, Harik IE, Choo CC (2006) Structural behavior of FRP composite bridge deck panels. J Bridg Eng 11:384–393
13. Farhey D (2005) Long-term performance monitoring of the Tech 21 all-composite bridge. J Compos Constr 9:255–262
14. Turner MK, Harries KA, Petrou MF, Rizos D (2004) In situ structural evaluation of a GFRP bridge deck system. Compos Struct 65:157–165
15. Liu Z, Majumdar PK, Cousins TE, Lesko JJ (2008) Development and evaluation of an adhesively bonded panel-to-panel joint for a FRP bridge deck system. J Compos Constr 12:224–233
16. Berman J, Brown D (2010) Field monitoring and repair of a glass fiber-reinforced polymer bridge deck. J Perform Constr Facil 24:215–222
17. Hong T, Hastak M (2006) Construction, inspection, and maintenance of FRP deck panels. J Compos Constr 10:561–572
18. Alampalli S (2006) Field performance of an FRP slab bridge. Compos Struct 72:494–502
19. Triandafilou LN, O'Connor JS (2010) Field issues associated with the use of fiber-reinforced polymer composite bridge decks and superstructures in harsh environments. Struct Eng Int 20:409–413
20. Park S-Z, Hong K-J, Lee S-W (2014) Behavior of an adhesive joint under weak-axis bending in a pultruded GFRP bridge deck. Compos B Eng 63:123–140
21. Keller T, Bai Y, Vallée T (2007) Long-term performance of a glass fiber-reinforced polymer truss bridge. J Compos Constr 11:99–108
22. Moon FL, Eckel DA, Gillespie JJW (2002) Shear stud connections for the development of composite action between steel girders and fiber-reinforced polymer bridge decks. J Struct Eng 128:762–770
23. Davalos JF, Chen A, Zou B (2011) Stiffness and strength evaluations of a shear connection system for FRP bridge decks to steel girders. J Compos Constr 15:441–450
24. Sotiropoulos SN, GangaRao HVS, Mongi ANK (1994) Theoretical and experimental evaluation of FRP components and systems. J Struct Eng 120:464–485
25. Gao Y, Chen J, Zhang Z, Fox D (2013) An advanced FRP floor panel system in buildings. Compos Struct 96:683–690
26. Hutchinson JA, Singleton MJ (2007) Startlink composite housing. Advanced Composites for Construction (ACIC 2007). University of Bath
27. Evernden M, Mottram J (2012) A case for houses to be constructed of fibre reinforced polymer components. Proceedings of the ICE - Construction Materials 3–13
28. Keller T, Tracy C, Hugi E (2006) Fire endurance of loaded and liquid-cooled GFRP slabs for construction. Compos A Appl Sci Manuf 37:1055–1067
29. Correia JR, Branco FA, Ferreira JG, Bai Y, Keller T (2010) Fire protection systems for building floors made of pultruded GFRP profiles: Part 1: experimental investigations. Compos B Eng 41:617–629
30. Satasivam S, Bai Y, Zhao X-L (2014) Adhesively bonded modular GFRP web–flange sandwich for building floor construction. Compos Struct 111:381–392
31. Satasivam S, Bai Y (2014) Mechanical performance of bolted modular GFRP composite sandwich structures using standard and blind bolts. Compos Struct 117:59–70
32. Wu C, Feng P, Bai Y (2014) Comparative study on static and fatigue performances of pultruded GFRP joints using ordinary and blind bolts. J Compos Constr 19(4):04014065
33. Luo F, Bai Y, Yang X, Lu Y (2015) Bolted sleeve joints for connecting pultruded FRP tubular components. J Compos Constr 04015024

34. ASTM. D3171 (2011) Standard test methods for constituent content of composite materials. West Conshohocken, United States
35. ASTM. D3039 (2000) Standard test method for tensile properties of polymer matrix composite materials. West Conshohocken, United States
36. Fawzia S, Zhao X-L, Al-Mahaidi R (2010) Bond-slip models for double strap joints strengthened by CFRP. Compos Struct 92:2137–2145
37. Clarke JL (1996) Structural design of polymer composites: EUROCOMP design code and handbook: E & FN Spon
38. Moses JP, Harries KA, Earls CJ, Yulismana W (2006) Evaluation of effective width and distribution factors for GFRP bridge decks supported on steel girders. J Bridg Eng 11:401–409
39. Keller T, Gürtler H (2006) In-plane compression and shear performance of FRP bridge decks acting as top chord of bridge girders. Compos Struct 72:151–162
40. Ahn I-S, Chiewanichakorn M, Chen SS, Aref AJ (2004) Effective flange width provisions for composite steel bridges. Eng Struct 26:1843–1851
41. Li G-Q, Li J-J (2007) Elastic stiffness equation of composite beam element. Advanced analysis and design of steel frames. John Wiley & Sons, Ltd, pp 35–52
42. Wu YF, Oehlers DJ, Griffith MC (2002) Partial-interaction analysis of composite beam/column members. Mech Struct Mach 30:309–332
43. Park K-T, Kim S-H, Lee Y-H, Hwang Y-K (2006) Degree of composite action verification of bolted GFRP bridge deck-to-girder connection system. Compos Struct 72:393–400
44. Wu Y-F, Griffith MC, Oehlers DJ (2004) Numerical simulation of steel plated RC columns. Comput Struct 82:359–371

Chapter 5
Composite Actions of Steel-Fibre Reinforced Polymer Composite Beams

Sindu Satasivam, Peng Feng, Yu Bai, and Colin Caprani

Abstract The shear stiffness of proposed blind bolts was experimentally investigated in this chapter through steel-FRP joints loaded in tension, for shear connections within the steel-fibre reinforced polymer (FRP) composite beam systems. The number of bolt rows (either one or two bolt rows) and the effect of the pultrusion orientation of the FRP web-flange sandwich slab on the joint stiffness and joint capacity were examined. It was found that joint capacity and failure modes were dependent on the pultrusion configuration of the FRP slab and the number of bolt rows. A unidirectional configuration consisting of an FRP slab with box-profiles (i.e. square hollow sections) parallel to flat panels exhibited shear-out failure, whereas a bidirectional orientation consisting of an FRP slab with box-profiles perpendicular to flat panels exhibited both shear-out and net-tension failure in the FRP component. The experimentally derived shear connector stiffness was then used in a proposed design formulation to predict the bending stiffness of modular steel-FRP composite beam systems, considering two kinds of partial composite actions. These were the composite action provided by the blind bolt shear connector and the composite action provided by transversely-oriented webs within the slab. Agreement was observed between the experimental results and the proposed design formulation.

Reprinted from Engineering Structures, 138, Sindu Satasivam, Peng Feng, Yu Bai, Colin Caprani, Composite actions within steel-FRP composite beam systems with novel blind bolt shear connections, 63–73, Copyright 2017, with permission from Elsevier.

S. Satasivam · Y. Bai (✉) · C. Caprani
Department of Civil Engineering, Monash University, Clayton, Australia
e-mail: yu.bai@monash.edu

P. Feng
Department of Civil Engineering, Tsinghua University, Beijing, China

5.1 Introduction

Fibre reinforced polymers (FRP) have been seen to offer lightness in weight, high strength and corrosion resistance, and as such have been used in structural applications [1]. Also FRP composites with glass fibres (GFRP) are economical due to the relatively low-cost of glass fibres and the automated manufacturing process known as pultrusion [2]. GFRP also exhibits low thermal conductivity [3], lower life cycle costs [4], and lower embodied energies [5]. Further cost savings are found during construction of FRP structures because of rapid on-site installation. Many studies have been conducted on steel-FRP composite systems in bridge superstructure construction, where FRP sandwich decks are connected to supporting steel girders with adhesive [6, 7], which provides full composite action; or mechanical shear connectors [8–10], which provide varying levels of composite action. Grouted shear studs that are pre-welded onto the steel beam or girder may be the most commonly used mechanical shear connections in steel-FRP composite systems [10, 11]. However, installation is a time consuming process, and grout can easily leak out of the connection if not controlled appropriately [12].

One-sided bolts have been used successfully in steel structures [13, 14], providing easy installation since access to only one face of the structure is required. However, the use of one-sided bolts as shear connectors in steel-FRP composite beams is limited. One-sided bolts such as BOM fasteners (from Huck International) were used in conjunction with adhesive bonding to connect FRP bridge decks to steel girders in the Wickwire Run and Laurel Hill Creek Bridges in West Virginia, USA [15, 16]. These fasteners contain a sleeve surrounding the bolt, which then collapses upon tightening of the bolt, clamping the components together [14]. The thread is then cut off with a specialised tool. Furthermore, these one-sided bolts are non-removable [17].

The one-sided bolts with retractable anchors (see Fig. 5.1) was introduced in the previous chapter and they do not require specialised tools and therefore allow for easier assembly and disassembly. Such one-sided bolts have been investigated in FRP double-lap joints [18], in bolted sleeve connections for tubular FRP frame structures [19], and as shear connections in built-up modular FRP one-way spanning sandwich structures [20]. A recent study introduced in the previous chapter used one-sided bolts with retractable anchors as shear connections in steel-FRP composite beams [21]. In this study, the FRP slabs were fabricated by adhesively bonding pultruded box profiles between FRP flat panels. Two pultrusion configurations were investigated: flat panel pultrusion directions placed parallel (unidirectional) or perpendicular (bidirectional) to the box profiles. The latter scenario was investigated in order to improve the

Fig. 5.1 M10 one-sided bolt with locking anchor

structural performance of the slab in the transverse beam direction by enhancing the junction between the face panels and the web-core box profiles to prevent premature cracking. This is a common occurrence seen in existing FRP bridge decks where the pultrusion orientation of the flat panels and webs lie transverse to the steel beam [22, 23].

Depending on the bolt spacing and the material and geometric properties of the steel-FRP composite system, one-sided bolt shear connectors may provide either full or partial composite action. The latter results from slip between the top face of the steel beam and the lower face of the FRP slab, therefore resulting in a reduced bending stiffness. The degree of composite action is affected by the shear stiffness of the one-sided bolt connection, and hence this is required for the design of such composite systems to ensure sufficient load transfer between the FRP slab and steel beam. However, the shear stiffness properties of these one-sided bolts have not yet been well characterised for applications as shear connectors in steel-FRP composite systems.

The shear stiffness is a design value that is experimentally determined by the load-slip relationship of the shear connection, i.e. the transmitted shear force causing slip at the interface between the slab and steel beam [24]. The standard push-out test is commonly used to determine the load-slip response in steel-concrete composite systems [25]. In this testing procedure, two concrete slabs are connected with shear connectors onto the flanges of a short-length steel beam. The slabs sit on a flat platen and loading is applied onto the steel beam. The longitudinal slip is then measured between each slab and the steel beam. Load-slip curves, where the load per connector is plotted against the bolt slip, are generally non-linear and hence the shear stiffness of the connector, also known as the slip modulus k, is defined as the secant stiffness at half the ultimate load [24]. Push-out tests are also used for timber-concrete composite systems [26, 27]. However, the standard test method outlined in ASTM D5652 [28] determines the shear connector stiffness from bolted timber joints loaded in tension.

In addition to the degree of composite action present between the steel beam and lower flat panel of the FRP slab, the degree of composite action between the lower and upper flat panels of the FRP slab must also be considered in design. Trapezoidal configurations, such as the DuraSpan deck [6], and built-up modular FRP slabs [21] show a partial degree of composite action between the upper and lower flat panels due to the weak in-plane shear stiffness of the transverse webs.

Several analytical methods consider the degree of composite action provided by the shear connection between the slab and supporting girder in concrete-steel systems [29–31], concrete-FRP systems [32, 33], and in steel-FRP systems [34]. In addition, design procedures have been used to consider the partial composite action within the FRP slab by Keller and Gürtler [35], but in this case adhesive bonding provided full composite action between the FRP and steel beam. Design procedures that include the degree of composite action provided by both the mechanical shear connections and the in-plane shear stiffness of the FRP slab have not yet been considered. These effects will therefore be considered in this chapter.

This chapter commences with an experimental investigation into steel-FRP single strap joints consisting of the proposed one-sided bolts and FRP sandwich slabs. The effects of pultrusion orientation and the number of bolt rows on stiffness and joint strength are investigated, and the slip modulus of the connection is obtained. This method is used instead of the standard push-out test as it is a straightforward and economical means of testing. Shear stiffness results from the experimental investigation are then verified via theoretical analysis and are used in a theoretical formulation to predict bending stiffness by considering the degree of composite action of the one-sided bolts. In addition to this, the theoretical formulation also considers the partial composite action present within the FRP slab, which arises due to the weak in-plane shear stiffness provided by the transverse webs of the box profiles.

5.2 Description of Steel-FRP Composite Beams

An investigation into steel-FRP composite beams was conducted in previous chapter, where two steel-FRP composite beams with bolted connections were tested under four-point bending. The FRP slabs were built-up sandwich sections consisting of FRP box profiles (of size 50 × 50 × 6 mm) adhesively bonded between two 6 mm-thick flat panels, which were then connected onto a 150UB18.0 steel beam with one-sided bolts as shown in Fig. 5.2. Two pultrusion orientations were investigated: a unidirectional orientation, where the fibres of both the flat panels and box profiles were placed in the longitudinal beam direction (see Fig. 5.2a), and a bidirectional orientation, where the longitudinal fibre direction of the flat panels was placed in the longitudinal beam direction, perpendicular to that of the box profiles (see Fig. 5.2b). One-sided bolts were placed in pairs at spacing of 105 mm. The composite beams were labelled

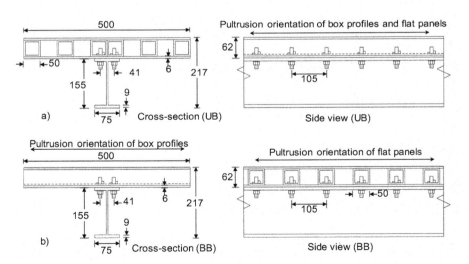

Fig. 5.2 Cross-section dimensions and bolt spacing of **a** composite beam UB and **b** composite beam BB (dimensions in mm)

as UB (unidirectional, bolted) and BB (bidirectional, bolted). All composite beams showed ductile load-deflection responses due to yielding of the steel beam, which occurred prior to material failure of the FRP slab.

In both the tested composite beams (UB and BB), the strain distribution along the depth of each cross-section indicated that partial composite action was present between the FRP slab and the steel beam. This partial composite action was provided by the one-sided bolt connection. Furthermore, the strain results also demonstrated that partial composite action was provided between the upper and lower flat panels in composite beam BB. It is necessary to quantify these two kinds of partial composite actions. Theoretical formulations also need to be developed to take into account these two kinds of partial composite actions in a composite beam system for predicting bending stiffness.

5.3 Experimental Investigation on Shear Stiffness of One-Sided Bolt Connectors

An experimental investigation is conducted on single-strap joints to determine the shear stiffness of one-sided bolt connectors. The single-strap joints consist of an FRP component to simulate the FRP slab. The FRP component is bolted on each end to two steel plates, which is then loaded in tension. Similar to a standard push-out test, the connectors in a single-strap joint configuration are loaded in shear and hence the slip of each connector at the steel-FRP interface can be measured.

5.3.1 Materials

FRP flat panels and box profiles consisting of E-glass fibres and polyester resin were the same materials as those used in the composite beam specimens in the previous chapter. The box profiles had longitudinal tensile modulus of 32 GPa and longitudinal tensile strength of 363 MPa [21]. The flat panels had longitudinal elastic modulus and tensile strength of 32 GPa and 393 MPa, and a transverse elastic modulus and tensile strength of 5 GPa and 22 MPa respectively. The volume fraction for the flat panels was 52% polyester resin and 48% fibres. The box profiles had similar fibre architecture to the flat panels, with two mat layers surrounding a unidirectional roving layer. The in-plane shear strength of the box profiles was 28 MPa, as determined in [36]. The in-plane shear properties of the GFRP flat panel were considered the same as that of the flat box profiles due to its similar fibre architecture. Steel plates were 9 mm thick with elastic modulus of 200 GPa. M10 one-sided bolts, supplied by the Blind Bolt Company, were used, the material properties of which are given in [21]. The one-sided bolts had a nominal tensile strength of 1000 MPa, a proof load stress of 900 MPa and shear capacity of 15.9 kN over slot and 23.2 kN over thread. Araldite

420 epoxy adhesive was utilised, with tensile strengths of 28.6 MPa, tensile modulus of 1.9 GPa and shear strength of 25 MPa.

5.3.2 Joint Specimens and Experimental Setup

Three joint types were fabricated and designated with the following naming convention: JB1, JU1 and JU2. The first letter (J) indicates it is a joint test, the second letter refers to the pultrusion orientation of the FRP component (B = bidirectional, U = unidirectional) and the numeral (1 or 2) refers to the number of bolt rows. Five specimens were tested for each joint type. Joint type JB1 is shown in Fig. 5.3a. It consisted of a bidirectional FRP component with two 50 × 50 × 6 mm box profiles adhesively bonded in a perpendicular direction to a 6 mm flat panel (to represent the bidirectional slab in composite beam BB tested in [21], see Sect. 5.2). The FRP component was then mechanically bolted to two steel plates 75 mm wide and 9 mm thick. Joint types JU1 and JU2 consisted of a unidirectional FRP component made of two 50 × 50 × 6 mm box profiles adhesively bonded in parallel to an FRP flat panel and mechanically bolted to steel plates to present the composite beam UB (unidirectional, bolted; see Sect. 5.2). Two bolt row configurations were investigated: one

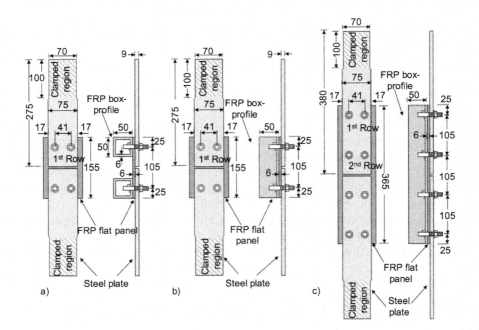

Fig. 5.3 Design and instrumentation of joint specimens. **a** JB1, **b** JU1 and **c** JU2 (dimensions in mm)

row (JU1, see Fig. 5.3b) or two rows (JU2, see Fig. 5.3c) to observe any differences in shear connector stiffness in joints with one or two rows of bolts.

The FRP flat panels were adhesively bonded to the box profiles following procedures outlined in [37]. The thickness of the adhesive was controlled by placing 0.5 mm-thick spacer wire at the adherend surfaces, in the same manner as that in the full-scale composite beams [21]. All joint specimens had bolt hole diameters of 10.5 mm and bolt spacings of 105 mm, and a tightening torque of 30 Nm was applied to each bolt, which was the same as that used for the composite beams. Similar to the one-sided bolts installed in the composite beams, the one-sided bolts in the joint specimens where placed so that the anchor was parallel to the direction of loading as this configuration exhibits the highest bolt strength [19]. The width and thickness of all the steel coupons were 75 mm and 9 mm respectively, the same as that of the steel flanges in the full-scale composite beams. The end distance for joint type JB1 was limited to 25 mm due to the geometry of the box profiles. Therefore, end distances of 25 mm were also used for joint types JU1 and JU2 in order to provide direct comparisons to JB1. The width of the steel plates at the clamping region was 70 mm to allow it to fit into the grips of the testing machine. Specimens were clamped at both ends over a 100 mm gripping length (see Fig. 5.1). In all cases, failure arose within the bolted region and not near the gripping area.

Specimens were loaded in tension using a Baldwin 500 kN testing machine. Loading was applied at a displacement control rate of 1 mm/min. Strain gauges were placed on the face of the lower steel plate and on the flanges of the FRP box profiles at positions next to the bolt holes. Aramis 3D photogrammetry was employed to measure bolt displacement, and hence a characteristic speckled pattern (i.e. black dots on a white background) was painted onto the steel plates and bolts, as shown in Fig. 5.4. Displacement within the bolt-slip region was also measured using two string potentiometers, which were clamped onto either side of the specimen as

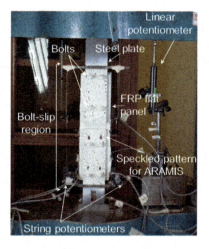

Fig. 5.4 Experimental setup of tension tests of joint specimens (shown for specimen JU2)

5.4 Results and Discussion

5.4.1 Joint Stiffness and Slip Modulus

The specimens tested in each joint type had similar load-displacement relationships and ultimate loads. Figure 5.5a therefore shows the typical load-displacement relationship within the bolted region for all joint types. For all specimens, in the initial stages of the load-slip behavior, there was a steep increase in loading with minimal displacement due to friction between the FRP component and the steel plates. The load at the end of this stage in JU2 is approximately double that of JU1 and JB1, proportional to the number of bolts in each joint type. The bolts started to come into contact with the bolt holes once the tensile load exceeded the friction force. This is shown in the decreased slope of the load-slip curve in Fig. 5.5a. Full contact was then made between the bolt and bolt hole, after which the slope of the load-slip curve increased and was linear until either FRP bearing failure in JU1 and JU2, or shear-out and net-tension failure of JB1 (see Sect. 5.4.2). Within the FRP component, there was no slip between the FRP flat panels and FRP box profiles due to the adhesive connection.

The average slip modulus is summarised in Table 5.1. These were found by obtaining the slip of each bolt using ARAMIS. The load per bolt is plotted against the bolt slip in Fig. 5.5b. The distribution of load per bolt row should be considered for JU2 specimens with two rows of bolts. According to [37], 57% of the total applied load is carried by the first bolt row (see Fig. 5.3) and the second bolt row carries 43% of the total applied load. The slip-modulus was taken as the slope of the linear portion from the load-slip curves in Fig. 5.5b. This region of the load-slip curve

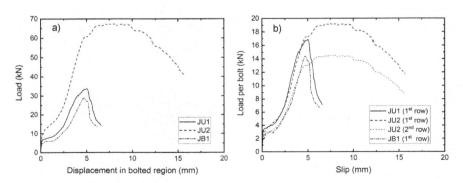

Fig. 5.5 a Typical load-slip curves within bolted region, b typical load-slip relationship per bolt for all joint types (shown for one typical specimen for each joint type)

Table 5.1 Geometry and results summary of joint tests (with standard deviation)

Joint type	JU1	JU2	JB1
Pultrusion orientation	Unidirectional	Unidirectional	Bidirectional
Number of rows	1	2	1
Spacing (mm)	105	105	105
Slip modulus 1st row bolts (kN/mm)	4.53 ± 1.02	4.96 ± 0.37	4.11 ± 0.60
Slip modulus 2nd row bolts (kN/mm)	–	3.83 ± 0.22	–
Average slip modulus (kN/mm)	4.53 ± 1.02	4.39 ± 0.29	4.11 ± 0.60
Experimental ultimate load $P_{u,e}$ (kN)	31.6 ± 1.3	67.8 ± 4.6	29.7 ± 4.1
Predicted ultimate load $P_{u,p}$ (kN)	34.9	61.1	24.7
$P_{u,p}/P_{u,e}$ (%)	110	90	83

represented the slip-modulus as it describes joint behaviour after the tensile forces exceeded the frictional forces between the FRP component and steel plates and after the bolts fully contacted the bolt holes. The slip modulus of JU1 was 4.53 kN/mm, whereas that of JU2 was 4.96 kN/mm in the first row of bolts and 3.83 kN/mm in the second row of bolts, each taken as the average from the five joint specimens. This difference in slip modulus per bolt row is due to the load level transmitted by each row, as observed by the differences in the load-slip relationships for each bolt row from Aramis experimental data. The average slip modulus across both bolt rows is 4.39 kN/mm, and this was only 3% lower than that of JU1.

The slip modulus of JB1 was 4.11 kN/mm, which was 10% lower than that of JU1 with the same number of bolt rows. This small difference is due to the pultrusion orientation of the FRP component. Other factors such as the bolt size, bolt spacing, applied torque, material properties of steel plates and friction at the steel and FRP panel interface remained the same in both JB1 and JU1. This indicated that the pultrusion orientation of the FRP slab had an effect on slip modulus.

5.4.2 Joint Capacity

Bearing and shear-out failures were observed in JU1. Bearing failure arose in joint types JU1 as localised crushing at the bolt holes in the FRP component of the joint and this was observed in the flat panel after removal of the steel plate once testing was completed. This bearing failure corresponded to a change in the load-displacement behaviour of Fig. 5.5a, whereupon non-linear load-displacement responses arose. JU1 reached an average ultimate load of 31.6 kN, which occurred due to shear-out failure of the FRP component (including the box profiles, flat panel and adhesive layer), as shown in Fig. 5.6a. This failure was associated with the sudden decline in load-carrying capacity shown in Fig. 5.5a. Average ultimate failure loads are summarised in Table 5.1.

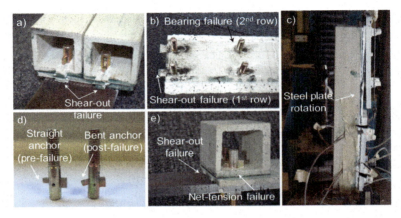

Fig. 5.6 a Shear-out failure of JU1, **b** shear-out failure and bearing failure of JU2, **c** rotation of steel plate in JU2, **d** bent anchors in bolts post failure in JU2, **e** shear-out and net-tension failure of JB1

A similar failure response was observed in JU2, which had the same unidirectional pultrusion orientation in the FRP component but with multiple bolt rows. Bearing failure arose as localised crushing of the FRP component of JU2. Progressive bearing failure continued until shear-out failure of the FRP component at an average ultimate load of 67.8 kN. Figure 5.6b shows shear-out failure in the first bolt row, which was the most heavily loaded row, and bearing failure in the second bolt row. However, unlike joint type JU1, joint type JU2 did not experience a sudden drop in load after shear-out failure, as shown in the load-displacement curves of Fig. 5.5a. Instead, the presence of multiple bolts introduced structural redundancy, causing a gradual decrease in load carrying capacity. Pull-out of the bolts was also observed in JU2. This arose due to bending moments that were created by eccentric forces at the single shear plane. The bolts were therefore loaded in tension, leading to rotation of the steel plate (as shown in Fig. 5.6c) and pull-out of the bolts. Bending of the bolt anchors was also observed in the post failure stage, as shown in comparison to a one-sided bolt with straight anchors prior to failure in Fig. 5.6d.

The joint type JB1, which had a bidirectional pultrusion orientation, showed a mixed-failure mode of shear-out in the flat panel and net-tension in the box profile, as shown in Fig. 5.6e. The shear-out failure arose within the flat panel, where the shear stresses exceeded the in-plane shear strength of the flat panel (28 MPa, see Sect. 5.3.1) in the direction of loading. However, shear-out failure did not arise in the box profile because the in-plane shear strength in the direction of loading was greater than that of the flat panel due to the presence of glass fibres in the transverse loading direction. The box-profile and adhesive layer instead failed via net-tension due to their low tensile strengths in the direction of loading. Shear-out and net-tension failures occurred simultaneously and was associated with a sudden drop in load-carrying capacity as shown in the load-displacement curve of Fig. 5.5a.

The joint capacity P_u can be predicted using Eq. 5.1 [19]:

$$P_u = \frac{N_i P_f}{\eta_i} \tag{5.1}$$

where N_i is the number of bolts within the ith row ($i = 1, 2$) and η_i is a load distribution coefficient. In joint types JU1 and JB1, $\eta_1 = 1.0$ as there is only one row of bolts. In joint type JU2, 57% of the total applied load is carried by the first bolt row ($\eta_1 = 0.57$) and 43% is carried by the second bolt row ($\eta_2 = 0.43$) for a steel-FRP joint [37]. P_f is the failure load depending on the failure mode. The failure load P_f for shear-out failure is obtained using Eq. 5.2 [2]:

$$P_f = 2 f_s t e \tag{5.2}$$

where f_s is the shear strength of the material, t is the thickness of the FRP component (flat panel, adhesive layer and flange of the box profile) and e is the end distance. The calculation includes both shear-out of the FRP material (with shear strength of 28 MPa, see Sect. 5.3.1) and shear-out of the 0.5 mm- thick adhesive layer (with shear strength of 25 MPa). From Eqs. 5.1 to 5.2, the joint capacities P_u of JU1 and JU2 were found to be 34.9 kN and 61.1 kN respectively. Compared to the experimental failure loads, the difference of the calculated values was +10%. Discrepancies between the experimental and predicted values could be due to initial bearing failures, which were not taken into account in the calculations. Predicted ultimate loads are summarised in Table 5.1.

Both shear-out and net-tension failures were observed in joint type JB1: shear-out arose within the FRP flat panel and net-tension arose in the box profile, as shown in Fig. 5.6e. The failure load P_f for net-tension failure is calculated using Eq. 5.3 [2]:

$$P_u = f_t(w - N_i d_h) t \tag{5.3}$$

where f_t is the tensile strength of the FRP in the direction perpendicular to the loading direction (22 MPa for the FRP box profile, 363 MPa for the FRP flat panel), w is the width of the FRP and d_h is the bolt holt diameter. The failure load P_f of JB1 was taken as the sum of the shear-out failure load of the flat panel (calculated using Eq. 5.2) and the net-tension failure load of the box-profile with the adhesive layer (calculated using Eq. 5.3). Ultimate joint capacity P_u was calculated as 24.7 kN, which gives a percentage underestimation of 17% compared to the experimental value.

Based on the joint tests, only joint type JU2 exhibited ductile failure. However, in the present study and based on the range of specimens tested, the failure mode of the joint type cannot be directly applied to the failure analysis of the composite beam system because of the two different structural forms. As presented in the previous chapter, composite beam BB failed via yielding of the steel beam followed by in-plane shearing of the web-flange junctions of the FRP slab; and composite beam UB failed via yielding of the steel beam followed by longitudinal shear of the upper and lower flat panels. No failure was observed in the bolted regions of the composite beams.

5.5 Theoretical Formulation of Partial Composite Actions

5.5.1 Bending Stiffness Considering Partial Composite Action at Steel/FRP Interface

The slip modulus experimentally determined from the bolt tests in Sect. 5.4.1 can be used to calculate the bending stiffness of the steel-FRP composite beam UB introduced in the previous chapter. This beam showed only one kind of partial composite action at the steel-FRP interface, which was provided by the one-sided bolt shear connectors.

Analytical equations considering shear stiffness of the connectors were derived by Nguyen et al. [32] for composite beams under four-point bending. This method only considers the partial composite action between the slab and steel girder that is provided by the one-sided bolts, and assumes that the curvatures of both the slab and steel portions are equal. However, in a composite beam system, the midspan and shear regions of the beam have different bending stiffness values due to different shear lag effects. Therefore, for the composite beam UB, the bending stiffnesses of the midspan and shear regions were calculated separately to consider different effective slab widths (summarised in Table 5.2). The curvatures ϕ_{mid} (midspan region) and ϕ_{shear} (shear region) are given in Eqs. 5.4 and 5.5 [32]:

Table 5.2 Experimental [21] and calculated stiffness properties for composite beams

Composite beam specimen	BB	UB
Pultrusion orientation	Bidirectional	Unidirectional
Effective width FRP deck b_e (mm)	455 (midspan) 445 (shear)	430 (midspan) 375 (shear)
Effective width box profiles $b_{e,box}$ (mm)	–	258 (midspan) 225 (shear)
Experimental yield load P_y (kN)	46	54
Experimental ultimate load P_u (kN)	79	102
Slip modulus k of connection k (kN/mm)	4.11	4.39
Theoretical bending stiffness ($\times 10^{12}$ Nmm^2)	–	2.59 (midspan) 2.56 (shear)
EC5 bending stiffness	2.29 (midspan) 2.29 (shear)	2.56 (midspan) 2.52 (shear)

$$\phi_{\text{mid}} = \frac{1}{E_1 I_0}[Pc - D_N(C_1 e^{\alpha x} + C_2 e^{-\alpha x} + \xi Pc)] \tag{5.4}$$

$$\phi_{\text{shear}} = \frac{1}{E_1 I_0}\left[\frac{PL}{2} - Px - D_N\left(C_3 e^{\alpha x} + C_4 e^{-\alpha x} + \frac{\xi PL}{2} - \xi Px\right)\right] \tag{5.5}$$

where P is the applied load, L is the span length, c is the length of the shear region (945 mm in the composite beams), and x is the distance measured from the centre of the span (see Fig. 5.7a). D_N is the distance between the geometric centroids of the FRP slab and steel beam and E_1 is the elastic modulus of the FRP slab, as shown in Fig. 5.7b for composite beam UB. Furthermore, I_0 is the second moment of area of the section assuming no composite action and is given in Eq. 5.6, α is given in Eq. 5.7, and factors C_1, C_2, C_3 and C_4 are given in Eqs. 5.8–5.10:

$$I_0 = I_1 + n I_2 \tag{5.6}$$

$$\alpha = \sqrt{\frac{k}{pE_1 I_0 A'}} \tag{5.7}$$

$$C_1 = C_2 = C_4 + \frac{\xi P}{2\alpha e^{-\alpha b}} \tag{5.8}$$

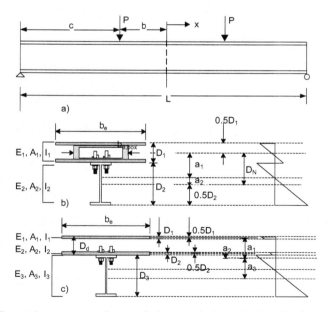

Fig. 5.7 **a** Geometric parameters of composite beam under four-point bending for theoretical analysis, **b** cross-section of unidirectional composite beam UB for theoretical analysis and the EC5 method, **c** cross-section of bidirectional composite beam BB for the EC5 method

$$C_3 = -C_4 e^{-\alpha L} \tag{5.9}$$

$$C_4 = -\frac{\xi P \cos h\alpha b}{\alpha(1 + e^{-\alpha L})} \tag{5.10}$$

where n is the modular ratio (E_2/E_1, and E_2 is the elastic modulus of the steel beam), I_1 and I_2 are the second moment of areas of the FRP slab and steel beam respectively (see Fig. 5.7b), k is the slip modulus of the bolts determined in Sect. 5.4.1, p is the longitudinal spacing of the bolts and b is half the flexural span length (420 mm in this study, see Fig. 5.7a). In addition, ξ and A' are given in Eqs. 5.11 and 5.12 respectively:

$$\xi = A' D_N \tag{5.11}$$

$$\frac{1}{A'} = D_N^2 + \frac{I_0}{A_0} \tag{5.12}$$

where $\frac{1}{A_0} = \frac{1}{A_1} + \frac{1}{NA_2}$, and A_1 and A_2 are the cross-sectional areas of the slab and steel beam respectively. The bending stiffness is directly calculated from the curvature (Eqs. 5.4 or 5.5) using Eq. 5.13:

$$\text{EI} = \frac{M}{\phi} \tag{5.13}$$

The bending stiffness obtained from Eq. 5.13 was 2.59×10^{12} Nmm² in the midspan region and 2.56×10^{12} Nmm² in the shear region. This shows good agreement with the experimental bending stiffness from the previous chapter(see Sect. 5.2 for details), as shown in Fig. 5.8 comparing the experimental and theoretical deflected shapes of composite beam UB at the experimental yield load 54 kN (P_y). The theoretical deflected shapes were obtained via double integration of the curvatures in both the bending and shear regions. The theoretical deflections show a percentage difference of 4% or less, suggesting that the slip modulus k obtained from the single-strap joints in Sect. 5.4.1 can be used to describe the bending stiffness of a steel-FRP composite system.

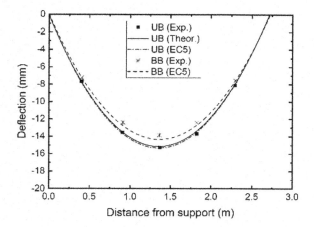

Fig. 5.8 Experimental, theoretical and EC5 deflected shapes for unidirectional composite beam UB at P_y (54 kN) and bidirectional composite beam BB at P_y (46 kN)

5.5.2 Bending Stiffness Considering Partial Composite Actions at Steel/FRP Interface and Within the FRP Slab

The analytical method presented in Sect. 5.5.1 can only consider partial composite action of the shear connector at the steel/FRP interface. However, composite beams with FRP deck as the top flange in a composite beam (such as the bidirectional modular sandwich slab [21] and the DuraSpan deck [6]) exhibit partial composite action within the slab, and hence design methods should consider this.

A design method was developed by Keller and Gürtler [35] based on Eurocode 5 (EC5) [38]. This method was applied to steel-FRP composite beams under four-point bending, with adhesive bonding as the shear connection between the FRP deck and steel girder. It took into account the full composite action provided by the adhesive connection . Only the partial composite action provided by the transverse webs within the deck is considered in the formulation of composite beams. However, this result may not be applicable in the case of bolted connections, where the stiffness of the shear connection is much less than that provided by adhesive bonding, and as such this partial composite action must be considered. Hence, this method is further developed here to include the degree of composite action of both the shear connection between the FRP and steel beam (as mechanical bolts) and that of the FRP sandwich deck. The results of this theoretical formulation can be validated by the experimental results from the bidirectional slab of composite beam BB tested in Satasivam and Bai [21], as one of the very limited experimental composite beam specimens with partial composite actions at both the steel-FRP interface and within the FRP slab.

The formulation presented in EC5 is for simply supported beams subjected to sinusoidal or parabolic bending moments [38]. However, the composite beams investigated in this study are subject to linearly varying bending moments. Hence, this method will first be verified with composite beam UB to observe any differences to the theoretical results obtained in Sect. 5.5.1. Composite beam UB is analysed using the two component cross-sections: the FRP slab ($i = 1$) and the steel beam ($i = 2$), as shown in Fig. 5.7b. Bending stiffness is calculated using Eq. 5.14 [38]:

$$\text{EI} = \sum_{i=1}^{3}(E_i I_i + \gamma_i E_i A_i \alpha_i^2) \quad (5.14)$$

where,

$$\gamma_1 = \left[1 + \frac{\pi^2 E_i A_i p}{kL^2}\right]^{-1} \quad (5.15)$$

$$\gamma_2 = 1 \quad (5.16)$$

γ_1 indicates the degree of composite action provided by the shear connectors at the steel-FRP interface, where a value of 1 for γ_1 indicates full composite action and a value of 0 indicates no composite action. The parameter a_2 is the distance measured from the neutral axis to the geometric centroid of cross-section for $i = 2$, which in this case is the steel girder (see Fig. 5.7b), as given in Eq. 5.17. Eq. 5.18 gives a_1, which is the distance between the neutral axis of cross-section $i = 2$ (steel beam) to the geometric centroid of cross-section $i = 1$ (the FRP slab, see Fig. 5.7b).

$$a_1 = \frac{\gamma_1 E_1 A_1 (D_1 + D_2)}{2\sum_{i=1}^{2} \gamma_i E_i A_i} \quad (5.17)$$

$$a_1 = D_N - a_2 \quad (5.18)$$

The bending stiffness of the composite beam UB calculated using the EC5 method was 2.56×10^{12} Nmm^2 (midspan region) and 2.52×10^{12} Nmm^2 (shear region). Deflections obtained from these bending stiffness values showed a percentage difference of less than 3% compared to the experimental deflected shape, giving a 25% reduction in difference compared to the analytical formulation presented in Sect. 5.5.1. Figure 5.8 shows the deflected shape of composite UB calculated using these bending stiffness values. The EC5 method shows a good comparison with both the experimental and theoretical deflected shapes, indicating that this method is effective in predicting bending stiffness.

The EC5 method can be extended to consider the partial composite actions provided by the bolts and the transverse webs of the FRP deck within one composite beam system. Based on the approach in [35], a normalised in-plane shear stiffness $\hat{k}$ of the bidirectional deck is equivalent to a shear stress τ_{xz} caused by a unit displacement u of 1 mm between the upper and lower flat panels, as given in Eq. 5.19.

$$\hat{k} = \tau_{xz} = G_{xz}\frac{u}{D_d} \quad (5.19)$$

where G_{xz} is the in-plane shear modulus of the slab and D_d is the depth of the slab (see Fig. 5.7c). The in-plane shear stiffness of the deck k_d is then obtained by multiplying the normalised in-plane shear stiffness with the effective width b_e, as shown in Eq. 5.20 [35]:

$$k_d = \hat{k}\, b_e \tag{5.20}$$

The in-plane shear stiffness of the bidirectional decks for composite beam BB was found via finite element (FE) analysis. As shown in Fig. 5.9a, a portion of the deck with a length of 105 mm and a width of 500 mm was modelled with Ansys software using Shell181 elements with material properties as given in Sect. 5.3.1, aligned with the centre of the box and flat plate thicknesses. The element had four nodes, each with six degrees of freedom (translations and rotations in the x, y and z directions). Adhesive bonding between the flat panels and box profiles was modelled by coupling coincident nodes at the panel and profile interfaces, as shown in Fig. 5.9b. The nodes of the bottom face of the lower flat panel were restrained in all directions. A unit displacement ($u = 1$ mm, see Fig. 5.9b) was then applied onto the upper flat panel. In order to determine the in-plane shear stress τ_{xz} given in Eq. 5.19, the force which caused the unit displacement was obtained from the FE analysis and was divided by the area of the top surface of the upper flat panel. This gave a shear stress (i.e. normalised shear stiffness) of 300 kPa mm for the FRP slab. The shear stiffness k_d for the bidirectional deck was then calculated from Eq. 5.20 as 136 MPa (midspan region) and 133 MPa (shear region). This FE approach was also used to calculate the in-plane shear stiffness of the DuraSpan deck examined in reference [35, 39], showing a good comparison (within 4% difference) to the experimental results reported.

The bending stiffness of composite beam BB is analysed using three cross-sections as shown in Fig. 5.7c, where the box profiles are not included as they provide a negligible contribution to the overall bending stiffness. To account for the partial

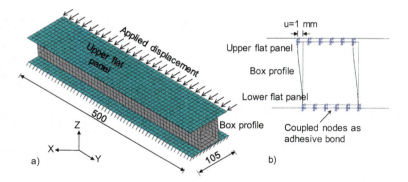

Fig. 5.9 **a** FE model to evaluate the in-plane shear stiffness of the box-profile in the transverse beam direction, **b** FE model with unit displacement ($u = 1$ mm) of upper flat panel

composite action within the deck, γ_1 from Eq. 5.15 has to be reformulated as Eq. 5.21 [35]:

$$\gamma_1 = \left[1 + \frac{\pi^2 E_1 A_1}{k_d L^2}\right]^{-1} \quad (5.21)$$

The level of partial composite action provided by the transverse webs between the upper and lower flat panels is indicated by γ_1, ranging from 0 (no composite action) to 1 (full composite action). The factor γ_2 remains the same as that given in Eq. 5.16, and γ_3 is given as Eq. 5.22 to take into account the degree of partial composite action of the bolt shear connection:

$$\gamma_3 = \left[1 + \frac{\pi^2 E_i A_i p}{k L^2}\right]^{-1} \quad (5.22)$$

For composite systems with two kinds of partial composite action, γ_3 indicates level of partial composite action at the steel-FRP interface provided by the shear connectors, ranging from 0 (no composite action) to 1 (full composite action).

The parameter a_2 from Eq. 5.23 is the distance measured from the neutral axis to the geometric centroid of cross-section i (= 2), which in the case of composite beam BB is the lower flat panel (see Fig. 5.7c). This is given in Eq. 5.23, adapted from [38] to consider the depth of the FRP slab:

$$a_2 = \frac{\gamma_1 E_1 A_1 (D_d) - \gamma_3 E_3 A_3 (\tau_f + D_3)}{2 \sum_{i=1}^{3} \gamma_i E_i A_i} \quad (5.23)$$

Furthermore, a_1 is the distance between the neutral axis of cross-section $i = 2$ (lower flat panel) to the geometric centroid of cross-section $i = 1$ (upper flat panel), and a_3 is the distance between the neutral axis of cross-section $i = 2$ (lower flat panel) to the geometric centroid of cross-section $i = 3$ (steel beam). a_1 and a_3 are given in Eqs. 5.24 and 5.25 respectively:

$$a_1 = -\left(D_d - \frac{D_1}{2} - \frac{D_2}{2} - a_2\right) \quad (5.24)$$

$$a_3 = \left(\frac{D_2}{2} + \frac{D_3}{2} + a_2\right) \quad (5.25)$$

The bending stiffness of composite beam BB was calculated using Eq. 5.14, and was found to be 2.29×10^{12} Nmm^2 in both the midspan and shear regions. The deflected shape obtained from the EC5 method at the experimental yield load of 46 kN [21] is shown in Fig. 5.8, showing a good agreement with the experimental deflected shape (with percentage difference of 5%).

In addition, it appears that the shear stiffness of the bolted connection has a greater effect than the shear-stiffness of the transverse webs of the FRP slab on the bending stiffness of the composite section. Assuming that full shear connection is provided by the transverse slabs within the FRP slab, the calculated bending stiffness would be 2.44×10^{12} Nmm^2 in the midspan region, which is only a difference of 6% compared to the scenario with partial shear connection between the upper and lower flat panels. However, if a full shear connection is provided between the FRP slab and steel beam, then the bending stiffness would be 2.95×10^{12} Nmm^2 in the midspan region, which is 30% greater than the bending stiffness of a composite beam with a partial shear connection between the slab and steel beam.

5.6 Concluding Remarks

This chapter presented an investigation into the shear stiffness of one-sided bolts used as shear connectors in steel-FRP composite systems. An experimental investigation was performed to quantify the shear stiffness, or slip modulus, of the shear connection by loading steel-FRP joints in tension. The shear stiffness of the connector was then used in a new design procedure to predict bending stiffness of the composite beam system with partial composite action at the steel-FRP interface. This proposed design procedure also considered the partial composite action between the upper and lower flat panels of the FRP slab caused by the weak in-plane shear stiffness of the transverse webs. The following conclusions can be drawn:

1. A single-strap joint configuration was used to determine the shear stiffness of novel one-sided bolts used as shear connectors in steel-FRP composite systems. There was a negligible difference in average slip modulus determined in the unidirectional joints with one (JU1) or two (JU2) bolt rows. The slip modulus of JU1 was only 3% greater than the average slip modulus of JU2. Furthermore, the effect of pultrusion orientation of the FRP slab showed a small difference in slip modulus, where the slip modulus of the unidirectional JU1 was 10% greater than that of the bidirectional JB1.
2. The joint types with unidirectional FRP slabs (JU1 and JU2) showed shear-out failure at the ultimate load. This failure arose after bearing failure was initiated. The shear-out failure in joint type JU1, with one row of bolts, resulted in a sudden drop in load-carrying capacity. However, the joint JU2 with two rows of bolts showed a gradual decrease in load-carrying capacity after shear-out as a result of structural redundancy that was introduced with the presence of multiple bolt rows. The prediction of joint capacity was reasonably accurate (with less than 10% difference) when considering the in-plane shear strength of the FRP materials and, in the case of the JU2, the load distribution per bolt row.
3. The joint type with the bidirectional FRP component, JB1, showed a mixed failure mode as simultaneous shear-out of the flat panel and net-tension of the transversely-placed box profile. Net-tension arose in the box profile due to the low transverse tensile strength in the direction of loading. The prediction of joint capacity was reasonable when taking into account these failure modes.

4. The experimental shear stiffness obtained from the single-strap joints was implemented into the theoretical formulation of composite beams considering the partial composite action resulted from the shear stiffness of the connectors for composite beams under four-point bending. This approach showed good agreement with the experimental bending stiffness of the composite beams, thus suggesting that the shear stiffness obtained from the joint tests can be used to describe the bending stiffness of steel-FRP composite systems. In addition, design procedures based on the EC5 method also showed good agreement with the experimental results for the unidirectional composite beam UB and the theoretical formulation, considering only the partial composite action provided by the one-sided bolt shear connection at the steel-FRP interface.

5. The EC5 method was further extended to incorporate the partial composite action within the FRP slab. This partial composite action was caused by the weak in-plane shear stiffness provided by the transverse webs. By quantifying the shear stiffness of the shear connectors and the shear stiffness of the FRP slab, the EC5 method can be used to describe the bending stiffness of a composite beam system with partial composite actions at the interface between the FRP slab and steel beam and between the upper and lower flat panels of the slab. The calculated results were compared with the experimental results from such a composite beam with partial composite action provided by the one-sided bolt connectors at the steel-FRP interface, and that provided by the transverse webs within the FRP slab. Improved agreement was found between the calculated and experimental results.

The proposed design formulation was shown to effectively describe the stiffness of certain FRP composite beam systems presented in this chapter. Further study may be conducted to fully develop this formulation for sufficient engineering design, especially in regards to additional types of connection methods, inelastic connection parameters and various material and geometric properties of FRP slabs.

References

1. Bakis CE, Bank LC, Brown VL, Cosenza E, Davalos JF, Lesko JJ et al (2002) Fiber-reinforced polymer composites for construction—State-of-the-art review. J Compos Constr 6:73–87
2. Bank LC (2007) Composites for construction: structural design with FRP materials. Wiley
3. Hollaway LC (2010) A review of the present and future utilisation of FRP composites in the civil infrastructure with reference to their important in-service properties. Constr Build Mater 24:2419–2445
4. Mara V, Haghani R, Harryson P (2014) Bridge decks of fibre reinforced polymer (FRP): a sustainable solution. Constr Build Mater 50:190–199
5. Halliwell S (2010) Technical papers: FRPs—The environmental agenda. Adv Struct Eng 13:783–791
6. Keller T, Gürtler H (2005) Composite action and adhesive bond between fiber-reinforced polymer bridge decks and main girders. J Compos Constr 9:360–368
7. Keller T, Gürtler H (2005) Quasi-static and fatigue performance of a cellular FRP bridge deck adhesively bonded to steel girders. Compos Struct 70:484–496

8. Keelor DC, Luo Y, Earls CJ, Yulismana W (2004) Service load effective compression flange width in fiber reinforced polymer deck systems acting compositely with steel stringers. J Compos Constr 8:289–897
9. Jeong J, Lee Y-H, Park K-T, Hwang Y-K (2007) Field and laboratory performance of a rectangular shaped glass fiber reinforced polymer deck. Compos Struct 81:622–628
10. Zhou A, Keller T (2005) Joining techniques for fiber reinforced polymer composite bridge deck systems. Compos Struct 69:336–345
11. Turner MK, Harries KA, Petrou MF, Rizos D (2004) In situ structural evaluation of a GFRP bridge deck system. Compos Struct 65:157–165
12. Davalos JF, Chen A, Zou B (2011) Stiffness and strength evaluations of a shear connection system for FRP bridge decks to steel girders. J Compos Construct 15:441–450
13. Lee J, Goldsworthy HM, Gad EF (2010) Blind bolted T-stub connections to unfilled hollow section columns in low rise structures. J Constr Steel Res 66:981–992
14. Barnett T, Tizani W, Nethercot D (2001) The practice of blind bolting connections to structural hollow sections: a review. Steel Compos Struct 1:1–16
15. Shekar V, Petro S, GangaRao H (2002) Construction of fiber-reinforced plastic modular decks for highway bridges. Transp Res Record: J Transp Res Board 1813:203–209
16. Hota GVS, Hota SRV (2002) Advances in fibre-reinforced polymer composite bridge decks. Prog Struct Mat Eng 4:161–168
17. Heslehurst R (2013) Design and analysis of structural joints with composite materials. DEStech Publications, Lancaster
18. Wu C, Feng P, Bai Y (2014) Comparative study on static and fatigue performances of pultruded GFRP joints using ordinary and blind bolts. J Compos Construct 19(4)
19. Luo F, Bai Y, Yang X, Lu Y (2015) Bolted sleeve joints for connecting pultruded FRP tubular components. J Compos Construct 20(1)
20. Satasivam S, Bai Y (2014) Mechanical performance of bolted modular GFRP composite sandwich structures using standard and blind bolts. Compos Struct 117:59–70
21. Satasivam S, Bai Y (2016) Mechanical performance of FRP-steel composite beams for building construction. Mater Struct 49(10):4113–4129
22. Hong T, Hastak M (2006) Construction, inspection, and maintenance of FRP deck panels. J Compos Constr 10:561–572
23. Berman J, Brown D (2010) Field monitoring and repair of a glass fiber-reinforced polymer bridge deck. J Perform Constr Facil 24:215–222
24. Johnson RP (2004) Composite structures of steel and concrete—Beams, slabs, columns, and frames for buildings, 3rd edn. Blackwell Publishing
25. Eurocode 4 (2004) Design of composite steel and concrete structures Part 1–1: general rules and rules for buildings. Brussels: European Committee for Standardisation (CEN). London, British
26. Carvalho EP, Carrasco EVM (2010) Influence of test specimen on experimental characterization of timber–concrete composite joints. Constr Build Mater 24:1313–1322
27. Clouston P, Bathon L, Schreyer A (2005) Shear and bending performance of a novel wood-concrete composite system. J Struct Eng 131:1404–1412
28. ASTM D5652 (1995) Standard test methods for bolted connections in wood and wood-based products. West Conshohocken, United States
29. Oehlers DJ, Nguyen NT, Ahmed M, Bradford MA (1997) Partial interaction in composite steel and concrete beams with full shear connection. J Constr Steel Res 41:235–248
30. Wang Y (1998) Deflection of steel-concrete composite beams with partial shear interaction. J Struct Eng 124:1159–1165
31. Nie J, Cai CS (2003) Steel-concrete composite beams considering shear slip effects. J Struct Eng 129:495–506
32. Nguyen H, Zatar W, Mutsuyoshi H (2015) Hybrid FRP–UHPFRC composite girders: part 2—Analytical approach. Compos Struct 125:653–671
33. Correia JR, Branco FA, Ferreira JG (2007) Flexural behaviour of GFRP-concrete hybrid beams with interconnection slip. Compos Struct 77:66–78

34. Chen A, Yossef M (2015) Analytical model for deck-on-girder composite beam system with partial composite action. J Eng Mech 142(2)
35. Keller T, Gürtler H (2006) Design of hybrid bridge girders with adhesively bonded and compositely acting FRP deck. Compos Struct 74:202–212
36. Satasivam S, Yu B, Zhao XL (2018) Mechanical performance of two-way modular FRP sandwich slabs. Thin-Walled Struct 184:904–916
37. Clarke JL (1996) Structural design of polymer composites: EUROCOMP design code and handbook. E & FN Spon
38. Eurocode 5 (2004) Design of timber structures. Brussels: European Committee for Standardisation (CEN). London, British
39. Keller T, Gürtler H (2006) In-plane compression and shear performance of FRP bridge decks acting as top chord of bridge girders. Compos Struct 72:151–162

Chapter 6
Fibre Reinforced Polymer Columns in Axial Compression

Lei Xie, Yu Bai, Yujun Qi, Colin Caprani, and Hao Wang

Abstract This chapter focuses on the mechanical performance of fibre reinforced polymer (FRP) columns with square hollow sections (SHS) in axial compression. Width-thickness ratio (b/t) is an important geometric parameter for the local buckling of plates and therefore also for such SHS columns. The effects of b/t on the failure modes and load-carrying capacities of pultruded glass fibre reinforced polymer (GFRP) SHS columns are investigated in this chapter. Two SHS with different b/t values of 10.7 and 15.9 respectively are examined under axial compression. Experimental results reveal that local buckling occurs in section B ($b/t = 15.9$) but not in section A ($b/t = 10.7$). From a theoretical analysis, a formulation of critical b/t values is established at the boundaries between the failure modes of such GFRP SHS columns under compression, considering the different boundary conditions of the SHS side plates. It is commonly understood that global buckling occurs in columns with higher non-dimensional slenderness λ. This is only true when the width-thickness ratio b/t is less than the derived critical value. Experimental results from this study and previous literature are consistent with the developed theoretical estimations of failure modes and load-carrying capacities for GFRP SHS columns, considering the effects of both non-dimensional slenderness, λ, and width-thickness ratio, b/t.

Originally published in Proceedings of the Institution of Civil Engineers—Structures and Buildings as "Xie L, Bai Y, Qi Y, Caprani C, Wang H. Effect of width-thickness ratio on capacity of pultruded square hollow polymer columns. Structures and Buildings 2018, 171(11): 842–854.", with permission from ICE Publishing.

L. Xie · Y. Bai (✉) · C. Caprani
Department of Civil Engineering, Monash University, Clayton, Australia
e-mail: yu.bai@monash.edu

Y. Qi
College of Civil Engineering, Nanjing Tech University, Nanjing, China

H. Wang
University of Southern Queensland, Toowoomba, Australia

6.1 Introduction

Glass fibre reinforced polymer (GFRP) composites hold promise for civil structural applications [1, 2], mainly due to their advantages of light weight and high corrosion resistance. The pultrusion process further allows structural GFRP members such as beams and columns to be supplied at affordable cost. In contrast to steel sections, however, the mechanical properties of pultruded GFRP differ in the longitudinal and transverse directions [1]. For example, Nunes et al. [3] reported that the tensile strength of a pultruded GFRP section was 308 MPa in the longitudinal direction but only 121 MPa in the transverse direction. The tensile and compressive strengths of pultruded GFRP in the longitudinal direction are comparable to those of steel, the interlaminar shear strength and elastic modulus in the same direction are low. The elastic modulus of common pultruded GFRP members is approximately 30 GPa, leading to a relatively low bending stiffness and more pronounced second-order effects for columns in compression [4]. Due to this low elastic modulus, global and local buckling may be critical for GFRP members in compression. Early experimental studies were carried out on I sections and square hollow sections (SHSs) by Barbero and Tomblin [5], Bank et al. [6], Tomblin and Barbero [7], Zureick and Scott [8], Hashem and Yuan [9] and Turvey and Zhang [10]. Later studies were conducted by Nguyen et al. [10, 11]. Table 6.1 provides more details of these experimental studies and the relevant theoretical and numerical work.

Table 6.1 Summary of global and local buckling studies of GFRP members from the literature

	Section type	Failure modes
Experimental studies		
Barbero and Tomblin [5]	I-sections	Global buckling
Zureick and Scott [8]	I-section and SHSs	Global buckling
Nguyen et al. [11, 12]	I- and C-sections	Lateral torsional buckling
Bank et al. [6]	I-sections	Local buckling
Tomblin and Barbero [7]	I-section and SHSs	Local buckling
Turvey and Zhang [10]	I-section	Local buckling
Hashem and Yuan [9]	SHSs and "universal" sections	Local buckling
Analytical studies		
Kollár [13] Kollár [14] Mottram [15] Qiao and Shan [16] Shan and Qiao [17] Cardoso et al. [18]	Composite plates with different boundary conditions	Local buckling
Numerical studies		
Turvey and Zhang [10] Laudiero et al. [19]	I-sections	Local buckling

Estimation of the load-carrying capacity of pultruded GFRP columns is built on the understanding of their failure modes. A prevailing method was proposed by Barbero and Tomblin [5], where interaction between local and global buckling for intermediate-length columns was considered through a load reduction factor based on an equation relating their non-dimensional slenderness, λ, and an empirical constant—the interaction factor, c. This method was further used [20, 21] with different values of c based on experimental results for more scenarios. Another design equation was proposed by Puente et al. [4] based on the approach used for steel structures in Eurocode 3. Again, the load reduction factor was related only to the geometric parameter of non-dimensional slenderness, λ. It is understood that local buckling depends highly on the width-thickness ratio (b/t), that is, the plate slenderness of the side plates of the GFRP members [22]. However, the effects of b/t on local buckling of GFRP plates were not well clarified in previous design approaches [4, 23]. Recently, the load-carrying capacity of pultruded GFRP SHS columns was investigated by Cardoso et al. [24] considering the effects of both column and plate slenderness. In that work, the load-carrying capacity was validated using experimental results from five different sections, where the failure modes of local buckling and compressive crushing were linked to the plate slenderness. A failure modes map was presented for GFRP columns of different slenderness. The boundaries between different failure modes were defined based on these experimental results.

In the present study, axial compression experiments were carried out on pultruded GFRP SHS columns with two different width-thickness ratios, b/t, and various non-dimensional slenderness, λ. The failure modes of column specimens with similar λ values but different b/t values were investigated through load-deformation and load-strain responses. The effects of b/t on global and local buckling and their interaction in GFRP SHS columns were determined. Based on orthotropic plate theory, a formulation was derived to determine the critical b/t values at the boundaries between failure modes. On that basis, a full failure modes map could be formed for GFRP SHS members with the boundaries between different failure modes defined by geometric parameters such as b/t and λ. This formulation was validated using the experimental results for GFRP columns from this study and the literature. Accordingly, the load-carrying capacity for such GFRP columns was formulated as a function of both non-dimensional slenderness λ and the width-thickness ratio b/t.

6.2 Experimental Study

6.2.1 Materials

The GFRP SHS columns used in this study had two section dimensions (columns A: 102 × 102 × 9.5 mm and columns B: 102 × 102 × 6.4 mm), as shown in Fig. 6.1. These section dimensions corresponded to b/t values of 10.7 and 15.9 respectively. The sections were made of polyester matrix reinforced with E-glass fibres, with

Fig. 6.1 Pultruded GFRP sections A and B

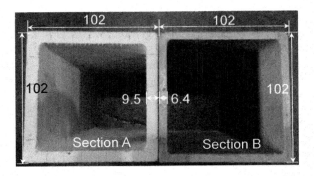

the fibre volume fraction V_f of 46.7% for A and 50.6% for B. The fibre architecture consisted of unidirectional roving in the centre and chopped strand mat on the surfaces. The material properties of these two sections were determined through coupon and short column tests and the results are given in Table 6.2. For example, the compressive strength, f_c, and elastic modulus in the longitudinal direction, E_1, were determined through short column axial compression tests with the setup shown in Fig. 6.2a. A strain gauge was applied on the specimen surface in the middle of the short column to measure the longitudinal strain. The measured axial loads during the experiments were converted to axial stress using the nominal cross-section area. A representative axial stress-strain curve for short column specimen A-100-1 is presented in Fig. 6.2b. The compressive strength was determined as the ultimate stress at failure (303 MPa), and the longitudinal elastic modulus as the slope of the linear curve, i.e. 31.7 GPa.

Table 6.2 Materials properties of GFRP sections

	Section A	Section B	Method
Longitudinal modulus, E_1 (GPa)	32.2	33.7	Short column test [24]
Transverse modulus, E_2 (GPa)	5.53	4.97	ASTM D3039
Longitudinal Poisson's ratio, υ_{12}	0.32	0.32	Short column test [24]
Transverse Poisson's ratio, υ_{21}	0.06	0.05	Classic Lamination Theory for orthotropic materials
Compressive strength, f_C (MPa)	290	274	Short column test [24]
In-plane shear modulus, G_{12} (GPa)	3.5	3.5	10° off-axis tensile test [33]
Interlaminar shear strength, f_{23} (MPa)	26.7	28.2	ASTM D2344

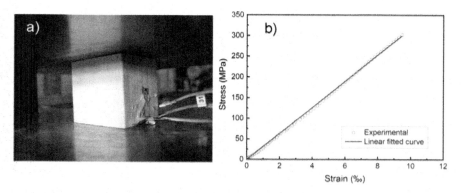

Fig. 6.2 a Short column compression test setup. b Axial stress-strain curve for specimen A-100-1

6.2.2 Specimens

In total, 20 column specimens were prepared and tested and the experimental parameters and results are summarized in Table 6.3, where specimen descriptors consist of section (A or B), length (mm), and specimen number (two of each). The short column specimens A-100 and B-100 were used to determine material compressive strengths for the sections. A-350 and B-200 were used to investigate short column performance. Intermediate and long column specimens for section A (1000, 1500, 2000 and 3000 mm) and for section B (1000 and 1500 mm) were used to examine local and global buckling and load-carrying capacities. Comparative studies between the two sections investigated the effects of *b/t* on the failure modes and load-carrying capacity. Non-dimensional slenderness λ was defined as [25]:

$$\lambda = \sqrt{\frac{P_C}{P_E}} \quad (6.1)$$

where P_C is the compressive failure load for a column (i.e. $P_C = A \cdot f_C$, where A is cross-section area and f_C is material compressive strength), and P_E is the global (Euler) buckling load.

The global (Euler) buckling load P_E was determined from the Euler equation ($P_E = \pi^2 EI/(kL)^2$), where E is the longitudinal elastic modulus, I is the moment of inertia of the section, kL is the effective column length.

6.2.3 Experimental Setup and Instrumentation

Axial compression tests were performed using an AMSLER testing machine with a capacity of 5000 kN. The typical experimental setup is shown in Fig. 6.3. The

Table 6.3 Experimental results for GFRP SHS columns under compression

Specimen ID	Length (mm)	Non-dimensional slenderness λ	Euler load P_E (kN)	Ultimate load P_U (kN)	Failure mode
A-100-1	100	0.08	/	1064	CF
A-100-2	100	0.08	/	1011	CF
A-350-1	350	0.29	/	1053	CF
A-350-2	350	0.29	/	952	CF
A-1000-1	1000	0.82	1555	1014	CF
A-1000-2	1000	0.82	1555	975	CF
A-1500-1	1500	1.24	691	511	GB/CF
A-1500-2	1500	1.24	691	544	GB/CF
A-2000-1	2000	1.62	389	353	GB
A-2000-2	2000	1.62	389	350	GB
A-3000-1	3000	2.43	173	165	GB
A-3000-2	3000	2.43	173	152	GB
B-100-1	100	0.08	/	669	CF
B-100-2	100	0.08	/	597	CF
B-200-1	200	0.16	/	636	LB
B-200-2	200	0.16	/	598	LB
B-1000-1	1000	0.80	988	462	LB
B-1000-2	1000	0.80	988	473	LB
B-1500-1	1500	1.19	439	452	GB/LB
B-1500-2	1500	1.19	439	411	GB/LB

boundary conditions on both bottom and top ends were pinned, using two rotational steel endplates attached to the rollers on the testing machine. Therefore, the effective length factor k was 1.0. The steel endplates had dimensions of 560 mm × 400 mm × 60 mm with a grid of tapped 20 mm diameter holes on their surfaces. Two angle plates were fixed on the steel endplates to constrain the specimens from lateral out-of-plane deflection. Three linear variable displacement transducers were placed at the centre and two edges of the bottom steel endplates to measure the axial displacement of the specimen end. Axial displacements at the two edges of the steel endplate were also used to determine any end rotation. A string potentiometer was used to measure the lateral displacement at mid-height of the specimens. Four strain gauges were attached to two adjacent sideplate surfaces of the section at mid-height: two vertical gauges SG1/SG3 measured the longitudinal strains, and two horizontal gauges SG2/SG4 measured the transverse strains. The compressive loading rate was 1 mm/min for all specimens.

Fig. 6.3 Experimental setup and instrumentation

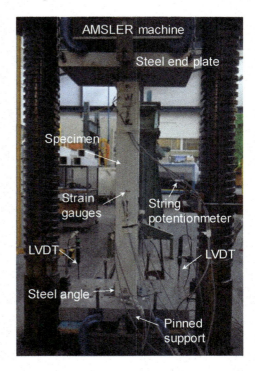

6.3 Experimental Results and Discussion

6.3.1 Failure Modes

Failures modes observed from the tested specimens are shown in Fig. 6.4 and are summarized in Table 6.3 together with the experimental ultimate loads, P_U. Three typical failure modes were categorized as follows: i) material compressive failure (CF), followed by junction separation in short column specimens; ii) local buckling (LB) of the sideplates, followed by junction separation in section B specimens; iii) global buckling (GB) of the columns, followed by junction separation in long column specimens.

Compressive failure was found in both sections A and B (A-100/350 and B-100), with sudden junction separation at the ultimate load (see Fig. 6.4a, b). This failure mode was also found in section A specimens of intermediate lengths (A-1000 and A-1500), as shown in Fig. 6.4c, d. Local buckling of the sideplates was found in section B specimens of short length (B-200) and intermediate lengths (B-1000 and B-1500), characterised by clear concave and convex deformations on adjacent plates, especially as the axial load approached the ultimate load. Figure 6.4e, f show such curved shapes after ultimate failure. Global buckling was initiated in slender section A specimens (A-2000 and A-3000) characterised by a constant axial load as the

Fig. 6.4 Typical failure modes of specimens. **a** A-100-1 ($\lambda = 0.08$, $b/t = 10.7$). **b** B-100-1 ($\lambda = 0.08$, $b/t = 15.9$). **c** A-1000-1 ($\lambda = 0.82$, $b/t = 10.7$). **d** A-1500-2 ($\lambda = 1.24$, $b/t = 10.7$). **e** B-1000-1 ($\lambda = 0.80$, $b/t = 15.9$). **f** B-1500-2 ($\lambda = 1.19$, $b/t = 15.9$). **g** A-2000-2 ($\lambda = 1.62$, $b/t = 10.7$). **h** A-3000-2 ($\lambda = 2.43$, $b/t = 10.7$)

lateral displacement increased (see Fig. 6.4g, h). It should be noted that LB did not occur during the post-buckling process of specimens A-2000 and A-3000, as shown in Fig. 6.4g, h.

The effects of non-dimensional slenderness λ on the failure modes of the GFRP columns were determined from the responses of section A specimens of different lengths. With increasing slenderness, the failure modes of GFRP members changed gradually from CF ($\lambda = 0.82, 1.24$) to GB ($\lambda = 1.62, 2.43$). The effects of width-thickness ratio b/t on the failure modes of GFRP columns were determined from comparisons between specimens of the same length but with different b/t values. For example, specimens B-1000 with the b/t value of 15.9 exhibited LB, whereas specimens A-1000 with the lower b/t value of 10.7 exhibited only CF. Specimens A-1500 showed CF in the middle during their approach to GB (see Fig. 6.4d). Interestingly, specimens B-1500 showed LB accompanying GB (see Fig. 6.4f), as discussed in greater detail with the load-strain and load-lateral displacement results.

6.3.2 Load-Strain Responses

The load-strain responses were used to validate the observed failure modes (local buckling in particular), and further to clarify the effects of *b/t* on pultruded GFRP SHS columns under compression. Figure 6.5 presents the load-strain response comparisons for specimens of the same length but with different *b/t* values, including A-1000-1 versus B-1000-1 in Fig. 6.5a, and A-1500-2 versus B-1500-1 in Fig. 6.5b. The strain results were measured from the longitudinal (SG1 and SG3) and transverse gauges (SG2 and SG4) as noted in Sect. 6.2.3. For section A, both the longitudinal and transverse strains measured from specimen A-1000-1 showed a linear increase with axial loading until the ultimate compressive failure at P_U of 1014 kN. Linear increase of compressive strains was also observed in specimens A-1500-2, with a further accelerated increase for the longitudinal strain (SG 3) on the compressive side when the ultimate load was approached. This phenomenon was the result of the additional compressive strains induced by the bending during global buckling.

The strains measured in section B specimens were linear at the beginning, as shown in Fig. 6.5a for B-1000-1 and b for B-1500-1. However, both showed gradual bifurcation in strain responses, i.e. the strains from SG1 and SG3 or SG2 and SG4 developed in opposite directions after the axial loads increased to about 450 kN for B-1000-1 and 400 kN for B-1500-1. Such bifurcated strain responses from section B specimens indicated that convex and concave deformations occurred on the adjacent GFRP plates of the SHS sections, clear evidence that local buckling occurred. The corresponding local buckling load could therefore be determined as the point at which the slope of the load-strain curve changed dramatically [6]. The difference in the load-strain responses between section A and B specimens of the same length further proved that the width-thickness ratio *b/t* played an important role in possible failure modes. Consequently, critical values of the *b/t* ratio needed to be determined to estimate failure mode for given relevant geometric parameters. This feature is addressed in Sect. 6.4.1 through a theoretical formulation.

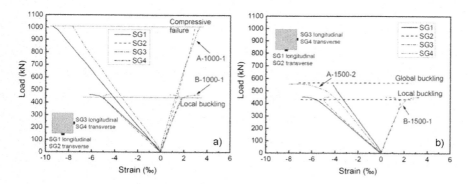

Fig. 6.5 Load-strain responses for sections with different *b/t* values. **a** A-1000-1 and B-1000-1. **b** A-1500-2 and B-1500-1

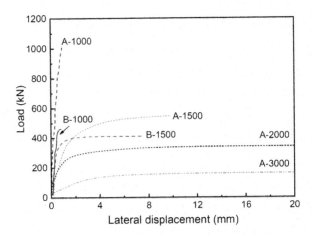

Fig. 6.6 Load-lateral displacement curves

6.3.3 Load-Lateral Displacement Curves

Figure 6.6 presents the load-lateral displacement curves for intemediate and slender specimens. Only one repeated specimen is presented here, as both results are similar. Generally, two response forms were found. Intermediate specimens A-1000 and B-1000 exhibited very limited lateral displacement (less than 1 mm) at ultimate loads, reflecting the absence of global buckling. Slender specimens A-2000 and A-3000 clearly showed global buckling, with lateral displacements of 66.3 mm and 134 mm respectively (out of range in Fig. 6.6) at the ultimate loads close to their Euler loads. In specimens B-1500, in addition to the local buckling failure mode identified from the load-strain responses, global buckling was also indicated, as shown in Fig. 6.6 where relatively large lateral displacements (7.6 mm at 411 kN) at the ultimate failure form a clear load plateau. The ultimate loads of specimens B-1500 were also close (−6.4%) to the Euler buckling load (439 kN). Lateral deformation (9.8 mm) was also measured in specimen A-1500 as its ultimate load (544 kN) approached the Euler load (691 kN, see Table 6.3).

6.3.4 Determination of Compressive Capacity P_C and Local Buckling Load P_{Local}

P_C is the compressive failure load and P_{Local} is the local buckling load of a GFRP column. As mentioned previously, short columns A-100 and B-100 were tested to determine the ultimate compressive failure loads (P_C) for the two sections. In general, both sections (A-100 and B-100) exhibited compressive failure in the form of crushing at mid-height (Fig. 6.4a, b). This occurred at average ultimate loads of 1038 kN and 633 kN respectively, giving compressive strengths of 290 MPa for section A and 274 MPa for section B. Specimens A-350 and B-200 were further

Fig. 6.7 Load-strain responses in short specimens A-350-1 and B-200-2

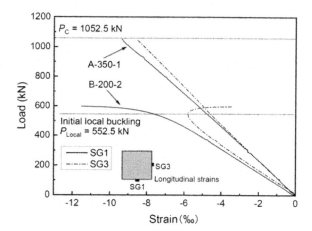

tested, and the differences in their load-strain responses were identified from Fig. 6.7. In a manner similar to specimens A-1000-1 in Fig. 6.5, specimens A-350-1 in Fig. 6.7 show a linear relationship between axial load and longitudinal strain until ultimate failure, suggesting a compressive failure. However, bifurcated strain responses were found in specimens B-200-2, indicating that local buckling began before the ultimate failure (for example at 552.5 kN in B-200-2).

It should be noted that the ultimate loads of specimens A-350 and B-200 were close to those measured from the shorter specimens, i.e. 101% of that of A-100 (1038 kN), and 94.5% of that of B-100 (633 kN). Therefore, it was important to understand the experimental ultimate loads in relation to the actual failure modes. As local buckling was observed in specimens B-200, orthotropic plate buckling theory [26] could be used to estimate the local buckling load P_{Local}. Considering the boundary condition of four simply supported edges (SSSS), the local buckling load $P_{\text{LOCAL}}^{\text{SSSS}}$ for a SHS coincided with the occurrence of local plate buckling, and so could be given as:

$$P_{\text{LOCAL}}^{\text{SSSS}} = \frac{\pi^2 A}{b^2 t} \left[D_{11} \left(\frac{mb}{a} \right)^2 + 2(D_{12} + 2D_{66}) + D_{22} \left(\frac{a}{mb} \right)^2 \right] \quad (6.2)$$

where a is the plate length (i.e. length of GFRP column), b and t are the plate width and thickness respectively, m is the number of buckled half-waves, A is the cross-section area and $D_{11}, D_{22}, D_{12}, D_{66}$ are bending stiffness coefficients of the composite plate relating to material properties. These values could be calculated according to classic orthotropic plate theory [14, 27].

In a SHS with the boundary condition of two loaded edges simply supported and two unloaded edges clamped (CCSS), the local buckling load $P_{\text{LOCAL}}^{\text{CCSS}}$ [16] is:

$$P_{\text{LOCAL}}^{\text{CCSS}} = \frac{\pi^2 A}{b^2 t}\left[D_{11}\left(\frac{mb}{a}\right)^2 + \frac{24}{\pi^2}(D_{12}+2D_{66}) + \frac{504}{\pi^4}D_{22}\left(\frac{a}{mb}\right)^2\right] \quad (6.3)$$

When m is greater than 3, the critical value of $P_{\text{LOCAL,cr}}^{\text{CCSS}}$ can be determined using the expression [17, 26]:

$$P_{\text{LOCAL,cr}}^{\text{CCSS}} = \frac{\pi^2 A}{b^2 t}\left[\frac{44.9}{\pi^2}\sqrt{D_{11}D_{22}} + \frac{24}{\pi^2}(D_{12}+2D_{66})\right] \quad (6.4)$$

The actual boundary condition at the two unloaded edges may be between simply-supported and clamped, depending on the stiffness of the junctions [17]. As a result, the boundary condition for a sideplate in a GFRP SHS may be between SSSS and CCSS, considering the boundary condition for the two loaded edges as simply supported (SS). Figure 6.8 compares the experimental local buckling loads of section B specimens with their theoretical local buckling loads under both SSSS and CCSS boundary conditions. In Fig. 6.8, the theoretical local buckling loads of section B are calculated using Eq. 6.2 and 6.3 as a function of the aspect ratio a/b. The relevant geometric parameters b and t used are nominal values of section B, and the bending stiffness coefficients $D_{11}, D_{22}, D_{12}, D_{66}$ are calculated using material properties (E_1, E_2, G_{12}) of section B listed in Table 6.2. The experimental local buckling loads of specimens B-200 and B-1000 with LB mode (determined through the load-strain responses as in Sect. 6.3.2) are also shown in Fig. 6.8. Satisfactory agreement (103–108%) with the theoretical curve based on boundary condition CCSS was found for specimen B-200. The experimental local buckling load was around 450 kN for specimens B-1000, which was between the theoretical curves of SSSS and CCSS boundary conditions, but closer to the latter curve of CCSS.

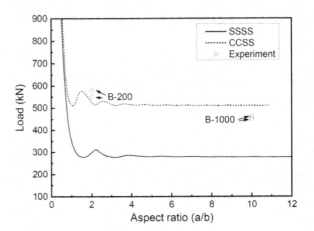

Fig. 6.8 Local buckling load comparison between theory and experiments for section B

6.4 Effects of Width-Thickness Ratio *b/t* on Failure Modes

6.4.1 Formulation of Critical b/t for Compressive Failure and Local Buckling

Although the width-thickness ratio *b/t* is recognized as an important parameter for plate local buckling, especially in steel SHS [28–30], its effect in GFRP SHS columns has not been well quantitatively investigated. The results from Sect. 6.3.2 showed that such GFRP SHSs with close slenderness values but different wall thickness (and therefore different *b/t* values) exhibited different compressive failure modes. It therefore became necessary to theoretically identify the critical *b/t* values which formed the boundaries between different failure modes.

Understandably, local buckling occurs first if the compressive capacity (P_C) of a SHS is higher than its local buckling load (P_{Local}); otherwise, compressive failure occurs first. The boundary between these failure modes is therefore $P_C = P_{Local}$. Substituting Eq. 6.2 of boundary condition SSSS, or Eq. 6.3 of CCSS, into this formulation for P_{Local} and Af_C for P_C, the formulation for the critical *b/t* values could be derived as:

$$(b/t)^{SSSS} = \sqrt{\frac{\pi^2}{fc}\left[\frac{E_1}{12(1-v_{12}v_{21})} \cdot \left(\frac{mb}{a}\right)^2 + 2\left(\frac{v_{12}E_2}{12(1-v_{12}v_{21})} + \frac{G_{12}}{6}\right) + \frac{E_2}{12(1-v_{12}v_{21})}\left(\frac{a}{mb}\right)^2\right]} \quad (6.5)$$

$$(b/t)^{CCSS} = \sqrt{\frac{\pi^2}{fc}\left[\frac{E_1}{12(1-v_{12}v_{21})} \cdot \left(\frac{mb}{a}\right)^2 + \frac{24}{\pi^2}\left(\frac{v_{12}E_2}{12(1-v_{12}v_{21})} + \frac{G_{12}}{6}\right) + \frac{504}{\pi^4} \cdot \frac{E_2}{12(1-v_{12}v_{21})}\left(\frac{a}{mb}\right)^2\right]} \quad (6.6)$$

As $(b/t)^{CCSS}$ becomes independent of *a/b* when *m* is greater than 3, its critical value could therefore be identified as:

$$(b/t)^{CCSS}_{cr} = \sqrt{\frac{\pi^2}{fc}\left[\frac{44.9}{\pi^2} \cdot \frac{\sqrt{E_1 E_2}}{12(1-v_{12}v_{21})} + \frac{24}{\pi^2}\left(\frac{v_{12}E_2}{12(1-v_{12}v_{21})} + \frac{G_{12}}{6}\right)\right]} \quad (6.7)$$

These formulas relate the critical values of the width-thickness ratio *b/t* to the material and geometric properties of the GFRP SHS columns. Therefore, such values could be determined during design, and the failure modes of either local buckling or compressive failure could be predicted accordingly.

6.4.2 Comparison with Experimental Results

Figure 6.9 compares the experimental failure modes of the GFRP SHS columns from the present work and a reference [24] to the critical width-thickness ratio $(b/t)_{cr}$

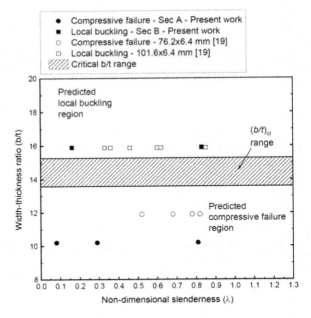

Fig. 6.9 Dependence of failure modes of local buckling and compressive failure on width-thickness ratio b/t and non-dimensional slenderness λ

according to Eq. 6.7. The boundary condition used here for the SHS sideplates was CCSS, based on the results in Sect. 6.3.4, and the material properties were as per Table 6.1 or the reference. The resulting $(b/t)_{cr}$ values were.

- 14.6 for section A and 14.96 for section B in the present work;
- 13.6 for SHS 76.2 × 6.4 mm and 15.3 for 101.6 × 6.4 mm in Cardoso et al. [24].

The range of critical width-thickness ratios (13.6–15.3) is shown in Fig. 6.9. The experimental failure modes (compressive failure and local buckling) of the SHS column specimens from the present work and Cardoso et al. [24] are marked in Fig. 6.9 with their non-dimensional slenderness λ and nominal width-thickness ratio b/t values. As estimated, local buckling occurred in the SHS specimens with nominal width-thickness ratios above this $(b/t)_{cr}$ range because the corresponding b/t values were greater than the critical values in the region; compressive failure occurred below this $(b/t)_{cr}$ range because the b/t values were less than $(b/t)_{cr}$.

6.4.3 Full Failure Modes Map

The effects of the width-thickness ratio b/t on the failure modes of local buckling and compressive failure were illustrated in Sect. 6.4.2. It was previously suggested that GFRP columns with high non-dimensional slenderness λ would fail in global buckling [24]. This might not be the case for sections with high width-thickness ratios because local buckling may occur prior to global buckling. Therefore, if the effects of

6 Fibre Reinforced Polymer Columns in Axial Compression

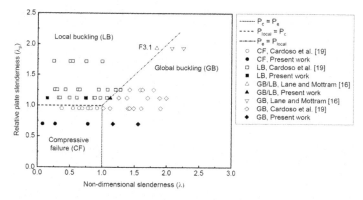

Fig. 6.10 Full failure modes map in relation to non-dimensional slenderness λ and plate relative slenderness λ_P

width-thickness ratio on local buckling and global buckling could be quantified, a full failure modes map, considering the effects of both non-dimensional slenderness λ and width-thickness ratio b/t, could be established. To take into account more section forms (such as I section and SHS) and the failure mode results from references, the plate relative slenderness, λ_P, proposed in Cardoso et al. [24] was used to represent effects from width-thickness ratios, defined as:

$$\lambda_p = \sqrt{\frac{P_C}{P_{\text{Local}}}} \qquad (6.8)$$

where P_C and P_{Local} are the compressive failure load and local buckling load of GFRP columns. Clearly, the critical width-thickness ratio b/t at the boundary between the failure modes of compressive failure and local buckling modes could be calculated, corresponding to the case when the plate relative slenderness λ_P equals 1.

The boundary between the failure modes of global buckling and compressive failure occurred when global buckling load P_E equalled compressive load P_C, i.e. $P_E = P_C$. Non-dimensional slenderness λ could be calculated using Eq. 6.1.

It should be noted that the compressive failure load P_C was used in Eq. 6.1 to define non-dimensional slenderness λ, rather than the local buckling load P_{Local} used in previous GFRP column studies [4, 20, 21]. This is because, as shown in Sect. 6.3.4, in sections with a low width-thickness ratio b/t value, compressive failure (not local buckling) would occur and the use of P_{Local} might result in a reduction factor χ greater than 1 for such cases. Clearly, when the non-dimensional slenderness λ equals 1, balanced failure of global buckling and compressive failure occurs. Similarly, balanced failure of local buckling and global buckling occurs when $P_{\text{Local}} = P_E$, and this balance could be expressed as $\lambda_p = \lambda$ based on the definitions of plate relative slenderness λ_p in Eq. 6.8 and non-dimensional slenderness λ in Eq. 6.1.

Based on the previous considerations, Fig. 6.10 presents a full failure modes map for compressive failure, local buckling and global buckling, along with their corresponding regions formed by three boundary lines: $\lambda = 1$ (for balanced failure of global buckling and compressive failure); $\lambda_P = 1$ (for balanced failure of compressive failure and local buckling), and; $\lambda_P = \lambda$ (for balanced failure of local buckling and global buckling). Accordingly, compressive failure may occur when both non-dimensional slenderness λ and plate relative slenderness λ_P are less than 1; local buckling occurs when λ_P is greater than 1 and λ_P is greater than λ; and global buckling occurs only when λ is greater than λ_P and λ is greater than 1. Failure modes of GFRP columns from experiments in the present work and references [21, 24] are included in Fig. 6.10, together with their non-dimensional slenderness λ and plate relative slenderness λ_P values. Non-dimensional slenderness λ is calculated by Eq. 6.1, where the compressive loads P_C are taken from references [21, 24]; global buckling loads P_E are calculated from the Euler equation using experimental results for elastic modulus E_1. Plate relative slenderness λ_P is calculated by Eq. 6.8, where the compressive loads P_C are from references [21, 24]; local buckling load P_{Local} is calculated from the critical flange buckling stress for wide-flange section in Mottram [15]; λ_P values for the SHS section were obtained from the reference [24].

Statistically, 94.3% of the points fall into the predicted three regions based on the proposed failure modes map presented in Fig. 6.10, and only a few points on the verge of the boundary line ($\lambda = 1$) exhibit a mode (global buckling) different from that predicted (compressive failure). Therefore, good agreement between the experimental and theoretical results is achieved. For example, circular points represent columns with observed compressive failure, and they are all well within the region formed by the two boundary lines ($\lambda_P < 1$ and $\lambda < 1$); square points represent columns with local buckling failure, and they are also within the region formed by two boundary lines ($\lambda_P > 1$ and $\lambda_P > \lambda$); finally, diamond points represent columns with global buckling failure, and they are within the region formed by two boundary lines ($\lambda > 1$ and $\lambda > \lambda_P$). It should be noted that columns with close λ_P and λ values (i.e. close to boundary line $\lambda_P = \lambda$) are marked as triangle points, and were reported with failure modes of both local and global buckling. Cardoso et al. [24] suggested that slender SHS columns with λ values greater than 1.3 would show the failure mode of global buckling only. However, as discussed above, it may be possible that slender columns having sections with large λ_P show local buckling (due to their thin sections and therefore low local buckling loads). That would also be consistent with the experimental results of Lane and Mottram [21] on wide-flange sections, where combined local and global buckling occurred for a specimen with calculated λ_P greater than λ; and only global buckling occurred for specimens with λ values greater than λ_P. It was reported by Lane and Mottram [21] that the shorter specimen F3.1 showed combined modes of local and global buckling whereas the longer specimens F3.5 and F3.8 showed only global buckling. The corresponding non-dimensional slenderness λ and plate relative slenderness λ_P of those specimens were calculated using Eq. 6.1 and 6.8, and are marked in Fig. 6.10. It was found that the hollow upward triangle point (specimen F3.1) was on the verge of the boundary line of global buckling and local buckling (i.e. line $\lambda = \lambda_P$) and within the local buckling prediction region. It is reasonable to affirm that the combined modes from specimen F3.1 and the results in Fig. 6.10 remain valid. It should be noted that the determination of λ_P and λ requires reliable representations of critical local buckling stress (i.e. P_{Local}) and compressive strength (i.e. P_c). The former requires four half wavelengths with the specimen length [15] and the latter requires the measurement of compressive strength through experiments.

6.5 Load-Carrying Capacity Considering Λ and b/t for SHS Sections

6.5.1 Formulation

The width-thickness ratio b/t had a dominant effect on the failure modes as quantified above. However, it seemed that the approaches for GFRP columns under compression in Barbero and Tomblin [7] and Puente et al. [4] were linked to non-dimensional slenderness λ without inclusion of the width-thickness ratio b/t. In this section, the load-carrying capacity of a GFRP SHS column, in the form of a reduction factor χ (defined as the ratio between the load-carrying capacity and P_C), was formulated as a function of both λ and b/t. For non-dimensional slenderness λ less than 1 and width-thickness ratio b/t value less than the critical value $(b/t)_{cr}$ (i.e. where the failure mode is compressive failure), the reduction factor χ is:

$$\chi = 1.0 \tag{6.9}$$

For non-dimensional slenderness λ less than 1 and width-thickness ratio b/t value greater than the critical value $(b/t)_{cr}$, according to Eq. 6.7, or for non-dimensional slenderness λ greater than 1 and width-thickness ratio b/t value greater than $\lambda \cdot (b/t)_{cr}$ (i.e. local buckling), the expression for the reduction factor χ can be derived from Eqs. 6.4 and 6.7, as:

$$\chi = \frac{P_{Local}}{P_C} = \frac{\frac{A \cdot f_C}{(b/t)^2} \cdot (b/t)_{cr}^2}{A \cdot f_C} = \frac{(b/t)_{cr}^2}{(b/t)^2} \tag{6.10}$$

where the critical width-thickness ratio value $(b/t)_{cr}$ can be calculated according to Eq. 6.7.

For non-dimensional slenderness λ greater than 1 and width-thickness ratio b/t value less than $\lambda \cdot (b/t)_{cr}$ (i.e. global buckling), the reduction factor χ can be given from the definition of non-dimensional slenderness λ (Eq. 6.1) as:

$$\chi = \frac{P_E}{P_C} = \frac{1}{\left(\frac{P_C}{P_E}\right)} = \frac{1}{\lambda^2} \tag{6.11}$$

Fig. 6.11 Change of reduction factor χ as a function of both λ and b/t, in comparison to existing design curves and experimental results

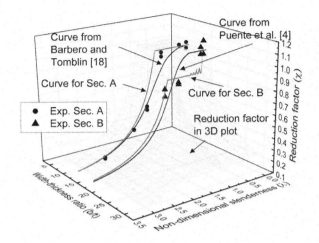

From Eqs. 6.9, 6.10 and 6.11 it can be seen that the load-carrying capacity of a GFRP column in the form of the reduction factor χ is a function of non-dimensional slenderness λ and width-thickness ratio b/t. Therefore, a three-dimensional graph can be established to illustrate the effects of both λ and b/t on the load-carrying capacity of GFRP columns, as shown in Fig. 6.11. The reduction factor χ remains at 1, forming a plateau for λ less than 1 and b/t below $(b/t)_{cr}$. This represents compressive failure only for short and thick sections. With an increase in non-dimensional slenderness λ or width-thickness ratio b/t, the reduction factor χ reduces due to other possible failure mechanisms of local or global buckling modes. Given a constant non-dimensional slenderness λ, the reduction factor χ remains unchanged for low width-thickness ratio b/t values until $(b/t)_{cr}$ is reached, i.e. the boundary between the failure modes of local buckling and compressive failure is reached. Similarly, for a constant width-thickness ratio b/t, the reduction factor χ remains constant when non-dimensional slenderness λ is less than 1. However, this constant value of the reduction factor χ might not always be 1 but could decrease depending on the width-thickness ratio b/t.

6.5.2 Comparisons

Load-carrying capacities were estimated for section A (A-350, A-1000, A-1500, A-2000 and A-3000) and section B (B-200, B-1000 and B-1500) based on their width-thickness ratios of 10.7 and 15.9 respectively as shown in Fig. 6.11. Design curves from previous references [4, 7] are also shown in Fig. 6.11, with satisfactory comparisons of reduction factors χ for specimens with non-dimensional slenderness λ of small values ($\lambda < 0.5$) and large values ($\lambda > 1.5$). However, for intermediate columns of non-dimensional slenderness $0.5 < \lambda < 1.5$ [31], the estimations based on previous design curves show interesting differences. For example, 18% underestimation was found for the reduction factor χ of specimens A-1000 with λ of 0.82, using the design curve of Barbero and Tomblin [7], while better underestimation (12%) was found for specimens B-1000 with λ of 0.80. Although only a limited number of results were used, these findings may be related to different failure modes associated with A-1000 and B-1000. The former was a thick section A corresponding to a low b/t value of 10.7 and failed in compression (see in Fig. 6.10); the latter, with a thinner section B, due to its large b/t value of 15.9, experienced local buckling failure mode. In previous studies [7, 20], mainly local and global buckling failure modes were used to develop the design curves, thereby yielding better comparisons with B-1000 than with A-1000. More experimental results for this aspect are necessary for further validation. In addition, geometrical imperfections have shown sensitive effects on the buckling modes of GFRP columns [21] and GFRP beams [32]. However, it appears in Fig. 6.11 that the buckling of SHS specimens from this study was less sensitive to geometrical imperfections than previous studies [5] using wide-flange open sections. It would be interesting in future study to determine the different sensitivities of section shapes to the effect of geometrical imperfections.

The design formulations (Eqs. 6.9–6.11) and Fig. 6.11, by taking into account both effects of non-dimensional slenderness λ and width-thickness ratio b/t, presented the estimation of load capacity, in particular for SHS, with a full range of geometric characteristics. The reduction factor χ remained 1 for specimens A-1000 with non-dimensional slenderness λ of 0.82 due to compressive failure associated with the width-thickness ratio b/t value of 10.7, corresponding to their experimental results of 0.98 (A-1000-1) and 0.94 (A-1000-2) respectively. For specimens B-1000 with λ of 0.80, where local buckling was observed because of the relatively high width-thickness ratio b/t value of 15.9, the calculated χ was 0.79, corresponding to their experimental results of 0.73 (B-1000-1) and 0.75 (B-1000-2).

6.6 Conclusions

Axial compression experiments were carried out on two pultruded GFRP SHS with different values of width-thickness ratio b/t and non-dimensional slenderness λ. The effects of width-thickness ratio b/t on the failure modes and load-carrying capacities of GFRP SHS columns under compression were therefore examined experimentally and further developed through theoretical formulations. A full map of failure modes including compressive failure, local and global buckling, as well as a three-dimensional graph of load-carrying capacity of GFRP columns under compression, were established considering the effects of both width-thickness ratio b/t and non-dimensional slenderness λ. From this work, the following conclusions can be drawn:

1. Experimental results for GFRP SHS columns with close non-dimensional slenderness λ but different width-thickness ratios b/t showed different failure modes. Under non-dimensional slenderness λ less than 1, the sections with the width-thickness ratio b/t of 15.9 would be more likely to fail in local buckling. For

example, for section B specimens with the length of 200 mm, this local buckling was indicated by bifurcation in load-strain responses and concave and convex deformed shapes of the two adjacent sideplates. Specimens with the width-thickness ratio b/t of 10.7, such as section A specimens with the length of 350 mm, could possibly fail through compressive failure. In the experimental results from specimens with the length of 1500 mm, the sections with the width-thickness ratio b/t of 15.9 showed combined modes of both local and global buckling, whereas those with the width-thickness ratio b/t of 10.7 showed only global buckling mode.

2. A theoretical formulation was developed to quantify the critical b/t values for the balanced failure of compressive failure and local buckling of GFRP SHS columns. If the actual b/t value is greater than the critical value, local buckling may occur; otherwise, compressive failure is more likely to occur. The failure modes predicted according to these critical b/t values match well with experimental results obtained from the present work and those from references. Such formulation requires only the inputs of material properties and hence can be used in the design stage for the determination of failure modes and thus capacity.

3. A theoretical failure modes map was further established to quantify the critical non-dimensional slenderness λ and width-thickness ratio b/t values for the boundary between failure modes of local and global buckling. Local buckling occurs when the plate relative slenderness λ_P is greater than 1 and also greater than non-dimensional slenderness λ; global buckling occurs when non-dimensional slenderness λ is greater than 1 and also greater than plate relative slenderness λ_P. With this formulation, a full map of failure modes can be theoretically formed, where the failure mode regions are defined by the boundary lines determined from the critical non-dimensional slenderness λ and plate relative slenderness λ_P values. A large number of previous experimental results, together with those from the present work, support the full failure modes map developed here. The map also highlights that the common understanding of global buckling associated with higher values of non-dimensional slenderness λ is accurate only when the width-thickness ratio b/t is below a critical value.

4. Once potential failure modes can be determined from the consideration of both non-dimensional slenderness λ and width-thickness ratio b/t, the corresponding load-carrying capacity can be estimated as a function of both λ and b/t. In this way, a three-dimensional design graph of reduction factor χ was established, replacing previous proposed design curves in which the reduction factor χ was only a function of non-dimensional slenderness λ. In such a graph, the reduction factor remains constant when non-dimensional slenderness λ is less than 1 and width-thickness ratio b/t is less than the critical value $(b/t)_{cr}$. However, the constant value of the reduction factor is not always 1, as it was in previous design curves, but may decrease depending on the width-thickness ratio b/t. Experimental results from a wide range of geometric characteristics of λ and b/t showed good agreement with the estimated load-carrying capacity for such GFRP SHS columns.

References

1. Bakis CE, Bank LC, Brown VL, Cosenza E, Davalos JF, Lesko JJ, Machida A, Rizkalla SH, Triantafillou TC (2002) Fiber-reinforced polymer composites for construction-state-of-the-art review. J Compos Constr 6:73–87
2. Hollaway LC (2010) A review of the present and future utilisation of FRP composites in the civil infrastructure with reference to their important in-service properties. Constr Build Mater 24:2419–2445
3. Nunes F, Correia M, Correia JR, Silvestre N, Moreira A (2013) Experimental and numerical study on the structural behavior of eccentrically loaded GFRP columns. Thin-Walled Struct 72:175–187
4. Puente I, Insausti A, Azkune M (2006) Buckling of GFRP Columns: An Empirical Approach to Design. Journal of Composites for Construction 10, 529-537
5. Barbero E, Tomblin J (1993) Euler buckling of thin-walled composite columns. Thin-Walled Struct 17:237–258
6. Bank LC, Nadipelli M, Gentry TR (1994) Local buckling and failure of pultruded fiber-reinforced plastic beams. J Eng Mater Technol Trans Asme 116:233–237
7. Tomblin J, Barbero E (1994) Local buckling experiments on FRP columns. Thin-Walled Struct 18:97–116
8. Zureick A, Scott D (1997) Short-term behavior and design of fiber-reinforced polymeric slender members under axial compression. J Compos Constr 1:140–149
9. Hashem ZA, Yuan RL (2000) Experimental and analytical investigations on short GFRP composite compression members. Compos B Eng 31:611–618
10. Turvey GJ, Zhang Y (2006) A computational and experimental analysis of the buckling, postbuckling and initial failure of pultruded GRP columns. Comput Struct 84:1527–1537
11. Nguyen TT, Chan TM, Mottram JT (2014) Lateral-torsional buckling resistance by testing for pultruded FRP beams under different loading and displacement boundary conditions. Compos B Eng 60:306–318
12. Nguyen TT, Chan TM, Mottram JT (2015) Lateral-torsional buckling design for pultruded FRP beams. Compos Struct 133:782–793
13. Kollár L (2002) Buckling of unidirectionally loaded composite plates with one free and one rotationally restrained unloaded edge. J Struct Eng 128:1202–1211
14. Kollár L (2003) Local buckling of fiber reinforced plastic composite structural members with open and closed cross sections. J Struct Eng 129:1503–1513
15. Mottram JT (2004) Determination of critical load for flange buckling in concentrically loaded pultruded columns. Compos Part B: Eng 35:35–47
16. Qiao P, Shan L (2005) Explicit local buckling analysis and design of fiber-reinforced plastic composite structural shapes. Compos Struct 70:468–483
17. Shan L, Qiao P (2008) Explicit local buckling analysis of rotationally restrained composite plates under uniaxial compression. Eng Struct 30:126–140
18. Cardoso DCT, Harries KA, Batista EDM (2014) Closed-form equations for compressive local buckling of pultruded thin-walled sections. Thin-Walled Struct 79:16–22
19. Laudiero F, Minghini F, Tullini N (2014) Buckling and postbuckling finite-element analysis of pultruded FRP profiles under pure compression. J Compos Constr 18(1)
20. Barbero E, Devivo L (1999) Beam-column design equations for wide-flange pultruded structural shapes. J Compos Constr 3:185–191
21. Lane A, Mottram JT (2002) Influence of modal coupling on the buckling of concentrically loaded pultruded fibre-reinforced plastic columns. Proc Inst Mech Eng Part L J Mater Des Appl 216:133–144
22. Pecce M, Cosenza E (2000) Local buckling curves for the design of FRP profiles. Thin-Walled Struct 37:207–222
23. Barbero E, Tomblin J (1994) A phenomenological design equation for FRP columns with interaction between local and global buckling. Thin-Walled Struct 18:117–131

24. Cardoso DCT, Harries KA, Batista EDM (2014) Compressive strength equation for GFRP square tube columns. Compos B Eng 59:1–11
25. Zahn JJ (1992) Re-examination of Ylinen and other column equations. J Struct Eng 118:2716–2728
26. Leissa AW (1987) A review of laminated composite plate buckling. Appl Mech Rev 40:575–591
27. Bank LC, Yin JS (1996) Buckling of orthotropic plates with free and rotationally restrained unloaded edges. Thin-Walled Struct 24:83–96
28. Craveiro HD, Paulo J, Rodrigues C, Laím L (2016) Buckling resistance of axially loaded cold-formed steel columns. Thin-Walled Struct 106:358–375
29. Shi G, Zhou W, Bai Y, Lin C (2014) Local buckling of 460 MPa high strength steel welded section stub columns under axial compression. J Constr Steel Res 100:60–70
30. Wang W, Kodur V, Yang X, Li G (2014) Experimental study on local buckling of axially compressed steel stub columns at elevated temperatures. Thin-Walled Struct 82:33–45
31. Bank LC (2006) Chapter 14—Pultruded axial members. Compos Constr (Wiley)
32. Bai Y, Keller T, Wu C (2013) Pre-buckling and post-buckling failure at web-flange junction of pultruded GFRP beams. Mater Struct 46:1143–1154
33. Odegard G, Kumosa M (2000) Determination of shear strength of unidirectional composite materials with the Iosipescu and 10° off-axis shear tests. Compos Sci Technol 60:2917–2943

Chapter 7
Fibre Reinforced Polymer Wall Assemblies in Axial Compression

Lei Xie, Yujun Qi, Yu Bai, Chengyu Qiu, Hao Wang, Hai Fang, and Xiao-Ling Zhao

Abstract Web-flange sandwich structures were built up by two glass fibre reinforced polymer (GFRP) panels and square hollow sections (SHS) in between through adhesive bonding or mechanical bolting. Experiments in compression were conducted in order to understand the failure modes including global and local buckling, load-bearing capacities, load–displacement curves and load-strain responses. Accordingly the effects of different connection methods and different spacing values between the SHS were clarified. Sudden debonding failure between GFRP panels and inner SHS columns was found on adhesively bonded specimens; while mechanically bolted specimens showed evident lateral deformation and progressive failure until the ultimate junction separation failure on the GFRP SHS columns. Local buckling was found on GFRP panels of specimens with a larger spacing between the two SHS. Finite element analysis and analytical modelling were performed to estimate the load–displacement curves and the critical stress for the local buckling on GFRP panels, where consistent agreements with experimental results were received.

Reprinted from Thin-Walled Structures, 145, Lei Xie, Yujun Qi, Yu Bai, Chengyu Qiu, Hao Wang, Hai Fang, Xiao-Ling Zhao, Sandwich assemblies of composites square hollow sections and thin-walled panels in compression, 106412, Copyright 2019, with permission from Elsevier

---

L. Xie · Y. Bai (✉) · C. Qiu · X.-L. Zhao
Department of Civil Engineering, Monash University, Clayton, Australia
e-mail: yu.bai@monash.edu

H. Wang
University of Sourthern Queensland, Toowoomba, Australia

Y. Qi · H. Fang
College of Civil Engineering, Nanjing Tech University, Nanjing, China

7.1 Introduction

Sandwich structures generally consist of two facesheets and a core where the facesheets can be made of glass fibre reinforced polymer (GFRP) composites and the core can be lightweight materials such as polymer foams and natural woods [1, 2]. Such sandwich structures with GFRP composites are featured with light weight and improved mechanical performance, and therefore presenting successful structural applications in many fields such as aerospace, aeronautical and marine industries [3]. In civil engineering applications, such sandwich structures [4–6] have attracted increasing interests in structural applications especially in modular buildings [7], e.g. beams [8–12], floors [13, 14], cladding [15, 16] and bridge decks [17–20].

Numerous studies were conducted for GFRP sandwich structures under bending [8–12, 18, 20], the work to investigate of their behaviour under compression for column or wall applications are relatively limited although it is understood that GFRP composites may be more critical in compression due to low elastic modulus and shear strength [21, 22]. The compressive behaviours of GFRP wall panels in sandwich structures using glass/polypropylene (PP) composites as the facesheet and polystyrene (PS) foam as the core were examined by Mousa and Uddin [23, 24]. Concentric and eccentric compression experiments were carried out on such sandwich structures for application of wall panels; and debonding between the facesheets and core was found as the main failure mode for such sandwich wall panels under compression. Furthermore, experimental results including the global buckling loads, lateral deflections and wrinkling stress were analysed and analytical formulations and finite element (FE) analysis were well validated accordingly. Slender sandwich panels with different rib configurations using GFRP laminates as facesheets and polyurethane (PU) foam as the core were developed by Mathieson and Fam [25] and experimental and numerical work was conducted to investigate their performance under axial compression. The effects of slenderness ratio on the failure modes and load-bearing capacities of such sandwich wall panels were analyzed and clarified in [26]. Another sandwich wall panels using hand lay-up GFRP as facesheets and PU foam as the core were developed by Abdolpour et al. [27]. Experiments were performed to study the mechanical performance of two scenarios, i.e. single sandwich unit and jointed sandwich units. It is found that the global buckling and GFRP local wrinkling are the dominant failure modes and further the axial load capacities of single and joint wall panels can be evaluated using proposed analytical formulas. The effects of foam density and slenderness ratios were examined by CoDyre and Fam [28] to understand the compressive strength of sandwich panels made of GFRP facesheets and polyisocyanurate foam core. Other studies on columns made of GFRP fabrics (as facesheets) and different cores including polyvinyl chloride (PVC) foam [29] or PU foam [30] are also conducted for structural applications.

One of the typical connection methods to build a sandwich structure using composites is adhesive bonding, which bonds the two facesheets and the core together [31–33]. Alternatively, mechanical bolting [34] has been used to form various GFRP structures including sandwich assemblies as reported in [35–39]. In comparing to adhesive bonding, mechanical bolting has characteristics of easy assembly and disassembly [40] and quasi-ductile behaviour [41]; while may be associated with higher stress concentration [42–45] at the bolted region. GFRP web-flange sandwich structures were developed using standard and one-sided bolts [46] and experimentally examined four point bending. It was shown that such GFRP web-flange sandwich structures [14, 46], also known as multi-cellular structures [47], have improved flexural stiffness over individual GFRP sections and been used for bridge deck and building floor applications [48, 49]. However, studies on such web-flange sandwich structures formed with adhesive bonding or mechanical bolting under compression are still needed especially for their potential column or wall applications.

This chapter presents an experimental investigation on the axial compressive performance of GFRP web-flange sandwich assemblies with bonded or bolted connections. The specimens were fabricated by two pultruded GFRP flat panels as face sheets sandwiching square hollow sections (SHS) using adhesive bonding or mechanical bolting. Experimental results on the failure modes, load-bearing capacities, load–displacement curves and load-strain responses were obtained and compared with numerical and theoretical modelling. The effects of connection types (adhesive bonding or mechanical bolting) and spacing between the core SHS on the overall load-bearing capacities and failure modes are discussed. Finite element (FE) analysis was also carried out to validate the experimental results in terms of the load–displacement curves with consideration of second order effects.

7.2 Experimental Investigation

7.2.1 Materials

Pultruded GFRP flat panels and SHS are used in this study to fabricate sandwich specimens. The nominal length of the used GFRP flat panels and SHS columns are 3000 mm to present the full scale for one-story column or wall applications. The nominal thickness of GFRP flat panels is 8 mm and the section dimension of GFRP SHS is 102 × 102 × 9.5 mm. Pultruded GFRP materials have unidirectional fibres in the longitudinal direction and are regarded as an orthotropic material. The material properties of used pultruded GFRP flat panels and SHS columns in the longitudinal and transverse directions were determined by coupon tests. Tensile properties of GFRP flat panels were determined as the average of five coupon specimens in accordance with ASTM D3039. The material properties of GFRP SHS in the longitudinal direction were determined through 100 mm height short column compression tests [50]. All the resulting properties are summarized in Table 7.1, where the elastic modulus of GFRP panels in the longitudinal direction (fibre orientation) and transverse direction is 30.5 GPa and 9.5 GPa, respectively; and the elastic modulus of GFRP SHS in the longitudinal direction is 28.1 GPa.

Table 7.1 Material properties of pultruded GFRP materials

		Direction	Elastic modulus (GPa)	Ultimate strength (MPa)
	Face sheet	Longitudinal	30.5	387.9
		Transverse	9.5	59.3
	SHS	Longitudinal	28.1	220.7

7.2.2 Specimens

Four GFRP sandwich specimens were prepared and experimentally studied. Those specimens were assembled by connecting two pultruded GFRP panels (as facesheets) and two pultruded GFRP SHS columns (as core) in between. Two different connection methods, i.e. adhesive bonding and mechanical bolting, are employed to assembly those sandwich specimens. Specimens are labelled based on their connection methods, e.g. "AB" represents adhesive bonding, and "MB" represents mechanical bolting. The configuration of those sandwich specimens after assembly is presented in Fig. 7.1. The spacing (c) between two SHS sections is an experimental parameter to investigate its effect on the mechanical performance of the sandwich specimens, with consideration of two spacing values of 96 and 246 mm. The nominal height of all specimens is 3000 mm, and the detailed dimensions of all specimens are measured and summarized in Table 7.2.

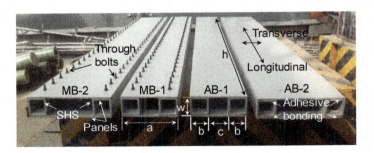

Fig. 7.1 Configurations of GFRP sandwich specimens

Table 7.2 Dimensions of GFRP sandwich specimens (unit in mm)

Specimen	a	c	w	h	T_c	T_p
AB-1	299.6	97.5	117.6	3005	10.3 ± 0.3	7.9 ± 0.2
AB-2	450.6	249.7	117.9	3005	10.2 ± 0.5	8.0 ± 0.3
MB-1	301.3	98.8	116.5	3005	10.2 ± 0.4	7.8 ± 0.3
MB-2	451.5	250.6	116.3	3005	10.1 ± 0.5	7.9 ± 0.2

Note a = specimen width; c = spacing between SHS sections; w = specimen depth; h = specimen height; T_c = thickness of SHS sections; T_p = thickness of panels

For all sandwich specimens, the longitudinal direction is the pultrusion fibre orientation, corresponding to the material properties provided in Table 7.1. AB series specimens were assembled using MA310, a two-part methacrylate adhesive with the nominal glue thickness of 0.6 mm. This adhesive has a tensile strength range of 24.1–31.0 MPa, a tensile modulus range of 1.0–1.2 GPa, and a shear strength of 20.7–24.1 MPa [51]. The surfaces of GFRP panels and SHS columns were firstly roughed using angle grinders and then cleaned using acetone solvent. After that, part A and B of MA310 adhesive were mixed together at the weight ratio of 1:1. The mixed MA310 adhesive was applied on one GFRP panel, and two SHS columns were bonded on the GFRP panel according to the geometric spacing defined in Table 7.2. To maintain consistent thickness of the adhesive layer, chopped steel wires with the diameter of 0.6 mm were used as spacer at the interfaces of GFRP SHS and face sheets. Heavy steel blocks were placed on the SHS columns to ensure the full contact of adhesive between the SHS columns and panels. The adhesive was cured at room temperature for 72 h. After that, the specimen was flipped and the GFRP panel on the other side was bonded to the inner SHS columns following the same procedures. MB series specimens were assembled by tightening M10 bolts (class 10.9) through the central line of the GFRP SHS, as shown in Fig. 7.1. The pitch between each bolt row is 100 mm. The overall length of the used M10 bolts is 150 mm, and the thread length is 35 mm. The nominal tensile strength of the bolts is 1000 MPa. The nominal diameter is 12.5 mm for of all bolt holes, and the nominal clearance distances are 51 mm (to the side edge) and 100 mm (to the end edge). For each MB series specimen, in total 58 through bolts are used for assembly.

7.2.3 Setup and Instrumentation

Testing machine with a loading capacity of 10,000 kN is used to apply the compression loading on the GFRP sandwich specimens. The overall setup is shown in Fig. 7.2a, b. The bottom end of the specimen sat on the steel base ground between two pieces of steel angles and was fixed by tightening using two through bolts. One side of each steel angle was welded on the steel base ground to ensure no displacement and rotation there. For the top end setup, the two pieces of steel angles were mounted by bolts to the loading end plate (see Fig. 7.2a). The top loading plate of the loading machine has a sphere inside, resulting in the pinned boundary condition for the top end.

The compressive loading rate was 1 mm/min in a displacement control mode for all specimens. The axial shortening displacement is measured by a linear variable displacement transducer (LVDT) installed at the central axis of the top loading plate. Lateral displacements on both front and back sides of the specimens are measured at the positions of 1/4, 1/2 and 3/4 height of the specimens as illustrated in Fig. 7.3. Strain gauges were adhered along the central line of specimens at 1/4, 1/2 and 3/4 height on both the front and back sides, as shown in Fig. 7.3. At each height position, strain gauges were placed in both longitudinal and transverse directions. In total 12

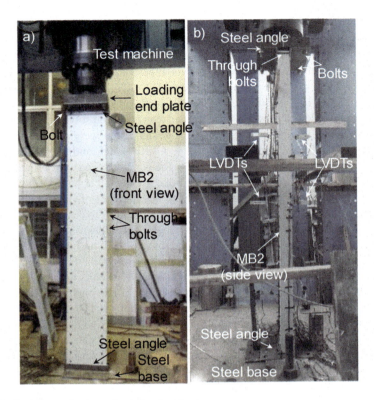

Fig. 7.2 Experimental setup **a** front view **b** side view

Fig. 7.3 Experimental instrumentation **a** Front view **b** Side view

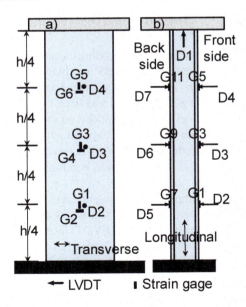

7 Fibre Reinforced Polymer Wall Assemblies in Axial Compression

strain gauges were used for each specimen. During the experiments, failure modes were monitored and recorded by several video cameras.

7.3 Finite Element Analysis

Finite element (FE) analysis was performed to evaluate the mechanical performance of GFRP sandwich specimens using Ansys. Both GFRP SHS and panels were modelled using element Solid45 which is an 8-node 3D solid structural element with three translational degrees of freedom. Element size was set as 20 mm to receive fine meshing results for these GFRP components. The material properties used in FE modelling were defined according to the experimentally acquired values listed in Table 7.1.

Figure 7.4a show the cross-section of AB specimens, in which the adhesive bonding between the inner GFRP SHS and outer GFRP panels were also modelled using Solid45 elements and the thickness was set as 0.6 mm. The bonding between GFRP panels and SHS was achieved by node coupling method, i.e. the coincident nodes between adhesive layers and GFRP panels/SHS were coupled for all degrees of freedom. For MB specimens assembled using through bolts, the bolted connection was modelled by coupling the coincident nodes only at the bolted regions between the GFRP SHS and GFRP panels, as shown in Fig. 7.4a. For each bolt position, the coincident nodes were coupled for all degrees of freedom. All the contact areas of MB specimens between the outer surface areas of the GFRP SHS sections and the

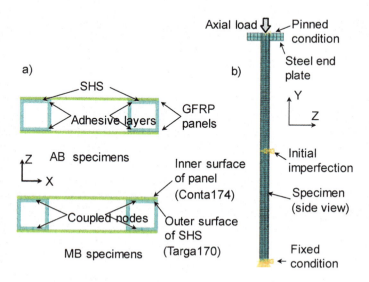

Fig. 7.4 FE modelling with **a** cross-sectional views of AB and MB specimens **b** side view and boundary conditions

corresponding inner areas of the GFRP panels were modelled using a pair of contact elements, i.e. Conta174 and Targa170. As shown in Fig. 7.4a for MB specimens, the outer surface areas of the two GFRP SHS was defined as target element Targa170, and the inner surface areas of GFRP panels was defined as contact element Conta174. This contact behavior allows free separation in the normal direction but defines the contact stiffness as the surface material with consideration of minor penetration (10% of the depth of the underlying element Conta174 [52]). In the tangent direction, the friction coefficient was set as 0.42 [53].

Regarding the boundary conditions, all the nodes on the bottom end of the specimens were constrained for all degrees of freedoms to represent the fixed boundary condition as shown in Fig. 7.4b. For the top end, a steel plate with dimensions of 500 × 500 × 50 mm was modelled for the actual loading plate during the experiments. The coincident nodes at the interface between the steel loading plate and the top end of the GFRP sandwich specimen were coupled as well to simulate the constraining effects of the steel angles. The translational degrees of freedom of the nodes in the central line of the top end of the steel plate were restrained in X and Z directions, leaving Y direction free for rotation to represent the pinned boundary condition for the top end, as shown in Fig. 7.4b.

An initial lateral deflection of 1/500 of the specimen height in the Z direction was defined in the middle section of the specimen as the initial imperfection to introduce geometric nonlinearity. After that, axial displacement loading was applied on the nodes of the central line of the top end on the steel plate and geometric non-linear analysis was further performed.

7.4 Experimental Results

7.4.1 Failure Modes

Detailed observations on the failure modes and progressions of each specimen were made during the experiments. For specimen AB-1, at the early stage of loading, adhesive bond cracking sounds were heard occasionally after the load increased to above 300 kN. No obvious lateral deformation was visually observed to indicate overall buckling. When the axial load reached around 1075 kN, a sudden loud sound were heard due to the separation of the panel on the back side from the inner SHS columns as shown in Fig. 7.5. The face sheet on the back side was severely buckled, but the inner SHS columns and the face sheet on the front side remained without noticeable buckling as no obvious lateral deformations were observed. Observation on the debonding area indicated that the debonding started from the middle of the specimen and extended to the ends (see Fig. 7.5). On the tip region of both ends, it was found that some adhesive was still in contact and not debonded. After the separation between the back face sheet and inner SHS, the sandwich specimen was considered as entirely failed and the experiment was stopped.

7 Fibre Reinforced Polymer Wall Assemblies in Axial Compression

Fig. 7.5 Failure mode of specimen AB-1

For specimen AB-2, the failure progression was similar to AB-1. With the increase in loading, adhesive cracking sounds were heard after the load became over 400 kN. The ultimate debonding failure occurred when the axial load reached 1038 kN, with a loud sound. As shown in Fig. 7.6, for this specimen both the front and back face sheets were severely buckled and separated with the inner SHS components. Again there was no obvious lateral deformation observed on the SHS components; and the experiment was stopped after the ultimate debonding failure.

For specimen MB-1, with the increase in the loading, overall buckling was visually observed in the early loading step (see Fig. 7.7a) and the lateral deformation continued increasing afterwards. When the axial load was approaching the peak load (1126 kN), several cracking sounds were heard while no cracking due to buckling was visually observed from the specimen surface. After the peak load, the load slightly dropped to 1082 kN and end crushing in association with the delamination of the back side GFRP face sheet (under compression) and the junction separation of SHS column were observed, as shown in Fig. 7.7b, c. After this failure, the axial load dropped substantially and the experiment was stopped thereafter.

Overall buckling initiation and increasing lateral deformation were also observed for specimen MB-2 at early loading stage. When the axial load reached 977 kN, wrinkling was found from the 1/3 height region (between 10 and 11th rows of bolts) of the back face sheet (see Fig. 7.8a), and the load dropped substantially to 780 kN. After that, the axial load continued to increase. At the load of 1048 kN, wrinkling

Fig. 7.6 Failure mode of specimen AB-2

was also found from the upper region (around at the height of 2.7 m, see Fig. 7.8b) of the front face sheet. The load then continued to increase and the ultimate failure was found from the back face sheet around at the height of 1.2 m, see Fig. 7.8c, where both face sheet wrinkling and also junction separation of the SHS occurred and this time this specimen lost its load capacity. It should be noted that no bending or shear failure on the through bolts were found. At the regions where the ultimate junction separation of GFRP SHS occurred, several longitudinal and transverse dents on the GFRP face sheet were found near the bolt hole regions as shown in Fig. 7.8d likely due to the local effects of bolting.

7.4.2 Load-Axial Displacement Curves

Figure 7.9 presents the load-axial displacement relationships for all specimens. It shows that for AB specimens, the load–displacement curves are highly linear until a sudden drop when the ultimate failure (debonding) occurs. For specimen MB-1, it has a slight drop at a peak load of 1126 kN, and keep carrying the loading until

7 Fibre Reinforced Polymer Wall Assemblies in Axial Compression 153

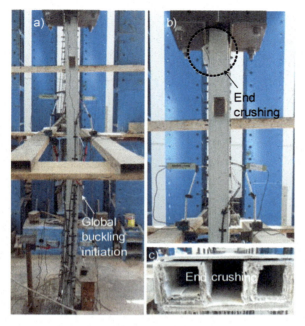

Fig. 7.7 Failure modes of specimen MB-1 **a** global buckling initiation **b** side view of end crushing **c** cross-sectional view of end crushing

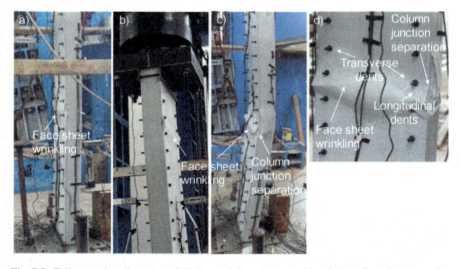

Fig. 7.8 Failure modes of specimen MB-2 **a** back face sheet wrinkling **b** front face sheet wrinkling **c** further back face sheet wrinkling and column junction separation **d** longitudinal and transverse local damages near bolt hole regions

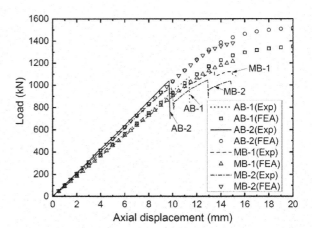

Fig. 7.9 Load and axial displacement curves for all specimens from both experiments and FE modelling

the ultimate failure (see Fig. 7.9) at load 1082 kN. For specimen MB-2, three load drops can be seen corresponding to the face sheet wrinkling observed in Fig. 7.8. Based on the comparison of the load-axial displacement curves of specimens with the same sectional width (AB-1 and MB-1, or AB-2 and MB-2), it is found that AB specimens have a slightly higher stiffness (slope of the curves) than MB specimens. This may be due to the composite action offered by different connection approaches. Adhesive bonding is considered with full composite action [14] while the mechanical bolting with partial composite action depending on the number and spacings of the mechanical bolts and their mechanical properties etc [46]. Also, the difference of effective widths between AB and MB specimens may lead to the difference in the axial stiffness. As further presented in Table 7.4 and Fig. 7.15, for the specimens with the same width but different connection method, AB specimens have a shorter effective width comparing to MB specimens; this further results in a higher critical local buckling stress for the face sheets and also a higher axial stiffness of the specimen. Ultimate axial displacements of each specimen at its ultimate load are summarized in Table 7.3. The ultimate displacements of specimen AB-1 and AB-2 are 11.4 and 9.7 mm and they are less than the corresponding values of specimens MB-1 (15.2 mm) and MB-2 (14.9 mm). It is understandable as the AB specimens experienced sudden debonding failure without the global buckling initiation while MB specimens showed progressive failure after the global buckling initiation.

FE modeling results are included in Fig. 7.9 to compare with the experimental results. Overall, the load–displacement curves from FE analysis showed good agreement with the experimental results especially for the linear developments at the initial stages. It should be noted that the FE results did not well describe the load–displacement behaviours for MB specimens after wrinkling of face sheets as such failure mode were not considered in the FE modelling.

7.4.3 Load-Lateral Displacement Curves

The load-lateral displacement curves are presented in Fig. 7.10. LVDT results measured from the 1/4 (D2 and D5), 1/2 (D3 and D6) and 3/4 (D4 and D7) height positions on both front and back sides of each specimen at the ultimate loads are summarized in Table 7.3. For specimen AB-1, the readings from D2, D3, D4 on the front side are negative values and D5, D6, D7 on the back side are positive values, suggesting that the specimen is under an overall lateral deformation towards to the front side. Prior to the adhesive debonding occurred on the back side at the axial load of 1075 kN, the largest lateral displacements from LVDTs D3 and D6 in the middle of the specimen are only 7.6 and 8.1 mm (see Table 7.3) respectively. After the sudden adhesive debonding failure, the lateral displacements D5, D6 and D7 on the back side panels increased substantially. The lateral deformations on specimen AB-2 are similar to AB-1. All the readings are relatively small (less than 6 mm) until the adhesive debonding failure occurred at ultimate load of 1038 kN. After the sudden adhesive debonding failure, all the lateral displacements increased substantially. It should be noted that from the lateral displacements of AB series specimens are low before the ultimate failure as evidenced in Fig. 7.10a, b, suggesting insignificant global buckling.

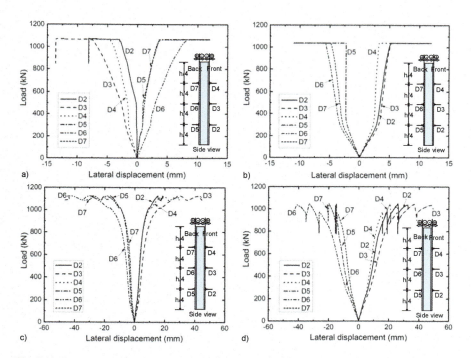

Fig. 7.10 Load and lateral displacement curves **a** AB-1 **b** AB-2 **c** MB-1 **d** MB-2

Table 7.3 Major experimental results of specimens

Specimen	Peak load P (kN)	Ultimate load P_u (kN)	Axial displacement at P_u (mm)	Lateral displacement at P_u (mm) (Front side)			Lateral displacement at P_u (mm) (Back side)		
				D2	D3	D4	D5	D6	D7
AB-1	1075	1075	11.4	3.0	7.6	4.5	3.7	8.1	3.6
AB-2	1038	1038	9.7	5.1	5.2	3.3	2.3	5.7	4.9
MB-1	1126	1082	15.2	17.4	45.0	30.6	22.1	47.6	33.3
MB-2	1048	1001	14.9	32.7	49.3	25.8	20.9	42.2	15.5

7 Fibre Reinforced Polymer Wall Assemblies in Axial Compression

Table 7.4 Comparisons between analytical and experimental results of ultimate compressive stress

Specimen	b_{eff} (mm)	Analytical results (MPa)		Experimental results (MPa)		Comparisons		
		σ_e	σ_{LB}	σ_u	$\sigma_{LB,\exp}$	σ_u/σ_e	σ_u/σ_{LB}	$\sigma_{LB,\exp}/\sigma_{LB}$
AB-1	96	110.0	552.1	88.2	/	0.80	0.16	/
AB-2	246	114.8	81.5	70.3	60.9	0.61	0.86	0.75
MB-1	196	109.9	149.8	94.9	/	0.86	0.63	/
MB-2	346	114.8	48.1	73.6	52.7	0.64	1.53	1.10

Note b_{eff} = effective plate width; σ_e = global buckling stress; σ_{LB} = local buckling stress; σ_u = experimental ultimate compressive stress; $\sigma_{LB,\exp}$ = experimental local buckling stress

For MB specimens, the load-lateral displacement curves presented in Fig. 7.10c, d show that lateral displacements developed continuously with relatively large values, unlike the sudden increase in AB series specimens. The largest lateral displacement readings before failure (D3 or D6, see Table 7.3) at the middle height of GFRP face sheets are 45.0 or 47.6 mm from specimen MB-1, and 49.3 or 42.2 mm from specimen MB-2. These large lateral deformations on MB series specimens made the bucked deformation visible during the experiments, and further suggesting that the global buckling may be initiated for the specimens (Table 7.4).

7.4.4 Load-Strain Responses

The load-strain responses of each specimen are presented from Figs. 7.11, 7.12, 7.13 and 7.14, with the corresponding positions of strain gauges marked in Fig. 7.3. On the front side of the specimens, strain gauges G1, G3, G5 represent the longitudinal

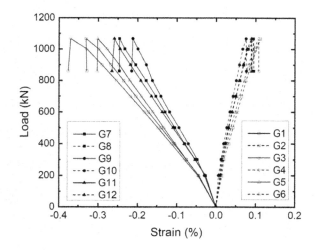

Fig. 7.11 Load-strain responses of specimen AB-1

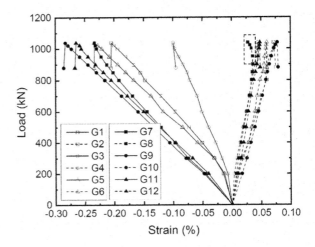

Fig. 7.12 Load-strain responses of specimen AB-2

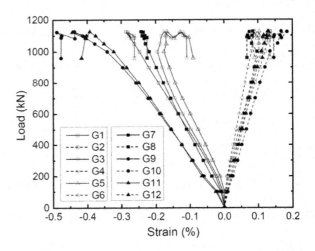

Fig. 7.13 Load-strain responses of specimen MB-1

strains on the 1/4, 1/2 and 3/4 height positions; and G2, G4, G6 correspond to the transverse strains on the same positions. On the back side of the specimens, strain gauges G7, G9, G11 represent the longitudinal strains on the 1/4, 1/2 and 3/4 height position; and G8, G10, G12 correspond to the transverse strains on the same positions. Figure 7.11 presents the load-strain responses on specimen AB-1, where both longitudinal and transverse strain responses appear linear. It is also shown that the slopes of longitudinal strains (G7, G9, G11) on the back side are higher than those (G1, G3, G5) on the front side. This is because the overall lateral deformation of the specimen as presented in Fig. 7.10a, in addition to the axial loading, caused the different compression on the front and back sides. The maximum compressive strain is -0.37% on G3 gauge located on the middle height of the front face sheet

Fig. 7.14 Load-strain responses of specimen MB-2

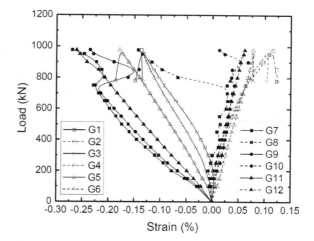

at the ultimate load of 1075 kN. The maximum transverse strain is 0.11% on G2 located on the 1/4 height position of the front face sheet at the ultimate load.

Figure 7.12 presents the load-strain responses of specimen AB-2. Again at the initial loading stage all the strain responses are linear. However different to specimen AB-1, the transverse strain from G8 showed a clear turning of direction after the compressive load of about 900 kN. This sudden change of strain values indicates that local deformation (buckling) at this location of the GFRP face sheet. The longitudinal strain development from G7 (on the same position of G8) also show change in slope, in support of the observation of the local buckling behaviour. The maximum compressive strain at the ultimate load of 1038 kN is -0.27% on gauge G9 located on the middle of the back side; and the maximum transverse strain is 0.08% on G10 located on the middle height position on the front side.

Figure 7.13 presents the load-strain responses of specimen MB-1. When the compressive load increased over 1000 kN, strain behaviours from G3, G5, G9, G11 on the 1/2 and 3/4 height positions gradually changed their development trends. This suggests the overall bending because of the initiation of global buckling and this is in accordance with the observed lateral deformation as shown in Figs. 7.7a and 7.10c. Figure 7.14 presents the load-strain responses of specimen MB-2 where more obvious change in the strain developments with loading are found. Clear turning points in the responses from gauges G7, G8 and G9 at load of 750 kN indicated local buckling occurred on the 1/4 and 1/2 height on the back face sheet. Although this local buckling was not observed visually during experiment, this region where the local buckling occurred was in general consistent with the further observed face sheet wrinkling (see Fig. 7.8a) on the lower area of the back face sheet. Other changes in the direction of strain developments are from gauges G7, G9 and G10 at the load of 830 kN, further confirming the occurrence of local buckling there on the back face sheet.

7.5 Discussions

7.5.1 Load-Bearing Capacity

The global buckling load of the sandwich specimens can be estimated using Eq. 7.1:

$$P_E = \frac{\pi^2 (EI)_{eq}}{(kL)^2} \quad (7.1)$$

where $(EI)_{eq}$ is the equivalent flexural stiffness of the section, k is the effective length ratio and L is the length of the specimen.

In this study, the boundary condition on the ends corresponds to one end pinned and the other fixed, therefore k is 0.7. The equivalent flexural stiffness of the section can be calculated using the below equation:

$$(EI)_{eq} = (EI)_{panels} + (EI)_{SHS}$$
$$= 2E_p \left(\frac{b_p t_p^3}{12} + b_p t_p d^2 \right) + 2E_c \left[\frac{b_c^4}{12} - \frac{(b_c - t_c)^4}{12} \right] \quad (7.2)$$

where E_p and E_c are the elastic moduli of GFRP panels and SHS columns, b_p and t_p are the width and thickness of GFRP panels, and b_c and t_c are the width and thickness of GFRP SHS; d is the distance between the central axis of the SHS sections and the central axis of the GFRP panels, i.e. $(b_c + t_p)/2$.

It should be noted that full composite action is assumed in Eq. 7.2 for the connection between the GFRP face sheets and inner SHS. However, only the bonded connection may provide full composite action [49]; while partial composite action was found on bolting connected web-flange sandwich systems [46]. The critical global buckling stress of all sandwich specimens are calculated using Eq. 7.1 and summarized in Table 7.4 for further comparison. It also can be found that the critical global buckling stress of MB specimens are overestimated indicating the partial composite action for the sandwich specimens with bolted connections.

For specimens AB-2 and MB-2 with a larger spacing c (246 mm) between the two SHS, local buckling may more likely occur on the GFRP face sheets between the two SHS sections, due to the relative high width-thickness ratios [50, 54]. The critical local buckling load can be calculated, taking into consideration the boundary conditions of the GFRP face sheets of the AB or MB specimens with different connections. For the former, the boundary condition of the GFRP face sheets may be close to simply supported at the top and bottom edges while as clamped due to the adhesive bonding at the two sides, i.e. CCSS as shown in Fig. 7.15. The equivalent plate width is therefore the clear spacing between the two SHS. The critical local buckling load for orthotropic GFRP plate with boundary condition of CCSS [55–57] can be calculated using the below equation:

7 Fibre Reinforced Polymer Wall Assemblies in Axial Compression

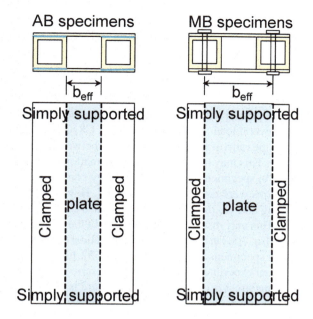

Fig. 7.15 Boundary conditions for GFRP face sheets in theoretical estimation of local buckling stress **a** cross-sectional view **b** front view

$$N_{Local,cr}^{CCSS} = \frac{\pi^2}{b_{eff}} \left[\frac{44.9}{\pi^2} \sqrt{D_{11} D_{22}} + \frac{24}{\pi^2} (D_{12} + 2D_{66}) \right] \quad (7.3)$$

where b_{eff} is the width of the plate, and D_{11}, D_{22}, D_{12}, D_{66} are the bending stiffness coefficients of the plate relating to material properties [57, 58]. Equation 7.3 was further used for MB series specimens. However the effective plate width b_{eff} may be considered as the distance between the two through bolts in the two SHS sections from the same row (as indicated in Fig. 7.15). Using Eq. 7.3, the critical local buckling stress of AB and MB series specimens are calculated and summarized in Table 7.4.

In Table 7.4, the experimental ultimate compressive stress σ_u of specimens AB-1 and AB-2 are 88.2 MPa and 70.3 MPa, corresponding to the stress ratio of 0.80 and 0.61 to the calculated global buckling stress σ_e. This is in support of the observed failure mode of AB series specimens, i.e. sudden debonding without global buckling. For MB specimens, the experimental ultimate compressive stress σ_u are 94.9 MPa and 73.6 MPa, corresponding to the stress ratio of 0.86 and 0.64 to the global buckling stress σ_e; while the calculated values can be overestimated because of the assumption of full composite action as discussed above. In addition, for specimen MB-1, the stress ratio 0.86 can already reasonably suggest the global buckling initiation, as observed during the experiment (Fig. 7.7a). For specimen MB-2, the stress ratio is 0.64 and much lower than the theoretical global buckling stress. This is due to the occurrence of local buckling on MB-2 specimen associated with a larger width between the two through bolts from the same row (see Fig. 7.8a, b). The calculated local buckling stress σ_{LB} is 48.1 MPa and less than the calculated global buckling stress σ_e for specimen MB-2.

The comparisons of local buckling stress between the calculated and experimental values in Table 7.4 show satisfactory agreements for specimens AB-2 and MB-2. For these two specimens AB-2 and MB-2 with a larger spacing between the two SHS sections, the calculated local buckling stress σ_{LB} are 81.5 MPa and 48.1 MPa, and lower than the global buckling stress σ_e. σ_{LB} of specimen MB-2 is close to the experimental local buckling stress $\sigma_{LB,exp}$ of 52.7 MPa and local buckling was evidently observed from the failure mode (face sheet wrinkling, see Fig. 7.8a, b) and from several load-strain responses (G7, G8, G9 and G10 in Fig. 7.14). For specimen AB-2, the ratio of local buckling stress between experimental and theoretical results is 0.75. This may be due to the approximation of boundary conditions of GFRP face sheets in specimen AB-2. The boundary conditions of two sides in the longitudinal directions were considered as clamped. However, the constrains from the adhesive bonding and the thin thickness of the face sheets may still allow rotation at the side ends. This may result in overestimation of the critical theoretical local buckling stress σ_{LB} in Eq. 7.3. On the other hand, theoretical calculations on the local buckling stress σ_{LB} for specimens AB-1 and MB-1 are 552.1 MPa and 149.8 MPa and they are higher than their experimental ultimate compressive stress σ_u as well as the theoretical global buckling stress σ_e This explains that there is no local buckling observed from these two specimens.

7.5.2 Effects of Spacing Between SHS Sections

As evidenced in load-strain responses presented in Figs. 7.11, 7.12, 7.13 and 7.14, the spacing value c shows clear effect on the local buckling of the GFRP face sheets between the two SHS sections. Specimens with a larger spacing value (AB-2 and MB-2) exhibited local buckling mode (as shown in Figs. 7.12 and 7.14); while the ones with a smaller spacing (AB-1 and MB-1) did not (see Figs. 7.11 and 7.13). Also the actual spacing between the two clamped side edges in MB specimens is larger than that between the two SHS components due to the inner positions of bolts. It also modified the boundary conditions and therefore affected the critical local buckling stress for the GFRP face sheets between the two bolts as discussed previously. In the development of such GFRP web-flange sandwich assemblies for column or wall applications, the spacing between SHS sections should be carefully determined as it affects the failure mode and overall load-bearing capacities.

7.5.3 Comparison of Bonded and Bolted Connections

From the observed failure modes, adhesively bonded specimens AB-1/2 exhibited sudden debonding. The bolted specimens MB-1/2 exhibited progressive failure associated with the evident global buckling initiation, and finally failed on the junction separation on the inner SHS. From the load-axial displacement curves, AB specimens showed a slightly higher axial stiffness than MB specimens. This reflects the difference in composite action offered by bonded or bolted connections. From the load-lateral displacement curves, evident lateral deformation observed from MB specimens (see Fig. 7.10c, d) while limited lateral deformation from AB specimens (lateral displacements less than 10 mm, see Fig. 7.10a, b). The differences in the load-bearing capacities of GFRP sandwich assemblies with these two types of connections are minor (within 5%).

7.6 Conclusions

This chapter investigated the performance of GFRP web-flange sandwich assemblies in compression for column or wall applications. The specimens were assembled by connecting two pultruded GFRP panels and two SHS using adhesive bonding or mechanical bolting. The failure modes, load-bearing capacities, load–displacement curves and load-strain responses were obtained from experiments and compared with theoretical and FE modelling. The effects of different connection methods and the spacing between SHS were discussed. The following conclusions can be drawn from this study:

1. Adhesively bonded specimens (AB-1/2) experienced sudden debonding failure between the GFRP face sheets and inner SHS sections without global buckling in the experimental study. Mechanical bolted sandwich specimens (MB-1/2) exhibited evident lateral deformation and global buckling initiation. Progressive failures were found from the bolted sandwich specimens. Specimen MB-1 with a smaller spacing value of 96 mm between two inner SHS showed end crushing and web-flange junction separation at the end, after the obvious lateral deformation without local buckling of the GFRP face sheets. Specimen MB-2 with a larger spacing of 246 mm exhibited clear local buckling of face sheet wrinkling during the loading and it ultimately failed by the junction separation of SHS columns at the middle height of the specimen.
2. Load and displacement curves showed slightly higher axial stiffness of AB specimens than the MB specimens with the same section width. This may be due to the different degrees of composite action between the adhesive bonding and mechanical bolting and difference in the effective widths between two GFRP SHS. Adhesive bonding provided full composite action between the GFRP face sheets and inner SHS; while mechanical bolting is associated with partial composite action. FE results on the load-axial displacement curves show good agreements with experimental results for both AB and MB specimens before failure. Load and lateral displacement curves of MB specimens showed more obvious lateral deformation than AB specimens. This also indicates global buckling initiated on MB specimens while not on AB specimens.

Load-strain responses evidenced local buckling occurred on the GFRP face sheets of specimens AB-2 and MB-2.

3. It can be therefore identified that the spacing between the two SHS sections play an important role on the local buckling of the GFRP face sheets. Specimens AB-1 and MB-1 with a smaller spacing value of 96 mm did not exhibit local buckling on the face sheets while the local buckling was clearly found from specimens AB-2 and MB-2 with a larger spacing between the two inner SHS sections. Analytical results based on orthotropic plate theory was able to estimate the local buckling stress for the face sheets with satisfactory agreement with the experimental results.

The results from this study may demonstrated the compressive performance of the developed GFRP sandwich structures for potential column and wall applications by assembly of GFRP flat panels as face sheets and SHS as core sections. Different connection methods such as adhesive bonding and mechanical bolting and spacing values between core SHS as major design parameters may result in different failure modes, composite action degrees and load-bearing capacities.

References

1. Allen HG (1969) Analysis and design of structural sandwich panels. Pergamon Press
2. Daniel IM (2000) Abot JL (2007) Fabrication, testing and analysis of composite sandwich beams. Compos Sci Technol 60:2455–2463
3. Mallick PK (2007) Fiber-reinforced composites: materials, manufacturing, and design. CRC Press
4. Keller T (2016) FRP sandwich structures in bridge and building construction. In: Proceedings of the eighth international conference on fibre-reinforced polymer (FRP) composites in civil engineering, pp 23–28
5. Manalo A, Aravinthan T, Fam A, Benmokrane B (2017) State-of-the-art review on FRP sandwich systems for lightweight civil infrastructure. J Compos Constr 21(1)
6. Fang H, Bai Y, Liu W, Qi Y, Wang J (2019) Connections and structural applications of fibre reinforced polymer composites for civil infrastructure in aggressive environments. Compos B Eng 164:129–143
7. Ferdous W, Bai Y, Duc T, Manalo A, Mendis P (2019) New advancements, challenges and opportunities of multi-storey modular buildings—a state-of-the-art review. Eng Struct 183:883–893
8. Fam A, Sharaf T (2010) Flexural performance of sandwich panels comprising polyurethane core and GFRP skins and ribs of various configurations. Compos Struct 92:2927–2935
9. Fang H, Sun H, Liu W, Wang L, Bai Y, Hui D (2015) Mechanical performance of innovative GFRP-bamboo-wood sandwich beams: experimental and modelling investigation. Compos B 79:182–196
10. McCracken A, Sadeghian P (2018) Partial-composite behavior of sandwich beams composed of fiberglass facesheets and woven fabric core. Thin-Walled Struct 131:805–815
11. Manalo AC, Aravinthan T, Karunasena W, Islam MM (2010) Flexural behaviour of structural fibre composite sandwich beams in flatwise and edgewise positions. Compos Struct 92:984–995
12. Manshadi BD, Vassilopoulos AP, De Castro J, Keller T (2012) Instability of thin-walled GFRP webs in cell-core sandwiches under combined bending and shear loads. Thin-Walled Struct 53:200–210

13. Zhu D, Shi H, Fang H, Liu W, Qi Y, Bai Y (2018) Fiber reinforced composites sandwich panels with web reinforced wood core for building floor applications. Compos B Eng 150:196–211
14. Satasivam S, Bai Y, Zhao XL (2014) Adhesively bonded modular GFRP web-flange sandwich for building floor construction. Compos Struct 111:381–392
15. Sharaf T, Fam A (2013) Analysis of large scale cladding sandwich panels composed of GFRP skins and ribs and polyurethane foam core. Thin-Walled Struct 71:91–101
16. Shen SY, Masters FJ, Upjohn HL, Ferraro CC (2013) Mechanical resistance properties of FRP/polyol-isocyanate foam sandwich panels. Compos Struct 99:419–432
17. Dey TK, Srivastava I, Khandelwal RP, Sharma UK, Chakrabarti A (2013) Optimum design of FRP rib core bridge deck. Compos B Eng 45:930–938
18. Liu Z, Majumdar PK, Cousins TE, Lesko JJ (2008) Development and evaluation of an adhesively bonded panel-to-panel joint for a FRP bridge deck system. J Compos Constr 12:224–233
19. Lombardi NJ, Liu J (2011) Glass fiber-reinforced polymer/steel hybrid honeycomb sandwich concept for bridge deck applications. Compos Struct 93:1275–7283
20. Keller T, Rothe J, de Castro J, Osei-Antwi M (2014) GFRP-balsa sandwich bridge deck: concept, design, and experimental validation. J Compos Constr 18(2)
21. Bai Y, Keller T (2009) Shear Failure of Pultruded Fiber-Reinforced Polymer Composites under Axial Compression. J Compos Constr 13:234–242
22. Puente I, Insausti A, Azkune M (2006) Buckling of GFRP Columns: An Empirical Approach to Design. J Compos Constr 10:529–537
23. Mousa MA, Uddin N (2011) Global buckling of composite structural insulated wall panels. Mater Des 32:766–772
24. Mousa MA, Uddin N (2012) Structural behavior and modeling of full-scale composite structural insulated wall panels. Eng Struct 41:320–334
25. Mathieson H, Fam A (2016) Numerical modeling and experimental validation of axially loaded slender sandwich panels with soft core and various rib configurations. Eng Struct 118:195–209
26. Mathieson H, Fam A (2014) Axial Loading Tests and Simplified Modeling of Sandwich Panels with GFRP Skins and Soft Core at Various Slenderness Ratios. J Compos Constr 19:1–13
27. Abdolpour H, Escusa G, Sena-Cruz JM, Valente IB, Barros JAO (2017) Axial performance of jointed sandwich wall panels. J Compos Constr 21:1–12
28. CoDyre L, Fam A (2016) The effect of foam core density at various slenderness ratios on axial strength of sandwich panels with glass-FRP skins. Compos B Eng 106:129–138
29. Tao J, Li F, Zhang D, Liu J, Zhao Z (2019) Manufacturing and mechanical performances of a novel foam core sandwich-walled hollow column reinforced by stiffeners. Thin-Walled Struct 139:1–8
30. Wang L, Liu W, Fang Y, Wan L, Huo R (2016) Axial crush behavior and energy absorption capability of foam-filled GFRP tubes manufactured through vacuum assisted resin infusion process. Thin-Walled Struct 98:263–273
31. Keller T, de Castro J, Schollmayer M (2004) Adhesively bonded and translucent glass fiber reinforced polymer sandwich girders. J Compos Constr 8(5):461–470
32. Keller T, Vallée T (2005) Adhesively bonded lap joints from pultruded GFRP profiles. Part II: joint strength prediction. Compos B Eng 36:341–350
33. Keller T, Vallée T (2005) Adhesively bonded lap joints from pultruded GFRP profiles. Part I: stress-strain analysis and failure modes. Compos B Eng 36:331–340
34. Bank LC (2006) Composites for construction. Wiley
35. Girão Coelho AM, Mottram JT (2015) A review of the behaviour and analysis of bolted connections and joints in pultruded fibre reinforced polymers. Mater Des 74:86–107
36. Dakhel M, Donchev T, Hadavinia H (2019) Behaviour of connections for hybrid FRP/steel shear walls. Thin-Walled Struct 134:52–60
37. Qureshi J, Mottram JT (2013) Behaviour of pultruded beam-to-column joints using steel web cleats. Thin-Walled Struct 73:48–56
38. Wu C, Zhang Z, Bai Y (2016) Connections of tubular GFRP wall studs to steel beams for building construction. Compos B 95:64–75

39. Luo FJ, Huang Y, He X, Qi Y, Bai Y (2019) Development of latticed structures with bolted steel sleeve and plate connection and hollow section GFRP members. Thin-Walled Struct 137:106–116
40. Mosallam A (2011) Design guide for FRP composite connections. American Society of Civil Engineers (ASCE)
41. The European Structural Polymeric Composites Group (1996) Structural design of polymer composites eurocomp design code and handbook, vol 35
42. Egan B, McCarthy CT, McCarthy MA, Frizzell RM (2012) Stress analysis of single-bolt, single-lap, countersunk composite joints with variable bolt-hole clearance. Compos Struct 94:1038–1051
43. Joseph APK, Davidson P, Waas AM (2018) Progressive damage and failure analysis of single lap shear and double lap shear bolted joints. Compos A Appl Sci Manuf 113:264–274
44. Xiang J, Zhao S, Li D, Wu Y (2017) An improved spring method for calculating the load distribution in multi-bolt composite joints. Compos B Eng 117:1–8
45. Xie L, Bai Y, Qi Y, Wang H (2019) Pultruded GFRP square hollow columns with bolted sleeve joints under eccentric compression. Compos B Eng 162:274–282
46. Satasivam S, Bai Y (2014) Mechanical performance of bolted modular GFRP composite sandwich structures using standard and blind bolts. Compos Struct 117:59–70
47. Zhou A, Keller T (2005) Joining techniques for fiber reinforced polymer composite bridge deck systems. Compos Struct 69:336–345
48. Keller T, Gürtler H (2005) Composite action and adhesive bond between fiber-reinforced polymer bridge decks and main girders. J Compos Constr 9:360–368
49. Coleman JT, Lesko JJ, Cousins TE, Temeles AB, Zhou A (2005) Laboratory and field performance of cellular fiber-reinforced polymer composite bridge deck systems. J Compos Constr 9:458–467
50. Xie L, Bai Y, Qi Y, Caprani C, Wang H (2018) Effect of width-thickness ratio on capacity of pultruded square hollow polymer columns. Proc Inst Civ Eng Struct Build 171:842–854
51. ITW Performance Polymers (2018) Plexus MA310. https://itwperformancepolymers.com/
52. Contact Technology Guide release 15.0 (2013) ANSYS Inc, Canonsburg PA
53. McCarthy CT, Gray PJ (2011) An analytical model for the prediction of load distribution in highly torqued multi-bolt composite joints. Compos Struct 93:287–298
54. Pecce M, Cosenza E (2000) Local buckling curves for the design of FRP profiles. Thin-Walled Struct 37:207–222
55. Shan L, Qiao P (2008) Explicit local buckling analysis of rotationally restrained composite plates under uniaxial compression. Eng Struct 30:126–140
56. Cardoso DCT, Harries KA, Batista EDM (2014) Closed-form equations for compressive local buckling of pultruded thin-walled sections. Thin-Walled Struct 79:16–22
57. Kollár L (2003) Local buckling of fiber reinforced plastic composite structural members with open and closed cross sections. J Struct Eng 129:1503–1513
58. Qiao P, Chen Q (2014) Post-local-buckling of fiber-reinforced plastic composite structural shapes using discrete plate analysis. Thin-Walled Struct 84:68–77

Chapter 8
Fibre Reinforced Polymer Columns with Bolted Sleeve Joints under Eccentric Compression

Lei Xie, Yu Bai, Yujun Qi, and Hao Wang

Abstract This chapter presents an investigation into the performance of pultruded glass fibre reinforced polymer (GFRP) square hollow columns under eccentric compression, i.e. subjected to both compression and bending. Eccentric compression experiments were performed on slender GFRP column specimens at different eccentricities. Bolted sleeve joint was employed to connect the GFRP column specimens and loading end plates. The relationship between the load-bearing capacities of GFRP columns and the eccentricities was received and discussed. The interaction curve between compression load and bending moment due to eccentricity (*P-M* curve) was obtained from experiments and compared with finite element (FE) and design approaches. Results revealed that the compression performance of GFRP columns was significantly affected by the eccentricity and the moment capacity of bolted sleeve joint. Splitting failure developed from the initiative longitudinal cracks in the bolted sleeve joint region at the end of the columns was found as the ultimate failure, after the large lateral deformation. FE analysis presented satisfactory agreements with experimental results; furthermore, the stress analysis in the critical bolted sleeve joint region indicated that the in-plane shear stress was the dominant component leading to the splitting failure.

Reprinted from Composites Part B: Engineering, 162, Lei Xie, Yu Bai, Yujun Qi, Hao Wang, Pultruded GFRP square hollow columns with bolted sleeve joints under eccentric compression, 274-282, Copyright 2019, with permission from Elsevier.

L. Xie · Y. Bai (✉)
Department of Civil Engineering, Monash University, Clayton, Australia
e-mail: yu.bai@monash.edu

Y. Qi
College of Civil Engineering, Nanjing Tech University, Nanjing, China

H. Wang
University of Sourthern Queensland, Toowoomba, Australia

8.1 Introduction

Pultruded glass fibre reinforced polymer (GFRP) members [1, 2], due to their advantages such as low density, high strength, excellent corrosion resistance and fatigue properties, have been regarded as promising members for structural applications. The utilization of such pultruded GFRP members into structural applications requires the comprehensive knowledge of their performances under different loading scenarios. Since 1990s numerous studies have been conducted in this area, particularly on GFRP beams under bending [3–11] and GFRP columns under axial compression [12–19]. Meanwhile, design approaches for GFRP members subjected to compression and bending are developed in guidelines and standards, e.g., Eurocomp Design Code and Handbook [20], Italian National Research Council CNR-DT-205 [21] and ASCE pre-standard for Load & Resistance Factor Design (LFRD) of Pultruded Fiber Reinforced Polymer (FRP) structures [22].

However, there are limited experimental results on the performance of GFRP columns under eccentric compression, i.e. GFRP columns under both axial compressive load and bending moment. Barbero and Turk [23] experimentally tested 22 pultruded GFRP I section beam-columns under eccentric loading with an eccentricity of 25.4 mm. Results showed that flange buckling was the major failure mode for I section beam-columns due to the combined action of axial load and bending moment. Comparing to concentrically loaded specimens, the ultimate loads of specimens with eccentricity reduced significantly up to 50%. Mottram et al. [24] investigated the buckling behaviours of ten slender GFRP wide-flange I section column specimens under three types of eccentric compression loading configurations including: i) symmetrical loading with both ends pinned; ii) symmetrical loading with one end pinned and one end fixed; and iii) unsymmetrical loading with both ends pinned. Furthermore, specimens were tested at four eccentricities individually for type i) and ii) loading, and at two eccentricities for type iii) loading. Results showed the increase in eccentricity decreased the maximum axial load, alongside the slight increase in the end-moment of column specimens. Nunes et al. [25] carried out experimental studies on the performance of GFRP I section columns loaded with the ratio of eccentricity to height of cross-section e/h at 0.00, 0.15 and 0.30. The results showed that the axial load capacity reduced up to 40% with the increase in eccentricity; and numerical analysis using finite element (FE) approach and generalized beam theory received good agreements with the experimental results.

For the investigations on eccentrically loaded GFRP columns as introduced above, it was found that I sections were studied and the covered eccentricities were relatively small. Therefore, this chapter intends to investigate the performance of GFRP column specimens with square hollow sections (SHS) under eccentric loading with large eccentricities. Such SHS members showed better resistance against local/torsional buckling [26] and shear failure at the web-flange junction region than open section members [27–30]. Connections for GFRP structures played an important role in their structural performances and in many cases dominated the capacities. Two main types of connections including bolted connection [31–33] and bonded connection [34–37] are investigated previously. The former has advantages including easy assembly and disassembly [38], and not sensitive to peeling loading; however has high stress concentration at the joint region [39]. The latter generally has high joint strength at room temperature while is inconvenient for applications on site and also difficult for disassembly [40] and sensitive to elevated and high temperatures [41, 42]. In addition, the effects of connection performance on the overall column behaviour subjected to both compression and bending are taken into account through the bolted sleeve joint (BSJ) proposed in [33, 43, 44] specifically for SHS members. GFRP SHS columns were tested under compression at three different eccentricities. The experimental results on axial load and bending moment capacities, i.e. *P-M* interaction diagram were therefore obtained and further compared with design approaches for pultruded GFRP members. Detailed FE analysis including the contact behaviours of bolts between GFRP columns and steel sleeve connectors was carried out to validate the experimental results including failure modes, load–displacement curves and load–strain responses, enabling an in-depth understanding of critical stress components for the ultimate failure.

8.2 Experimental Program

8.2.1 Materials and Specimens

Pultruded GFRP SHS columns used in this study are with a nominal section dimension of 102 × 102 × 9.5 mm. The nominal length of GFRP column specimens is 3000 mm. The GFRP composites are made of polyester matrix reinforced by E-glass fibres with a fibre volume fraction V_f of 46.7% [45]. The fibre architecture consists of unidirectional roving in the centre, and chopped strand mat on the surfaces. Material properties of such GFRP composites are determined through coupon tests as previously reported in [28, 46], and the material properties of steel tube, end plate and through bolts used in this study are determined previously in [44, 47, 48]. All material properties are summarized in Table 8.1. The non-dimensional slenderness of the column specimens examined in this study is 2.43 according to the results obtained in [46] and the definition in [49]. Previous understanding also indicated that slender columns in association with global buckling failure mode generally presented their non-dimensional slenderness greater than 1.3; therefore the GFRP column specimens in this study are considered as slender ones.

*Pultruded GFRP is regarded as an orthotropic material, in which the longitudinal direction is major fibre orientation; on the transverse plane, it is assumed as an isotropic material [47].

Table 8.1 Material properties of pultruded GFRP columns, steel tube and through bolts

Material	Direction*	E (GPa)	ν	F_t(MPa)	F_c(MPa)	F_y(MPa)	G (GPa)	S (MPa)
GFRP	Longitudinal	32.2	0.32	307	290	/	3.5	26.7
	Transverse	5.5	0.06	19.5	33.4	/	/	/
Steel tube		216.9	0.3	520	/	326.2	/	/
Steel endplate		218.8	0.3	511	/	332.4	/	/
Steel through bolts		235	0.42	596	/	1043	/	/

Note E = elastic modulus; $ν$ = poisson's ratio; F_t = tensile strength; F_c = compressive strength; F_y = yield strength; G = shear modulus; S = in-plane shear strength

8.2.2 Bolted Sleeve Joint (BSJ)

BSJ is designed and manufactured as shown in Fig. 8.1. First, a sleeve connector as shown in Fig. 8.1a was formed by welding a square hollow steel tube (section dimension of 83 × 83 × 10 mm and a height of 160 mm) with a steel end plate (dimension of 243 × 105 × 10 mm, and a thickness of 10 mm). Two rows of holes with a nominal diameter of 12.5 mm were drilled on the four side areas of the steel tubes and designed for further connecting to GFRP columns using through bolts. The vertical distance between two rows of holes is 50 mm, and the horizontal distance between two columns of holes is 43 mm (see Fig. 8.1b). After joining the sleeve connector and GFRP SHS columns using M12 steel through bolts as shown in Fig. 8.1b, the end plate of a BSJ was connected to the base plate of the testing machine as shown in Fig. 8.2a.

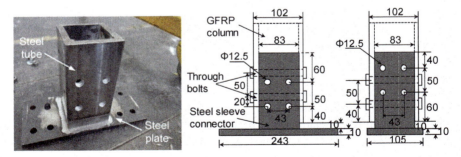

Fig. 8.1 Configuration of a bolted sleeve joint **a** steel sleeve connector **b** schematic diagram on joining the sleeve connector with GFRP column using through bolts

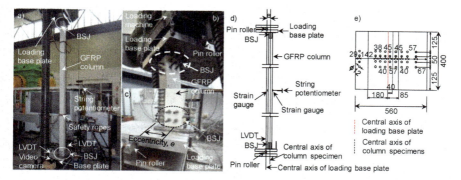

Fig. 8.2 Experimental setup of eccentric compression experiments **a** Overall setup **b** BSJ and the setup on the top **c** BSJ and the setup on the bottom **d** Schematic diagram of the setup **e** Top view diagram of the loading base plate

8.2.3 Experimental Setup and Instrumentation

A testing machine with a loading capacity of 5000 kN was used to apply the compression load on GFRP SHS specimens, and the overall setup is shown in Fig. 8.2a. To achieve the combined loading of compression and bending on the specimens, a steel base plate of 60 mm thickness was designed and manufactured as shown in Fig. 8.2b, c. A schematic diagram on the setup is shown in Fig. 8.2d. As the top view diagram of the loading base plate shown in Fig. 8.2e, it has an array of tapped holes with 12 mm diameter in parallel with the central axis. GFRP column specimens with BSJs on both ends can be mounted on the steel base plate at different rows of holes using M12 bolts, therefore an eccentricity (e) can be received between the central axis of the loading base plate (where the axial loading is applied) and the central axis of the GFRP column specimens, as indicated in Fig. 8.2b, c and e. By varying the different rows of holes to be connected with the BSJs, different values of eccentricity (e) can be introduced to the GFRP column specimens.

The boundary conditions on both ends of column specimens were pinned–pinned and this was achieved by placing the steel loading base plate on a pin roller. For the bottom setup, the loading base plate was positioned on top of the pin roller as shown in Fig. 8.2c; and the pinned condition for the top setup was similar. The spherical contact areas between the loading base plate and the pin roller enabled the base plate to rotate along the central axis of the pin roller. In addition, the connecting plates on the side of the top loading base plate (see Fig. 8.2b) were applied to connect the testing machine and the top loading base plate, and extra steel chains were also used to prevent the drop of the loading base plate during experiments. The maximum rotation capacity of this loading base plate was 25 degrees and deemed to be adequate for this study. Through this setup, an eccentric load was applied on the GFRP column specimen, i.e. the specimen end was subjected to both axial compressive loading and bending moment simultaneously. During the experiment process, safety ropes were applied around the outer surface of GFRP columns to prevent potential material bursting when the ultimate failure occurs.

In this experimental study, performances of GFRP SHS specimens under compressive loading with three different eccentricities (40 mm, 85 mm and 180 mm) are investigated. The results on axial load and bending moment capacities are compared with the one from axial compression only (i.e. $e = 0$). For all specimens, the compressive loading rate was 1 mm/min. The axial shortening between the two loading base plates were measured by two linear variable displacement transducers (LVDTs) positioned in the central axis of the bottom loading base plate (see Fig. 8.2a, d). One string potentiometer was applied to measure the lateral displacement at the mid-height of the column specimens (see Fig. 8.2d). Two strain gauges were adhered on the compressive and tensile surfaces of GFRP column specimens at the middle height to measure the axial strains, as shown in Fig. 8.2d. The applied loads, displacements and strains were recorded using a data logger. The failure modes were monitored and recorded using a video camera recorder.

8.3 FE Analysis

8.3.1 Model Description

FE analysis was performed to simulate the mechanical responses of the GFRP column specimens under eccentric compression using Ansys. This detailed model was built based on the experimental configuration, including the steel loading base plate, steel bolted sleeve joints, through bolts, GFRP columns, and contact areas between them, as shown in Fig. 8.3. Element Solid45, an 8-node 3D structural element with three translational degrees of freedom, was applied to model the GFRP columns, steel tubes and endplates, and through bolts. Mapped meshing method was applied to achieve fine meshing, and the element size was controlled to be less than 5 mm in the BSJ region. The coincident nodes between the loading base plate and the endplate

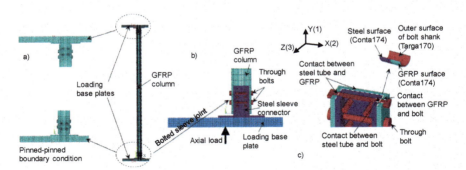

Fig. 8.3 Detailed FE model **a** overall model **b** cross-section of a BSJ **c** contact elements and areas

of the sleeve connector were coupled together. The translational degrees of freedom for the nodes in the central line of the top loading base plate were constrained in all directions; while those on the bottom loading base plate were constrained in x and z directions to simulate the pinned–pinned boundary condition, as shown in Fig. 8.3a. To simplify the model, the bolt head was modelled together with the washer in a cylinder with a diameter of 24 mm and the shank of the bolts was also modelled as a cylinder with a smaller diameter of 12 mm. The pretension of the through bolts was modelled using Pres179 pretension element in the middle of the bolt shank [47], and a 5 kN pretension load (equivalent to the torque value of 30 Nm) was applied on the bolt shank.

The contact/slip behaviours between the bolt shanks and holes, the washers and GFRP column, and the sleeve connector and GFRP column were modelled using a pair of contact elements, i.e. Conta174 and Targa170. As shown in Fig. 8.3c, the outer surface of bolt shanks was defined as target element Targa170, and the inner surface of holes on both steel sleeve connector and GFRP columns as contact element Conta174. The friction coefficient was set as 0.25 [47] between GFRP and steel (GFRP holes and bolt shanks, inner surface of GFRP column and outer surface of steel sleeve connector, outer surface of GFRP column and bolt washer) and 0.44 [50] between steel and steel (steel sleeve connector holes and bolt shanks). The material properties used in the FE analysis were defined according to those listed in Table 8.1. An axial displacement load was applied on the nodes in the central axis of the bottom base plate. To model the behaviours of GFRP column specimens under axial loading, a small initial imperfection with an eccentricity of 1 mm (approximately 1% of the section width) was introduced and a geometric non-linear analysis was further performed. No initial imperfection was introduced when modeling the behaviours of GFRP column specimens under eccentric loading because the applied eccentricity already introduced a geometric nonlinearity to the column specimens.

8.3.2 Failure Criterion

Tsai-Wu failure criterion [51] was applied to predict the failure of GFRP material. In this study, the inverse of Tsai-Wu strength ratio index I_F [52] was used and it is defined as:

$$I_F = 1.0 / \left(-\frac{B}{2A} + \sqrt{\left(\frac{B}{2A}\right)^2 + \frac{1}{A}} \right) \tag{8.1a}$$

and

$$A = \frac{\sigma_1^2}{F_{1t}F_{1c}} - \frac{\sigma_2^2}{F_{2t}F_{2c}} - \frac{\sigma_3^2}{F_{3t}F_{3c}} + \frac{\tau_{12}^2}{F_{12}^2} + \frac{\tau_{23}^2}{F_{23}^2} + \frac{\tau_{13}^2}{F_{13}^2}$$

$$+ c_{12} \frac{\sigma_1 \sigma_2}{\sqrt{F_{1t} F_{1c} F_{2t} F_{2c}}} + c_{23} \frac{\sigma_2 \sigma_3}{\sqrt{F_{2t} F_{2c} F_{3t} F_{3c}}} + c_{13} \frac{\sigma_1 \sigma_3}{\sqrt{F_{1t} F_{1c} F_{3t} F_{3c}}} \quad (8.1b)$$

$$\boldsymbol{B} = \left(\frac{1}{F_{1t}} + \frac{1}{F_{1c}}\right)\sigma_1 + \left(\frac{1}{F_{2t}} + \frac{1}{F_{2c}}\right)\sigma_2 + \left(\frac{1}{F_{3t}} + \frac{1}{F_{3c}}\right)\sigma_3 \quad (8.1c)$$

where σ_i is the stress in the i direction ($i = 1, 2, 3$), which corresponds to y, x, z in the global coordinating system, respectively, see Fig. 8.3c; τ_{ij} is the shear stress in the ij plane ($j = 1, 2, 3$); c_{ij}, the Tsai-Wu coupling coefficients are set as -1; F_{it} and F_{ic} are the tensile and compressive strengths of GFRP composite material in the i direction, which are given in Table 8.1 (where direction 1 corresponds to the longitudinal direction, and direction 2 corresponds to the transverse one); F_{ij} is the shear strength in the ij plane. In this study, the shear strength in different planes (F_{12}, F_{23}, F_{13}) is assumed to be the same and set as 26.7 MPa of the longitudinal shear strength (F_{12}) listed in Table 8.1. If the inverse of Tsai-Wu strength ratio index $I_F >$ 1, the GFRP composite material is considered failed.

8.4 Results and Discussion

8.4.1 Failure Modes

The observed failure modes for all specimens are similar. Firstly, considerable lateral flexural deformation (as shown in Fig. 8.4a) was observed, with displacements increased with the increment in eccentric compressive loads. It should be noted

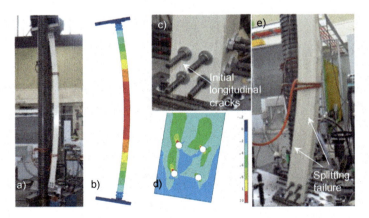

Fig. 8.4 Typical failure modes (on specimen e-85–1): lateral flexural deformation from **a** experimental results ($P = 65$ kN) and **b** FE results ($P = 65$ kN); initial longitudinal cracks from **c** experimental results ($P = 75$ kN) and **d** FE results ($P = 75$ kN); and **e** splitting failure at ultimate load ($P = 88$ kN)

that during the global flexural deformation, no local buckling of the column side plates was observed. In previous experimental studies on pultruded GFRP I section columns under axial compression [53–58] and eccentric compression [24], local buckling on the flange were widely reported. This was possibly due to the relatively larger width-thickness ratio of the GFRP section, as discussed in [46].

As shown in Fig. 8.4c, initial longitudinal cracks were observed as originated at the location between the two rows of bolt holes in the bottom BSJ region on the pressure side of GFRP column specimens. With the increase in loading, longitudinal cracks were also found on the tension side of the column specimens as well. Those longitudinal cracks gradually propagated from the bottom BSJ regions to the middle of the column specimens, and eventually formed the splitting failure as shown in Fig. 8.4e at the ultimate loads.

FE analysis shows consistent results on the failure modes of GFRP column specimens, as the flexural deformed shape shown in Fig. 8.4b. In Fig. 8.4d, Tsai-Wu failure criterion was applied on GFRP materials to indicate the potential failure, and the critical regions with the inverse of Tsai-Wu strength ratio index I_F larger than 1 were marked in red colour. Such critical regions were found near the bolt holes (on their pressed side) and also on the tensile side of the GFRP columns, in consistency with the locations where longitudinal cracks initiated during the experimental process (see Fig. 8.4c).

8.4.2 Load-Axial Shortening Curves

Figure 8.5 presents the load-axial shortening curves for all specimens loaded at different eccentricities. For specimen e-0 loaded without eccentricity, the load-axial shortening curve is linear until the compressive load reached over 160 kN (with the corresponding axial displacement of 6 mm). After that, it gradually remains in a

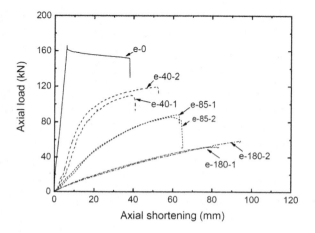

Fig. 8.5 Load-axial shortening curves

Table 8.2 Summary of experimental results

Specimens	Eccentricity (mm)	P_u(kN)	CoV (%)	M_u (kN.m)	CoV (%)	d (mm)	CoV (%)	δ (mm)	CoV (%)
e-0	0	155.9	4.9	0.0	/	34.0	10.3	187.8	7.2
e-40	40	114.8	5.7	4.6	5.7	46.7	17.9	159.8	16.5
e-85	85	87.0	2.4	7.4	2.4	64.5	3.1	153.0	2.0
e-180	180	55.2	9.1	10.0	9.2	88.7	8.4	126.4	13.5

Note P_u = average ultimate axial load; M_u = average ultimate bending moment; d = average maximum axial shortening; δ = average maximum lateral displacement at the mid height of column specimens; CoV = Coefficient of Variation

plateau between 150 to 160 kN alongside the prominent increase in axial shortening. The ultimate load at final splitting failure of specimen e-0 is 151.1 kN, and the maximum value of axial shortening is 38.1 mm. This acquired ultimate load is close to the ones (165 and 152 kN) received previously in [35] with the same section at the same length (3000 mm) under axial compression, and close to the theoretical Euler buckling load (173 kN). The average ultimate load for specimens under axial compression is therefore determined as 155.9 kN.

The load-axial shortening curves for eccentrically loaded specimens, unlike specimen e-0, show nonlinear behaviours from the much earlier stage of the loading process. The slope of those curves gradually decreases as the loading progresses; and also reduces with the increase in the eccentricity. It is also shown that repeated tests on two specimens with the same eccentricity receive close mechanical responses. The detailed experimental results are listed in Table 8.2. It should be noted that in Fig. 8.5, the axial shortening used were the ones measured on the central lines of the loading base plates and they may not be the actual axial shortening of the GFRP columns due to the eccentricity and rotations of the loading base plates. However such measurements are still able to present the geometric nonlinear characteristics of the eccentrically loaded specimens, as illustrated in Fig. 8.5. More accurate structural responses from the columns were taken into account through the measurements of their load and lateral displacement results and strain measurements. Based on the comparison of the ultimate axial loads and axial shortening for different specimens, the average ultimate loads of specimens with eccentricities of 40 mm, 85 mm and 180 mm are 114.8 kN, 87.0 kN and 55.2 kN, corresponding respectively to 73.6%, 55.8% and 35.4% of the average ultimate load of specimen e-0 (155.9 kN) without eccentricity. Meanwhile, the average maximum axial shortening of specimen e-40, e-85 and e-180 are 45.2 mm, 64.5 mm and 88.7 mm, corresponding respectively to 137.2%, 189.5% and 260.5% of the average counterpart value of specimen e-0 (34.0 mm).

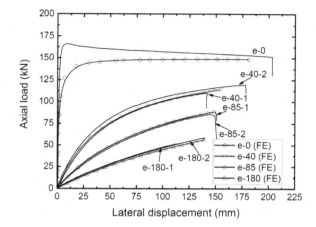

Fig. 8.6 Load-lateral displacement curves

8.4.3 Load-Lateral Displacement Curves

Figure 8.6 presents the load-lateral displacement curves for all specimens loaded at different eccentricities. For specimen e-0, similar to the previous load-axial displacement curve, a long and flat plateau of the axial load between 150 to160 kN was observed until the ultimate failure occurred when the maximum lateral displacement reached 187.8 mm. This again indicated the global buckling of the e-0 specimens. For specimens loaded with eccentricities, the load-lateral displacement curves exhibited nonlinear responses at early stage of the loading. The average maximum lateral displacements of specimens with eccentricities of 40 mm, 85 mm and 180 mm are 159.8 mm, 153.0 mm and 126.4 mm, corresponding respectively to 85.2%, 81.5% and 67.3% of the average counterpart value of specimen e-0 (187.8 mm). FE results on the load-lateral displacement curves are also obtained and compared with experiments in Fig. 8.6, showing good agreement.

8.4.4 Load-Axial Strain Responses

The axial strains measured by strain gauges G1 and G2 at the mid height on column specimens are presented in Fig. 8.7. Due to similarity, only one curve is presented for the two specimens with the same eccentricity. Axial strains on specimen e-0 remain negative values in the loading phase up to 80 kN, indicating both sides are under compression. The two strains from experimental measurements presented bifurcation after the axial load reached 80 kN, and G2 values gradually turn to positive values indicating this side turns to under tension. After the axial strains G1 and G2 significantly bifurcated at the maximum load of 165 kN, the axial load remained at a plateau between 150 and 160 kN, with only the increase in axial strains. The

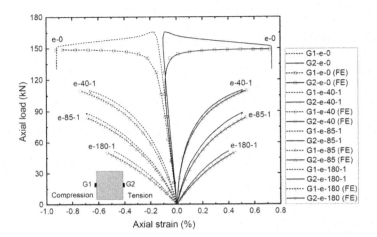

Fig. 8.7 Load-axial strain responses

ultimate axial strains G1 and G2 reached to -0.91% and 0.73% respectively when the specimen failed at the ultimate axial load of 151.1 kN.

For specimens loaded with eccentricities, the axial strains G1 and G2 bifurcated from the beginning of loading and the load-axial strain curves are nonlinear. Compared to the specimen e-0, the ultimate axial strains at the ultimate failure decreased with the increase in the eccentricity. The ultimate axial compressive strains of specimens e-40–1, e-85–1 and e-180–1 are −0.65%, −0.68% and −0.47%, respectively, corresponding to reduction of 28.6%, 25.2% and 48.3% to that of specimen e-0. As shown in Fig. 8.7, FE results on the load-axial strain responses generally compared well with the experimental results. For specimen e-85, the FE result of ultimate compressive strain at the axial load of 88 kN is -0.70%, showing 3% deviation with the experimental result (−0.68%). More obvious difference in axial strain responses was found for specimen e-0 when the strains approached to their bifurcation and this may be due to the effects of initial imperfections (in the FE modelling, only small initial eccentricity of 1 mm was considered as a form of initial imperfection).

8.4.5 FE Verification of Splitting Failure at BSJ Region

In Sect. 8.4.1, the failure initiation in the BSJ region is validated using Tsai-Wu failure criterion. However, the contribution of major stress components (compressive stress, transverse stress or shear stress) to the splitting failure is not identified or fully understood yet, in reference to the FE analysis from Nunes et al. [25] where the shear stress was indicated as the dominant component leading to the initial failure of GFRP I section columns under eccentric compression. In this section, the major stress component for the observed failure mode is investigated using FE analysis

8 Fibre Reinforced Polymer Columns with Bolted Sleeve Joints ... 179

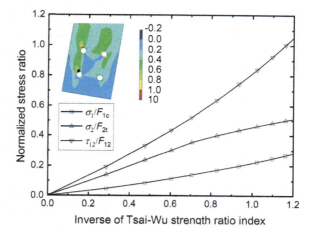

Fig. 8.8 FE results of normalized stress ratios on specimen e-85

and such results would assist in understanding of the ultimate failure and capacity, especially for the specimens with eccentricities.

Figure 8.8 presents the relationship between normalized stress ratios and the inverse of Tsai-Wu strength ratio index I_F in the critical region where longitudinal cracks initiated. Due to similarity, only the results from specimen e-85 are presented, where normalized stress ratios of longitudinal compressive stress (σ_1/F_{1c}), transverse tensile stress (σ_2/F_{2t}) and in-plane shear stress (τ_{12}/F_{12}) in the critical region marked in a black spot are plotted against the Tsai-Wu strength ratio index I_F. As shown, the normalized in-plane shear stress ratio (τ_{12}/F_{12}) is always larger than compressive stress (σ_1/F_{1c}) and tensile stress (σ_2/F_{2t}) ratios, especially when I_F approaches 1. When I_F reaches 1 which indicates the failure initiation of GFRP materials, the in-plane shear stress ratio τ_{12}/F_{12} is 0.82 for specimens e-85, which is significantly higher than the tensile stress ratio σ_2/F_{2t} of 0.46 and compressive stress ratio σ_1/F_{1c} of 0.22. This result well evidences that the in-plane shear stress has the dominant contribution for the splitting failure of GFRP columns with BSJs under eccentric compression.

8.4.6 Load-Bearing Capacities of GFRP Columns with BSJs

The design equation of nominal load-bearing capacity (without consideration of resistance factors) for pultruded GFRP columns under combined loading of compression and bending from guidelines [22, 59] is given as:

$$\frac{P}{P_u} + \frac{M}{M_u} = 1 \tag{8.2}$$

where P_u is the least value among the critical axial loads of the pultruded GFRP columns under different failure modes, i.e. local and global buckling, and material compressive failure; M_u is the least value of the critical bending moments which causes either local and global buckling or material compressive or tensile failure. This equation is built based on the linear interaction between axial loading and bending moment [23, 24]. In this study, the applied bending moment at the end of GFRP columns can be determined as:

$$M = P \cdot e \qquad (8.3)$$

where P is the applied axial load, and e is the eccentricity.

By substituting Eq. (8.3) into Eq. (8.2), the relationship between the axial load and P eccentricity e can be drawn as:

$$P\left(\frac{1}{P_u} + \frac{e}{M_u}\right) = 1 \qquad (8.4)$$

and it can be reformed as

$$\frac{1}{P} = \frac{1}{P_u} + \frac{1}{M_u} \cdot e \qquad (8.5)$$

In Eq. (8.5), P_u and M_u are constants for a pultruded GFRP column once its material property and geometry are given. In this study, the values of P_u (155.9 kN) and M_u (22.1 kN m) are received from this study and previous studies on the same GFRP section under axial compression [46] with 3000 mm height and pure bending [28] with a similar span length of 2700 mm. By applying the P_u and M_u values into Eq. (8.5), a design approach describing the relationship between the load-bearing capacity of the GFRP columns used in this study and eccentricities is obtained in Fig. 8.9. Experimental results of ultimate loads and eccentricities for the GFRP columns investigated in this study are also included in Fig. 8.9 as individual points.

Comparison between the experimental results and design guideline shows the ultimate load-bearing capacities from experiments for all eccentrically loaded specimens (e-40, e-85 and e-180) are lower than the values in the design guideline. The deviations increase with the increase in eccentricity, and the largest deviation is associated with the specimens e-180 being up to 17% and 27%. Such overestimations from design approaches that experimental results are due to the observed failure mode occurred in the BSJ region of column specimens rather than the member failure as considered in the design approaches. Because connections in GFRP structures may become critical components in the system, it is necessary to take into account the joint capacity in comparison to the member capacity. Therefore in the following the axial load and bending moment interaction of GFRP columns with consideration of the BSJ capacity is further presented in the following Sect. 8.4.7.

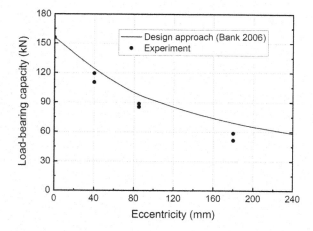

Fig. 8.9 Relationship between load-bearing capacity and eccentricity

8.4.7 P-M Interaction Curve

Figure 8.10 presents the *P-M* interaction curve for GFRP SHS columns with BSJs used in this study. As shown, experimental results from the GFRP column specimens investigated in this study are included as individual points. Experimental results on the ultimate bending moment of the same GFRP section (M_u = 22.1 kN.m, [28]) and the same BSJ (M_u = 11.5 kN.m, [44]) are also included. The bending capacity of the BSJ was experimentally investigated in [44], where the bolt holes dimensions and positions on the compression and tension sides of the BSJ under bending are in consistency with the ones used in this study, and the same batch of through bolts was used.

Experimental results from this study show that the moment capacity of GFRP columns with BSJs increases with the increase in eccentricity, along with the

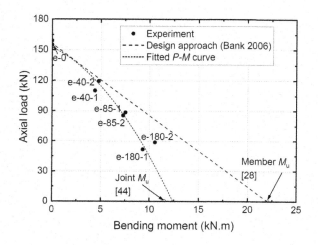

Fig. 8.10 *P-M* interaction curve and comparison with design approach

load-bearing capacities decreases substantially; this forms a nonlinear relationship between axial loads and bending moments. Specimen e-180–1 and 180–2 with the largest eccentricity of 180 mm presented their moment capacities of 9.3 kNm and 10.6 kNm, approaching the joint moment capacity of 11.5 kNm. The interaction curve from the design approaches [59] forms a linear relationship between the load-bearing capacity and moment capacity of members, according to Eq. (8.5). The nonlinear P-M interaction curve of GFRP SHS columns with BSJs behaves differently to the design approach especially when the curve is approaching to cases with larger bending moments (i.e. associated with larger eccentricities). Again this behavior is dominated by the joint capacity rather than the member capacity and due to the relatively low moment capacity of the BSJ in this study (only 52%) comparing to the GFRP member. This interaction curve indicates the joint capacity in developing GFRP structural applications is critical and should be carefully considered. It also implied that such interaction curves may also perform differently as member capacities of GFRP columns may vary with their diverse ultimate failure modes introduced by the material characteristics different from steel and concrete.

8.5 Conclusions

Experimental and numerical studies on pultruded GFRP SHS columns subjected to eccentric compression are presented in this chapter. Such column specimens are with bolted sleeve joints (BSJ) on both ends and are mounted on pined-pined end supports and loaded at three different eccentricities. Results on the failure modes, load-bearing capacities, axial and lateral displacements, and strain responses were obtained from the experiments. In addition, an interaction curve of the axial load and bending moment of GFRP SHS column specimens was obtained and compared with design approaches for such specimens. The main findings can be concluded as follows:

1. Lateral flexural deformation was observed for all specimens, and this is associated with the global buckling behaviour of such slender column specimens without local buckling during the loading. Longitudinal cracks are found subsequently initiated between two rows of bolt holes in the bottom BSJ region on the pressure side of GFRP column specimens, and gradually propagated to the middle of the column alongside the increasing loading, till ultimate splitting failure on the GFRP composites.
2. The increase in the eccentricity leads to a decrease in the stiffness of GFRP specimens from both the load versus axial shortening and lateral displacement curves. For eccentrically loaded specimens, the load–displacement curves become more nonlinear because of combined compressive loads and bending moments in association with second order effects. The ultimate compressive loads decreased substantially with the increase in eccentricity. The deviations of the compressive load at the ultimate state between the experiments and design

approaches increase when the eccentricity increases. This is because the found failure was in the BSJ region rather than the member failure suggested from design approaches.
3. A *P-M* interaction curve for GFRP SHS columns with BSJs is formed based on the experimental results. It shows a nonlinear behaviour between the axial loads and bending moments, instead of a linear relationship suggested from design approaches. The ultimate bending moments for specimen e-180 with the largest eccentricity are approaching the BSJ moment capacity but much less (48%) than the member moment capacity under pure bending. This nonlinear *P-M* interaction curve of GFRP columns with BSJ indicates the importance of joint design when developing GFRP members for structural applications.
4. FE results show good agreement with experimental results in terms of the failure modes, load-lateral displacement curves and load-strain responses. Predicting the splitting failure of GFRP composite materials in the BSJ region using Tsai-Wu failure criterion receives consistent results with experimental results. Further stress analysis indicates the shear stress is the dominant component which leads the splitting failure.

References

1. Bakis CE, Bank LC, Brown VL, Cosenza E et al (2002) Fiber-reinforced polymer composites for construction-state-of-the-art review. J Compos Constr 6(2):73–87
2. Hollaway LC (2010) A review of the present and future utilisation of FRP composites in the civil infrastructure with reference to their important in-service properties. Constr Build Mater 24(12):2419–2445
3. Bank LC, Bednarczyk PJ (1990) Deflection of Thin-Walled Fiber-Reinforced Composite Beams. J Reinf Plast Compos 9(2):118–126
4. Borowicz DT, Bank LC (2011) Behavior of Pultruded Fiber-Reinforced Polymer Beams Subjected to Concentrated Loads in the Plane of the Web. J Compos Constr 15(2):229–238
5. Correia JR, Branco F, Gonilha J, Silva N et al (2010) Glass Fibre Reinforced Polymer Pultruded Flexural Members: Assessment of Existing Design Methods. Struct Eng Int 20(4):362–369
6. Correia J.R., Branco F.A., Silva N.M.F., Camotim D., et al. (2011) First-order, buckling and post-buckling behaviour of GFRP pultruded beams. Part 1: Experimental study. Computers & Structures 89(21–22): p. 2052–2064.
7. Bank LC, Gentry TR, Nadipelli M (1996) Local buckling of pultruded FRP beams—analysis and design. J Reinf Plast Compos 15(3):283–294
8. Nguyen TT, Chan TM, Mottram JT (2014) Lateral-torsional buckling resistance by testing for pultruded FRP beams under different loading and displacement boundary conditions. Compos B Eng 60:306–318
9. Ascione L, Giordano A, Spadea S (2011) Lateral buckling of pultruded FRP beams. Compos B Eng 42(4):819–824
10. Turvey GJ (1996) Lateral buckling tests on rectangular crosssection pultruded GRP cantilever beams. Compos B Eng 27(1):35–42
11. Ascione F, Feo L, Lamberti M, Minghini F et al (2016) A closed-form equation for the local buckling moment of pultruded FRP I-beams in major-axis bending. Compos B Eng 97:292–299
12. Hashem ZA, Yuan RL (2000) Experimental and analytical investigations on short GFRP composite compression members. Compos B Eng 31(6–7):611–618

13. Hashem ZA, Yuan RL (2001) Short versus long column behavior of pultruded glass-fiber reinforced polymer composites. Constr Build Mater 15(8):369–378
14. Lane A, Mottram JT (2002) Influence of modal coupling on the buckling of concentrically loaded pultruded fibre-reinforced plastic columns. Proc Inst Mech Eng Part L-J Mater-Des Appl 216(L2):133–144
15. Zureick A, Scott D (1997) Short-term behavior and design of fiber-reinforced polymeric slender members under axial compression. J Compos Constr 1(4):140–149
16. Barbero E, Raftoyiannis I (1993) Local buckling of FRP beams and columns. J Mater Civ Eng 5(3):339–355
17. Bai Y, Keller T (2009) Shear failure of pultruded fiber-reinforced polymer composites under axial compression. J Compos Constr 13(3):234–242
18. Bai Y, Vallée T, Keller T (2009) Delamination of pultruded glass fiber-reinforced polymer composites subjected to axial compression. Compos Struct 91(1):66–73
19. Gliszczynski A, Kubiak T, Borkowski L (2018) Experimental investigation of pre-damaged thin-walled channel section column subjected to compression. Compos B Eng 147:56–68
20. Group TESPC (1996) Eurocomp design code and handbook, in structural design of polymer composites. CRC Press
21. CNR (National Research Council of Italy) (2008) Guide for the design and construction of structures made of FRP pultruded elements, in Technical Document CNR-DT 205/2007, Rome
22. Engineers ASoC (2010) Pre-Standard for Load & Resistance Factor Design (LRFD) of Pultruded Fiber Reinforced Polymer (FRP) Structures, American Composites Manufacturers Association
23. Barbero ET, Malek, (2000) Experimental investigation of beam-column behavior of pultruded structural shapes. J Reinf Plast Compos 19(3):249–265
24. Mottram JT, Brown ND, Anderson D (2003) Buckling characteristics of pultruded glass fibre reinforced plastic columns under moment gradient. Thin-Walled Struct 41(7):619–638
25. Nunes F (2013) Correia M., Correia J.R., Silvestre N., et al., Experimental and numerical study on the structural behavior of eccentrically loaded GFRP columns. Thin-Walled Struct 72:175–187
26. Kollár L (2003) Local buckling of fiber reinforced plastic composite structural members with open and closed cross sections. J Struct Eng 129(11):1503–1513
27. Bai Y, Keller T, Wu C (2013) Pre-buckling and post-buckling failure at web-flange junction of pultruded GFRP beams. Mater Struct 46(7):1143–1154
28. Satasivam S, Bai Y (2014) Mechanical performance of bolted modular GFRP composite sandwich structures using standard and blind bolts. Compos Struct 117:59–70
29. Feo L, Mosallam AS, Penna R (2013) Mechanical behavior of web–flange junctions of thin-walled pultruded I-profiles: an experimental and numerical evaluation. Compos B Eng 48:18–39
30. Turvey GJ, Zhang Y (2006) Shear failure strength of web–flange junctions in pultruded GRP WF profiles. Constr Build Mater 20(1–2):81–89
31. Mottram JT, Turvey GJ (2003) Physical test data for the appraisal of design procedures for bolted joints in pultruded FRP structural shapes and systems. Prog Struct Mat Eng 5(4):195–222
32. Girão Coelho AM, Mottram JT (2015) A review of the behaviour and analysis of bolted connections and joints in pultruded fibre reinforced polymers. Mater Des 74:86–107
33. Luo F, Bai Y, Yang X, Lu Y (2015) Bolted sleeve joints for connecting pultruded FRP tubular components. J Compos Constr 20(1)
34. Keller T, Vallée T (2005) Adhesively bonded lap joints from pultruded GFRP profiles. Part I: stress–strain analysis and failure modes. Compos Part B: Eng 36(4):331–340
35. Keller T, Vallée T (2005) Adhesively bonded lap joints from pultruded GFRP profiles. Part II: joint strength prediction. Compos Part B: Eng 36(4):341–350
36. Ascione F, Lamberti M, Razaqpur AG, Spadea S (2017) Strength and stiffness of adhesively bonded GFRP beam-column moment resisting connections. Compos Struct 160:1248–1257
37. Zhang ZJ, Bai Y, Xiao X (2018) Bonded sleeve connections for joining tubular glass fiber–reinforced polymer beams and columns: experimental and numerical studies. J Compos Constr 22(4)

38. Keller T, Bai Y, Vallée T (2007) Long-term performance of a glass fiber reinforced polymer truss bridge. J Compos Constr 11(1):99–108
39. Bai Y, Yang X (2013) Novel joint for assembly of all-composite space truss structures: conceptual design and preliminary study. J Compos Constr 17(1):130–138
40. Clarke JL (1996) eurocomp design code and handbook. CRC Press, Structural Design of Polymer Composites
41. Bai Y, Keller T (2011) Effects of thermal loading history on structural adhesive modulus across glass transition. Constr Build Mater 25:2162–2168
42. Nguyen TC, Bai Y, Zhao XL and Al-Mahaidi R (2011) Mechanical characterization of steel/CFRP double strap joints at elevated temperatures. Compos Struct 93:1604–1612
43. Luo F, Yang X, Bai Y (2015) Member capacity of pultruded gfrp tubular profile with bolted sleeve joints for assembly of latticed structures. J Compos Constr 20(3)
44. Wu C, Zhang Z, Bai Y (2016) Connections of tubular GFRP wall studs to steel beams for building construction. Compos B Eng 95:64–75
45. Satasivam S, Bai Y, Yang Y, Zhu L et al (2018) Mechanical performance of two-way modular FRP sandwich slabs. Compos Struct 184:904–916
46. Xie L, Bai Y, Qi Y, Caprani C et al (2018) Effect of width–thickness ratio on capacity of pultruded square hollow polymer columns. Proc Inst Civil Eng Struct Build 171(11):842–854
47. Zhang Z, Wu C, Nie X, Bai Y et al (2016) Bonded sleeve connections for joining tubular GFRP beam to steel member: numerical investigation with experimental validation. Compos Struct 157:51–61
48. Zhang Z, Bai Y, He X, Jin L et al (2018) Cyclic performance of bonded sleeve beam-column connections for FRP tubular sections. Compos B Eng 142:171–182
49. Cardoso DCT, Harries KA, Batista EDM (2014) Compressive strength equation for GFRP square tube columns. Compos B Eng 59:1–11
50. Shi G, Shi Y, Wang Y, Bradford MA (2008) Numerical simulation of steel pretensioned bolted end-plate connections of different types and details. Eng Struct 30(10):2677–2686
51. Tsai SW, Wu EM (1971) A general theory of strength for anisotropic materials. J Compos Mater 5(1):58–80
52. ANSYS Mechanical APDL Material Reference Release 15.0 (2013) ANSYS Inc.
53. Turvey GJ, Zhang Y (2006) A computational and experimental analysis of the buckling, post-buckling and initial failure of pultruded GRP columns. Comput Struct 84(22–23):1527–1537
54. Pecce M, Cosenza E (2000) Local buckling curves for the design of FRP profiles. Thin-Walled Struct 37(3):207–222
55. Tomblin J, Barbero E (1994) Local buckling experiments on FRP columns. Thin-Walled Struct 18(2):97–116
56. Cardoso D, Harries K, Batista E (2014) Compressive local buckling of pultruded GFRP I-sections: development and numerical/experimental evaluation of an explicit equation. J Compos Constr 19(2)
57. Mottram JT, Brown ND, Anderson D (2003) Physical testing for concentrically loaded columns of pultruded glass fibre reinforced plastic profile. Proc ICE Struct Build 156:205–219
58. Correia MM, Nunes F, Correia JR, Silvestre N (2013) Buckling behavior and failure of hybrid fiber-reinforced polymer pultruded short columns. J Compos Constr 17(4):463–475
59. Bank LC (2006) Composites for construction: structural design with FRP materials. Wiley, New York

Chapter 9
Connections of Fibre Reinforced Polymer to Steel Members: Experiments

Chao Wu, Zhujing Zhang, and Yu Bai

Abstract Timber and steel studs or posts are commonly used in wall constructions for buildings. In this context and with the results from previous chapters, pultruded glass fibre reinforced polymer (GFRP) studs may provide an alternative solution considering their light weight and improved durability. However, integrating the GFRP wall studs to a steel frame structure is challenging, as proper connection methods are required. A sleeve connection was proposed and examined in this chapter for wall studs to steel beams. Pultruded GFRP stud was fastened to the sleeve connector by one of three methods: ordinary bolt, one-sided bolt and adhesive bond. The connector was then fastened to the steel beam through ordinary bolts. Connections with conventional steel angles were also prepared for comparison purpose. A series of moment-rotation experiments were conducted on these stud-to-beam connections. In addition, two stud lengths were designed in order to study the connection behaviour under shear force dominant loading and moment dominant loading conditions. Experimental results were obtained including failure mode, moment-rotation response, shear-rotation response, joint rotational stiffness and capacity. It was found that the bonded sleeve connection outperformed all the other connections and was classified as a rigid and partial strength connection.

9.1 Introduction

Steel frame structures have become a major structural system for building construction in many countries. The cost of these steel frame buildings can be largely reduced by integrating light weight walls, as examined by many researchers [1, 2]. Traditional lightweight wall systems normally incorporate external facing sheets sandwiching

---

Reprinted from Composites Part B: Engineering, 95, Chao Wu, Zhujing Zhang, Yu Bai, Connections of tubular GFRP wall studs to steel beams for building construction, 64–75, Copyright 2016, with permission from Elsevier.

C. Wu · Z. Zhang · Y. Bai (✉)
Department of Civil Engineering, Monash University, Clayton, Australia
e-mail: yu.bai@monash.edu

a series of wall studs or posts. These wall studs could be tubular or open steel sections or solid timber sections [3, 4]. Pultruded glass fibre reinforced polymer (GFRP) profiles have great potential to offer an alternative solution as wall studs in building constructions [5–9]. These profiles exhibit desirable advantages including light weight, low thermal conductivity, and high resistance to corrosion and chemical attack [10–17]. Such advantages enable the design and construction of structural systems [18–23] with reduced maintenance cost and carbon dioxide emission [9, 10]. For example, a built-up web-flange sandwich configuration has been developed for beam/slab applications in buildings [25, 26] and introduced in previous chapters. This built-up section consists of a series of standard pultruded GFRP box or I profiles sandwiched between two pultruded GFRP flat panels. Such a sandwich configuration may be further incorporated as a wall in the frame structures. However reliable connection methods are necessary in order to transfer the load between the pultruded GFRP studs or posts and the surrounding steel beams.

Although the information on connecting GFRP wall studs to surrounding frame structure is limited in the literature, valuable experiences can be referred to in published work on pultruded GFRP beam-to-column connections [27–35]. Considering the low moduli of the pultruded GFRP profiles, semi-rigid connection is mandatory in order to increase the overall structural stiffness [28, 36]. The efforts started with emulating methods that had been widely used for connecting steel members. The pultruded seated angles were used as web and flange cleats and fastened by bolts to connect pultruded GFRP beams and columns by Bank et al. [33] (see Fig. 9.1a). Due to the inherent anisotropic nature of pultruded GFRP profiles (which is very different from the isotropic steel members), these steel-like connections leaded to premature failures such as web-flange separation of the pultruded column, and fracture of the pultruded cleat angles. Reinforcing elements were added later in [28] to the column web to avoid those premature failures. Through bolts were also used to engage both column flanges in carrying the load. However, those modified connections may become complicated and therefore may not be economical. After understanding the unique failure modes of these pultruded connectors, a prototype connecting element was then successfully proposed in [34] which was made by wrapping a pultruded GFRP angle with two layers of composites sheeting (Fig. 9.1b). Connection using

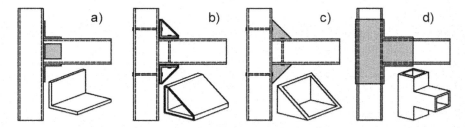

Fig. 9.1 a seated angle connection [23]. **b** wrapped angle connection [29]. **c** universal connector. [25] **d** cuff connection [32] (not to scale)

this prototype connector developed approximately twice the strength of the connection with steel-like connector (Fig. 9.1a), though only one-third of the initial stiffness was obtained.

A universal connector (UC) was proposed in [30, 36] as a step forward in the pultruded GFRP beam-to-column connections [27]. This UC connector is similar to the prototype wrapped angle connector in [34], but with the gusset plates more directly integrated into the seated angles (see Fig. 9.1c). An increase in the moment capacity by 280% was achieved in comparison to the connection in Fig. 9.1a, and the connection behaved as a rigid joint at the initial moment-rotation stage [30]. Since the UC connector was made of GFRP composites, limitations were observed with local failure in the UC connector and shear punching of the bolts through the column flange. These local connector failures are not desirable due to the brittleness and lack of energy dissipation capacity of GFRP composites. In the review article [27], steel connectors were proposed to ensure the connecting elements not being the weak link in the beam-to-column connections. In the experimental studies on the connections as shown in Fig. 9.1a [38, 39], it was found that the moment capacity and stiffness were increased by 70% and 78% respectively, when steel seated angles were used instead of the pre-preg GFRP seated angles [38, 39]. Such connectors continued resembling the geometry of those used in the steel constructions which may not be suitable for connecting the pultruded GFRP profiles. It was also noticed that the connection failure shifted from the steel angle connector to the GFRP beams, leading to a limited plasticity in the moment-rotation response of the connection.

A cuff connector was also proposed [37, 40] as a single monolithic unit made through vacuum assisted resin transfer molding (VARTM) of stitched E-glass fabric and resin (see Fig. 9.1d). It was intended for connecting pultruded GFRP members of tubular sections which can fit into the hollow section of the cuff. The fabrication and testing of the cuff connections were later conducted in [41]. It was found that the moment capacity doubled the value of the connection in Fig. 9.1a while the stiffness kept almost unchanged. However, the cuff connection is hardly economical when repetitive VARTM fabrication process is necessary for connecting beams and columns of different sizes. In addition, premature bond failure between beam and cuff was observed [31, 41] indicating a need for bond quality control. Similar connections were also employed when developing a truss structure assembled with FRP tubular sections [42, 43].

From the above studies, several points have to be considered in the design of the connections for pultruded GFRP studs. For example, simple mimicking practices in steel structures may lead to poor connection performance due to the anisotropic mechanical properties of GFRP composites. Also, semi-rigid connection is necessary for compensating the low moduli of the GFRP composites, so that maximizing the connection stiffness is important. Further the connector should be geometrically flexible and suitable for connecting a large variety of pultruded profiles of difference sizes. Moreover, the benefits of using steel components can be considered which ensure the improvement in the connection moment capacity and stiffness. Steel components also introduce plasticity and energy dissipation characteristics to the connection. In addition, connection with cuff geometry may be desirable through the integrated contributions from all the connected components with improved mechanical properties.

This chapter investigates the connection methods between pultruded GFRP wall studs to steel beam, with the expectation of incorporating the pultruded GFRP sandwich wall for light-weight steel frame construction. Inspired by the cuff configuration developed in [37, 40], a sleeve connector was designed suitable for connecting the GFRP wall stud to a steel beam. It was expected that the proposed connection system would have improved stiffness and strength, and ductile failure mechanisms. A series of moment-rotation experiments were conducted to evaluate the mechanical performance of the connections. Experimental results were reported in terms of the failure mode, moment-rotation response, shear-rotation response, rotational stiffness, moment and shear capacity. Comparisons with the conventional seated angle connection were also made. Finally, the proposed connection method was classified by the capacity and rigidity according to the existing design guidelines.

9.2 Experimental Program

9.2.1 Connection Design

A steel sleeve connector welded with an end plate is proposed for connecting the pultruded GFRP tubular wall stud to a steel I beam using bolts (see Fig. 9.2a). It is named as 'sleeve' because the connector is inserted in the GFRP tubular section, which may provide similar functions as the cuff connection proposed in [37, 40]. When connecting the stud and the beam, the steel tube is inserted in the stud hollow section, and the end plate is bolted to the flange of the I beam. In order to compare with the conventional steel angle connection as shown in Fig. 9.2b, only the flanges of the GFRP stud were fastened to the steel tube when connected using bolts.

Both sleeve and angle connections were designed with the same geometric parameters including the bolt locations, tube length and angle leg length, thickness of steel plates. This is to exclude effects of geometric differences on the performances of the two types of connections. Considering the lack of design standards for connecting GFRP profiles to steel members, the design requirements specified in steel standard (AS 4100 [44]) and GFRP composite guidelines [45], Italian guidelines CNR-DT 205/2007 [46], and ASCE pre-standard [47] were referred to when determining the connection dimensions in Fig. 9.2. The requirements in these standards on critical parameters are presented in Table 9.1. Side distance means the distance from the bolt centre to the side edge of the plate. End distance is the distance from the bolt centre to the plate end. Pitch distance is the distance between the centres of two bolts. As

9 Connections of Fibre Reinforced Polymer to Steel ...

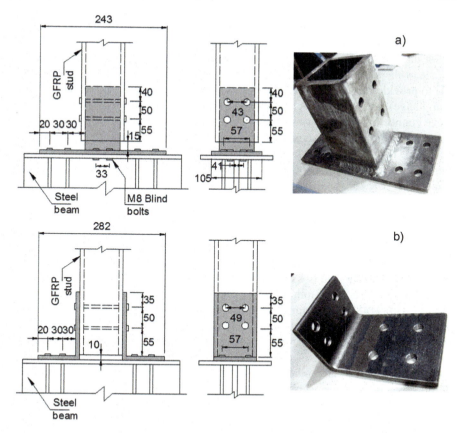

Fig. 9.2 Detailed dimensions of **a** sleeve connection; **b** seated angle connection (unit in mm, all bolts are M12 except clarified otherwise)

Table 9.1 Design requirements for steel connections and GFRP connections with parameters used in the current study

Design parameters	Steel	FRP			Current study
	[39]	[40]	[41]	[42]	
Hole clearance	≤2 mm	≤5%d*	≤1 mm	≤1.6 mm	0.5 mm
Bold diameter	N/A	$t^* \leq d \leq 1.5t$	$t \leq d \leq 1.5t$	$9.53 \leq d \leq 25.4$	12 mm
Side distance	≥1.5d	≥1.5d	≥0.5d	≥1.5d	1.5d
End distance	≥1.5d	≥3d	≥4d	≥2d	4d
Pitch distance	≥2.5d	≥3d	≥4d	≥4d	4d

*d is the bolt diameter; t is the plate thickness

can be seen in Table 9.1, the final design parameters in Fig. 9.2 satisfied all the steel and GFRP design requirements.

For the steel angle connection, the GFRP stud had a gap of 10 mm to the flange surface of the steel beam (Fig. 9.2b). This gap was selected according to AS 4100 [44] to provide rotational freedom during the loading process (stud not contacting beam flange), and to offer installation tolerance of the connection [27]. For the steel sleeve connection, the gap was 15 mm (from end of the stud to the surface of the end plate in Fig. 9.2a). This gap was also for rotational freedom of the stud and accommodated the welding size ranging from 5 to 10 mm. Leaving a gap between connecting components has been a common practice in similar work on the GFRP connections [27, 28, 30, 38].

9.2.2 Materials

The pultruded GFRP stud is a square hollow section (SHS) with a measured dimension of 102 × 102 × 9.5 mm. Tensile coupon tests were conducted according to ASTM D-3039 [48] to determine the tensile Young's modulus and tensile strength in the pultrusion direction. The interlaminar shear strength was obtained through short-beam tests according to ASTM D-2344 [49]. Five coupons were tested for each property and the results are listed in Table 9.2. The GFRP stud has E-glass fibres as reinforcements which are embedded in a polyester resin. The fibre architecture and the fibre volume fraction were investigated according to ASTM D-3171 [50]. The GFRP tube has symmetric and balanced fibre rovings in the centre sandwiched by two layers of CFM mats on each side. The average measured fibre volume fraction was 46.7% considering the density of E-glass fibre of 2.57 g/cm^3 [19].

The steel I beam has a flange width of 153 mm and the total height of the section is 158 mm. The flange is 9.4 mm thick and the web thickness is 6.6 mm. The root

Table 9.2 Mechanical properties of GFRP stud, steel beam, sleeve connector and steel angle

Components	Yield strength (MPa)	Yield strain (%)	Ultimate strength (MPa)	Tensile modulus (GPa)	Interlaminar shear strength (MPa)
GFRP stud	N/A	N/A	306.5	30.2	26.7
Steel beam flange	356.9	0.169	513.7	211.6	N/A
Steel beam web	366.7	0.168	527.3	221.1	N/A
Sleeve tube plate	326.2	0.154	519.7	216.9	N/A
Sleeve end plate	332.4	0.153	511.1	218.8	N/A
Steel angle	349.7	0.159	496.7	220.7	N/A

radius between the web and flange is 8.9 mm. Coupons were cut from the web and the flange and the tensile tests were conducted to determine the tensile properties of the steel beam according to ASTM E-8/E-8 M [51]. The mechanical properties of the steel beam are listed in Table 9.2.

For the sleeve connector, four steel plates of 10 mm thick were welded to form a tubular section (or SHS) which was then welded to a 10 mm thick end plate. The tube has a length of 160 mm. The external dimension of the tube was controlled so that a 0.5 mm gap between the external surface of the steel tube and the internal surface of the GFRP stud was achieved. This gap ensured the ease of inserting the tube into the GFRP stud. The thickness of the steel angle was also 10 mm thick for the purpose of comparison with the sleeve connector. One leg of the steel angle is 150 mm long and the other leg is 90 mm long with a root radius of 10 mm. Tensile coupon tests were conducted according to ASTM [51]. The mechanical properties of the steel tube, end plate, and steel angle are presented in Table 9.2. M12/8.8 all threaded rods with a length of 250 mm were used to fasten the connector (either sleeve or angle) to the steel beam. Both beam flanges were anchored using these rods so that the whole I section participated in the connection action. These rods had a minimum tensile proof load of 48.9 kN, and an ultimate tensile load of 67.4 kN.

When connecting the sleeve connector to the GFRP stud, one of the three methods was used: ordinary bolt, one-sided bolt and adhesive bond. The aim is to understand how different connection methods behave so that the optimal solution can be selected for future applications. The ordinary bolts were M12/8.8 with a length of 150 mm, a minimum tensile proof load of 48.9 kN, and an ultimate tensile load of 67.4 kN (the same as the all threaded rod of class 8.8). M12 one-sided bolts were with a length of 70 mm, a tensile capacity of 18.8 kN, and a shear capacity of 33.7 kN. The one-sided bolts are convenient especially for connecting tubular sections when the access is limited for tightening the ordinary bolts (as for the sleeve connector in this chapter). More detailed configuration and installation information of the one-sided bolts can be found in [32]. M8 one-sided bolts were used for connecting the end plate of the sleeve connector to the steel beam. The length was 50 mm, with a tensile capacity of 6.9 kN, and a shear capacity of 14.6 kN. Flat washers of grade ZY (ASTM F-436 [52]) were used which were 2.6 mm thick with external/internal diameters of 26.3/14.2 mm, respectively. Plain hex nuts of class 8 according to AS 1112.1 [53] were used which had a minimum proof load of 74.2 kN.

Araldite 420 epoxy was used as adhesive for bonding the GFRP stud and the sleeve connector. The properties of Araldite 420 were reported in [54]. The tensile strength and the modulus were 28.6 MPa/1.9 GPa, respectively, and the shear strength was 25.0 MPa. When bonding the GFRP stud and the sleeve connector, firstly, the resin layer of the internal surface of the stud was removed using a 60-grit sandpaper. This was to expose the internal fibres of the stud for better adhesion. Then the external surface of the sleeve connector was sand blasted to expose the fresh steel surface. A thick layer of adhesive was then applied on the internal surface of the stud and the external surface of the sleeve connector. The steel tube was then inserted in the stud and the extra adhesive was squeezed out. The connection was then held in position for curing for at least seven days before the subsequent testing.

When tightening the bolts, a clamping force of 100 Nm was applied to steel-to-steel components (i.e. connector to steel beam). This bolt torque was applied to eliminate the connection slip when adhesive bonding was not present [27, 38, 39]. When connecting GFRP stud, a lower torque of 30 Nm was applied to avoid any premature crush of the GFRP composites. This torque was also comparable to the similar practices on GFRP connections in the literature [30, 55, 56].

9.2.3 Specimens

Two groups of connections were prepared for the subsequent moment-rotation testing. Each group contained four specimens, including three sleeve connections (Fig. 9.2a) according to the three different methods (i.e. ordinary bolt, one-sided bolt and adhesive bond), and one steel angle connection (Fig. 9.2b). The two groups of connections had the same details except for the cantilever length of the stud. The stud was 600 mm long in the first group and 1700 mm long in the second. These two lengths were selected to investigate the connection response under the shear dominant loading (600 mm) and the moment dominant loading (1700 mm) conditions. Correspondingly, the specimens with 600 mm stud length are named as the shear dominant connections and those with 1700 mm stud length are named as the moment dominant connections. In addition, for the wall stud applications in buildings with a level height of 3–4 m, the stud has zero bending moment approximately at the mid height under in-plane shear loading with fix–fix ends constrains. Therefore, 1700 mm length stud under cantilever loading condition (see Fig. 9.3) in this chapter is also a representative scenario for such structural applications.

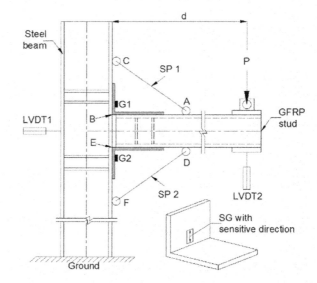

Fig. 9.3 Typical experimental setup and instrumentation (not to scale; 'SP1' and 'SP2': string pots used to measure the extension of 'AC' and 'DF'; 'G1' and 'G2': strain gauges; 'P': applied loading at the end of GFRP stud; 'd': the span length of the cantilever GFRP stud)

9 Connections of Fibre Reinforced Polymer to Steel … 195

Table 9.3 Specimens and the experimental results

Specimens	$K_{e,M}$ (kN.m/rad)	$K_{e,V}$ (kN/rad)	M_e (kN.m)	V_e (kN)	M_u (kN.m)	V_u (kN)
S-AG-OB	206.3	341.0	6.0	10.0	11.2	18.5
S-SL-OB	246.4	407.3	7.8	12.9	11.5	19.0
S-SL-BB	280.7	464.0	7.8	12.9	11.4	18.8
S-SL-AB	911.0	1505.8	8.3	13.7	19.4	32.1
M-AG-OB	189.1	110.6	6.3	3.7	9.7	5.7
M-SL-OB	278.0	162.6	8.2	4.8	11.1	6.5
M-SL-BB	320.8	187.6	7.9	4.6	9.9	5.8
M-SL-AB	1019.9	596.4	8.5	5.0	≥ 15.5*	≥ 9.1*

* The test was stopped due to the stretch limit of the load cell was reached. Higher ultimate moment capacity should be expected if the specimen were loaded further until failure

Eight specimens are prepared as listed in Table 9.3 (see Sect. 9.3.4). The label of each specimen contains three parts. In the first part, 'S' means the shear force dominant connections (with a stud length of 600 mm) while 'M' refers to the moment dominant connections (with a stud length of 1700 mm). The type of connector, 'angle' or 'sleeve', can be identified in the second part of the label as 'AG' or 'SL', respectively. The last part of the label refers to the method that is used for connecting the sleeve and the stud: 'OB' for ordinary bolt, 'BB' for one-sided bolt and 'AB' for adhesive bonding. These abbreviations will be used in the following discussions except specified otherwise.

9.2.4 *Experimental Set-Up and Instrumentation*

Figure 9.3 shows a typical cantilever set-up of an AG connection. The same set-up was used for all the other connections. The steel beam was in a vertical direction and fixed rigidly to the ground, while the GFRP stud was in a horizontal position. This set-up was designed for the ease of the load application through an Instron loading cell with 250 kN capacity. The loading speed was 1 mm per min. Each connection was loaded until a peak load was achieved and the load started to drop, or the maximum stretch of the load cell was reached (which is 200 mm).

In order to ensure the rotational freedom at the loading point of the GFRP stud, a special loading mechanism was designed as shown in Fig. 9.4. The GFRP stud was clamped with top and bottom steel plates of 20 mm thick and 100 mm long in the stud direction. The top and bottom plates were connected and held in place using four bolts. Two stoppers were welded underneath the top plate to avoid any side movement of the GFRP stud under loading. Then the top plate was welded to a hinge which was pinned to the Instron loading cell. A timber block was firmly inserted in

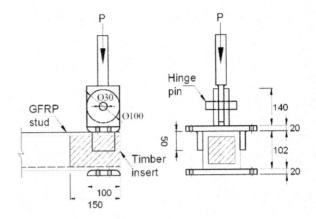

Fig. 9.4 Loading point mechanism (unit in mm)

the GFRP stud to avoid the crushing of the GFRP tubular stud at the loading point. The length of the timber block was 150 mm.

The rotation angle of the connection was measured through the changes of AC and DF in Fig. 9.3 by two string pots (SP1 and SP2). The two ends of each string pot were fixed at the predetermined locations. In other words, the locations of A, B, C, D, E and F in Fig. 9.3 were known and fixed under loading, while the changes in AC and DF were measured. Assuming the angle between BA and BC is α and the angle between ED and EF is β, they can be calculated by the following equations:

$$\cos \alpha = \frac{BA^2 + BC^2 - (AC + \delta_1)^2}{2\alpha \times BA \times BC} \quad (9.1)$$

$$\cos \beta = \frac{ED^2 + EF^2 - (DF + \delta_2)^2}{2 \times ED \times EF} \quad (9.2)$$

where δ_1 and δ_2 are changes in the lengths of AC and DF respectively. Since the stud is initially perpendicular to the beam, the rotation angles of the connection are simply

$$\Delta \alpha = \alpha - \frac{\pi}{2} \quad (9.3)$$

$$\Delta \beta = \beta - \frac{\pi}{2} \quad (9.4)$$

Similar measurement method of the rotation angle was adopted in [31, 37] where $\Delta \alpha$ was called opening angle and $\Delta \beta$ was named closing angle. The rotation angle of the connection, θ, is defined as the average of $\Delta \alpha$ and $\Delta \beta$. The shear force at the connection is equal to the applied load P, while the moment of the connection M can be calculated directly from the load P multiplied by its distance to the flange surface of the beam d (see Fig. 9.3). The moment-rotation stiffness (K_M) and shear-rotation stiffness (K_V) of the connection can be defined as

$$K_M = \frac{dM}{d\theta} \quad (9.5)$$

$$K_V = \frac{dP}{d\theta} \quad (9.6)$$

$$M = P \times d \quad (9.7)$$

$$\theta = \frac{\Delta\alpha + \Delta\beta}{2} \quad (9.8)$$

Strain gauges G1 and G2 were installed at the top and bottom corners of the AG and SL connectors to monitor whether the steel yielded or not at the critical locations during the test. LVDT1 was used to monitor the deformation of the beam under loading. This was mainly to secure the ground anchorage of the beam. The deformation of the beam will not affect the measurement of the rotation angles in Eqs. (9.1) and (9.2). Another LVDT2 was placed under the loading point to measure the deflection at the end of the stud. This was to ensure that the experiments could be safely stopped when the stretch of the load cell was reached. A video camera was used to record the progressive failure at the connection for all the specimens.

9.3 Experimental Results and Discussions

9.3.1 Failure Modes

The failure modes of all the connections were investigated after being dismounted from the experimental set-up. The detailed progressive failure mechanism of each specimen was analysed with the help of the video camera recording. The final failure modes of the shear force dominant specimens are shown in Fig. 9.5 and those for the moment dominant specimens are presented in Fig. 9.6. In addition to the close view at the connection zone after failure, the specimen was disassembled and the detailed failure pattern at the end of the GFRP stud was also presented in Figs. 9.5 and 9.6.

For the shear specimens shown in Fig. 9.5, all bolted connections (OB and BB in Fig. 9.5a, b, c) failed with damages observed at the end of the pultruded GFRP studs, and most of the damages concentrated at the top and bottom flanges of the GFRP section without any damage on the webs. This is because only the top and bottom flanges of GFRP stud were connected to the connector through bolts. To balance the external moment and shear force, the top and bottom flanges of the stud are subject to the bearing forces from the connector (AG or SL), which may cause flange damages (particularly at the four corners of the tubular section) as reported

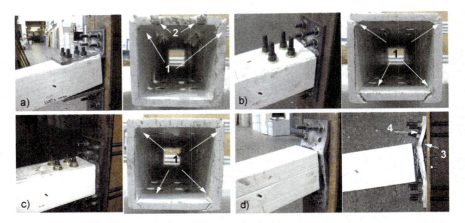

Fig. 9.5 Failure modes of shear specimens. **a** S–AG-OB, **b** S-SL-OB, **c** S-SL-BB, **d** S-SL-AB (the numbers indicate the damages of 1. web/flange separation; 2. shear out of top flange; 3. yielding of end plate; 4. bending and fracture of threaded rod)

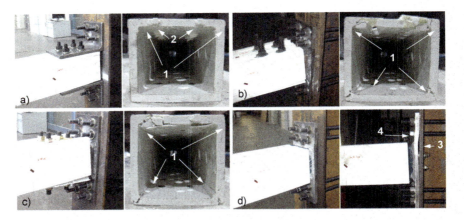

Fig. 9.6 Failure modes of moment specimens. **a** M-AG-OB, **b** M-SL-OB, **c** M-SL-BB, **d** M-SL-AB (the numbers indicate the damages of 1. web/flange separation; 2. shear out of top flange; 3. yielding of end plate; 4. bending of threaded rod)

in [16, 17]. In addition, the failure mode of the SL connection seems independent of the bolt type (i.e. OB or BB). Both OB and BB specimens exhibited similar failure modes as shown in Fig. 9.5b, c.

On the other hand, no damage on the GFRP stud was observed when it was adhesively bonded to the SL connector (S-SL-AB in Fig. 9.5d). This is because the GFRP stud was integrated with the SL connector through adhesive bonding. Therefore, the moment and the shear force are transferred through the whole composite section (GFRP stud and steel tube) to the steel beam. In addition, the stress concentration effects due to the bolts are avoided. This structural integrity of the bonded SL connection desirably enables the achievement of plasticity through the yielding of the steel end plate (see Fig. 9.5d) rather than the brittle failure of GFRP stud as experienced by bolted SL connections. The large yield deformation of the steel end plate leaded to the bending and fracture of the first row of the threaded rod which is close to the top surface of the steel tube (see Fig. 9.5d).

For moment dominant specimens in Fig. 9.6, the failure modes are similar to those of the corresponding shear specimens with the same connection method in Fig. 9.5. For example, all bolted specimens showed web-flange separations of the GFRP stud while the bonded SL connection exhibited no stud damage instead of extensive yielding of the steel end plate. However, the difference in the failure modes between the shear and moment specimens is the way in which their web/flange separated. The difference is illustrated in Fig. 9.7 along with explanations based on the shear and moment dominant mechanisms. The stud length was 1700 mm for the moment connection and 600 mm for the shear specimen. Under the same cantilever load, the moment at the moment connection at failure should be 2.8 times (1700/600) of the moment at the shear connection. Therefore, the moment connection was mainly subjected to relative rotation between the sleeve connector and the GFRP stud as shown in Fig. 9.7a. Under such a relative rotation, the top flange of the GFRP stud was subjected to concentrated loads at the bolt locations transferred from the bolts, and the bottom flange was subjected to a more uniformly distributed pressure transferred by the steel tube of the sleeve connector. The corresponding web/flange separations

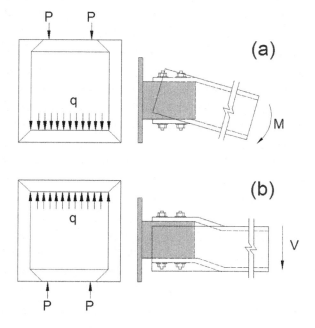

Fig. 9.7 Difference in failure mechanisms between **a** moment SL connection and **b** shear SL connection (*P* is the compression force due to bolts, *q* is the pressure from steel tube surface, *M* is moment and *V* is shear force at connection)

at the four corners are then represented on the left of Fig. 9.7a, as were further evidenced in Fig. 9.6c.

On the other hand, when the shear connection failed at a similar moment to that of the moment connection (see Table 9.3), the shear force at the shear connection should be around 2.8 times higher. Therefore, the connection was mainly subjected to shear force and a vertical movement relative to the GFRP stud (see the right part of Fig. 9.7b). Under such a relative shear movement, the bottom flange of the GFRP stud was subjected to concentrated loads at the bolt locations transferred from the bolts, and the top flange was subjected to a more uniformly distributed pressure transferred by the steel tube of the sleeve connector. The resulting web/flange separations at four corners are then presented on the left of Fig. 9.7b, as were also evidenced in Fig. 9.5c.

9.3.2 Moment-Rotation and Shear-Rotation Curves

According to the experimental setup and instrumentations in Fig. 9.3, the moment and rotation angle of the connection can be determined using Eqs. (9.7) and (9.8). The corresponding moment-rotation curves of all the shear specimens are plotted in Fig. 9.8a and those for the moment specimens are presented in Fig. 9.8b, respectively. The ultimate moment capacity (M_u) is defined as the moment value at the peak point of the corresponding moment-rotation curve in the following discussions.

It can be found for both AG and SL bolted connections (either shear or moment), the moment increased gradually with the rotation angle until the peak point was reached, after which the moment decreased quickly with the rotation angle. The decrease in moment was due to the damages to the GFRP stud as observed in Figs. 9.5 and 9.6. For bonded SL connections (either shear or moment), the moment linearly increased with the rotation angle until the extensive nonlinearity of the moment-rotation curve was observed. In this nonlinear stage, the slope of the moment-rotation curve reduced while the moment continued to increase with the rotation angle, which

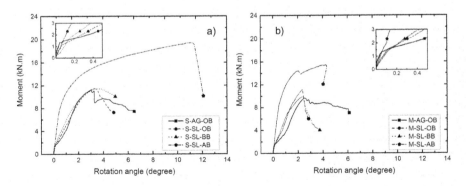

Fig. 9.8 Moment-rotation curves of **a** shear specimens, **b** moment specimens

is attributed to the strain-hardening behaviour of the steel connector. The nonlinearity and strain-hardening features of the bonded connections were mainly attributed to the yielding of the steel end plate of the SL connector (see Figs. 9.5d and 9.6d).

For bolted connections, the moment-rotation curves are similar independent of the bolt type (OB or BB) under either shear or moment dominant loading condition. In addition, except for a marginal decrease in the rotational stiffness, AG-OB connection (either shear or moment) also showed very similar moment-rotation response to those of the SL-OB and SL-BB counterparts. This is mainly attributed to the similar ultimate failure modes of bolted connections as shown in Figs. 9.5 and 9.6. On the other hand, it is obvious in Fig. 9.8 that the SL-AB connection (either shear or moment) performs the best in terms of rotational stiffness, moment capacity and energy dissipation capacity (the area under moment-rotation curve) than corresponding SL-OB and SL-BB connections. These improved performances are mainly because bonded SL connections only experienced connector yielding rather than accumulative damages in GFRP stud as observed in bolted connections. These nonlinearity and strain-hardening features are desirable, allowing the connection continues to carry the external loading by showing sufficient warning without sudden collapse.

To reveal more details of the initial stage, all moment-rotation curves in Fig. 9.8 are zoomed in with the rotation angle ranging from 0° to 0.5° and the moment ranging from 0 to 3 kN.m. It can be seen from the zoomed-in graphs that, there are two stages for AG and SL bolted connections (either shear or moment). Firstly, the moment linearly increased with the rotation angle up to a knee point. Then the second stage commenced with a decreased slope of the moment-rotation curve indicating a reduced rotational stiffness. The two-stage response of the bolted connections is mainly determined by the friction-slip mechanism between connecting components (e.g. bolts, washers, stud and AG/SL connectors). For the bolted connections, frictions between connecting components existed due to the tightening torque of bolts. The internal friction forces were exceeded by the external forces resulting in the reduction of the connection stiffness. This friction-slip phenomenon has been well documented for pultruded GFRP bolted connections [32]. On the contrary, the SL-AB (either shear or moment) connection kept a constant rotational stiffness throughout the whole initial stage. This is reasonable because the GFRP stud was bonded to the SL connector, and no slip can be developed under external moment and shear force.

The shear-rotation curves of all specimens are plotted in Fig. 9.9. The shear force dominant specimens are presented using different solid labels, while those moment dominant specimens are differentiated by various hollow labels. For either shear or moment connections, square label is for AG-OB; circle label represents SL-OB; triangular stands for SL-BB and pentagon label means SL-AB. As can be seen in Fig. 9.9, for either 'S' or 'M' specimens, SL-AB exhibited much higher shear-rotation stiffness and shear capacity than the other bolted specimens which was due to the improved integration between sleeve connector and GFRP stud by the adhesive bonding. In addition, the effects of the shear dominant and the moment dominant loading conditions on the shear-rotation curves can be obviously identified in Fig. 9.9. Comparing with the corresponding 'M' specimens, the 'S' specimens failed at a

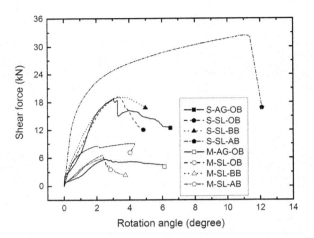

Fig. 9.9 Shear-rotation curves of all specimens (this figure is available in color online and in black and white in print)

much higher shear force and a similar moment (see Fig. 9.8). Consequently, the 'S' specimens yielded a larger rotation angel at failure due to the higher combination of shear and moment actions at the connection.

9.3.3 Strain Responses of Steel Connectors

Two strain gauges, G1 and G2 (see their positions in Fig. 9.3), were used to monitor the strain responses of each steel connection during the loading process. The results for all the shear specimens are plotted in Fig. 9.10, and those for the moment specimens are shown in Fig. 9.11. The curves of G1 and G2 are labelled with circle and

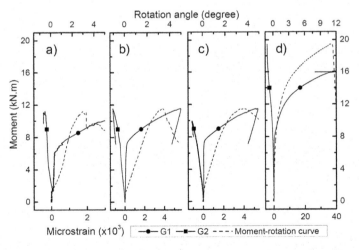

Fig. 9.10 Strain readings of shear specimens. **a** S-AG-OB, **b** S-SL-OB, **c** S-SL-BB, **d** S-SL-AB

9 Connections of Fibre Reinforced Polymer to Steel ...

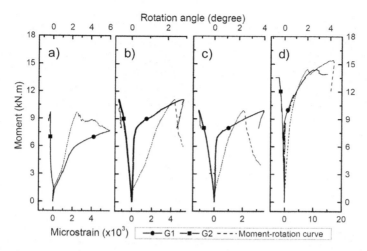

Fig. 9.11 Strain readings of moment specimens. **a** M-AG-OB, **b** M-SL-OB, **c** M-SL-BB, **d** M-SL-AB

square, respectively. For interpretation of the strain readings, the moment-rotation curve of the same specimen is also added in the corresponding G1 and G2 graph in Figs. 9.10 and 9.11.

As can be seen in Figs. 9.10 and 9.11, G1 is in tension with positive strain readings while G2 is under compression with negative readings. This is consistent with the cantilever loading setup (see Fig. 9.3). For all specimens, both G1 and G2 linearly increased until a specific moment value was reached. The G1 curves showed obvious yielding behaviour beyond this moment value, while G2 continued to increase in a relatively linear manner. Therefore, the moment value at the yielding point of G1 curve is defined as the elastic moment capacity (M_e) of the corresponding connection. The moment-rotation curve from the origin up to the elastic moment capacity is called elastic stage and the corresponding slope is defined as the elastic moment-rotation stiffness ($K_{e,M}$). Similarly, the corresponding shear force at M_e is defined as V_e, and the elastic shear-rotation stiffness is referred as $K_{e,V}$. It should be noted that $M_e = V_e \times d$, and $K_{e,M} = K_{e,V} \times d$, where d is the stud length in Fig. 9.3. This elastic rotational stiffness is important for the evaluation of the connection serviceability limit state [28].

It is interesting to see that, all connections yielded before the peak point of the moment-rotation curve was reached. The moment-rotation curves (see Figs. 9.10 and 9.11) however did not show much stiffness reduction after the yielding of G1 for specimens AG-OB, SL-OB and SL-BB. This is because their elastic moment M_e and ultimate moment M_u (see Table 9.3 in Sect. 9.3.4 for their values) are very close and the local yielding did not introduce stiffness reduction before the ultimate load was reached. For specimen SL-AB, M_u was much larger than M_e so that the stiffness reduction in the moment-rotation curve became more obvious due to the yielding of steel connector (see Fig. 9.8).

There is not much difference in terms of the strain responses when comparing Figs. 9.10 and 9.11. For AG connections (either shear or moment) in Fig. 9.10a and Fig. 9.11a, they have similar M_e at around 6 kN.m. For SL connections, M_e seems independent of the connection method (i.e. OB, BB or AB) as well as the loading conditions (either shear or moment). The M_e for all the SL connections falls at a value of about 8 kN.m. On the other hand, considering the difference in the stud length of the shear or moment connections, they are expected to have much different V_e values at the yielding of G1. The aim of Figs. 9.10 and 9.11 is to define the elastic stage and the corresponding moment and shear force parameters (M_e and V_e). More detailed comparisons on the stiffness and capacity values of the connections are presented in the following section.

9.3.4 Comparisons on the Connection Stiffness and Capacity

The values of $K_{e,M}$ and $K_{e,V}$, M_e and V_e, M_u and V_u of each connection specimen are listed in Table 9.3. For shear specimens, the S-AG-OB specimen showed the lowest elastic properties, i.e. $K_{e,M}$ (206.3 kN.m/rad), $K_{e,V}$ (341.0 kN/rad), M_e (6.0 kN.m) and V_e (10.0 kN). These parameters are lower by about 16% ($K_{e,M}$ and $K_{e,V}$), and about 23% (M_e and V_e), respectively, comparing to those of S-SL-OB specimen. These elastic connection properties indicate the advantage of the SL connector, which is more monolithic than the two single AG connectors.

Comparing to S-SL-OB, the M_e and V_e of S-SL-BB specimen remained the same (7.8 kN.m and 12.9 kN). However, both $K_{e,M}$ and $K_{e,V}$ increased by 14% (i.e. $K_{e,M}$ from 246.4 kN.m/rad to 280.7 kN.m/rad, and $K_{e,V}$ from 407.3 kN/rad to 464.0 kN/rad). The improved stiffness of S-SL-BB specimen mainly benefits from the fastening mechanism of the one-sided bolts for this particular SL connector. The comparison of the fastening mechanisms between S-SL-OB and S-SL-BB is illustrated in Fig. 9.12. It can be seen that, when pretensioned using the same torque, one-sided bolting tends to apply clamping force to the connecting components through both external and internal contact surfaces. This mechanism leads to a relatively rigid

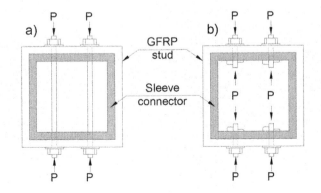

Fig. 9.12 Illustration of the fastening mechanisms of **a** ordinary bolting, **b** one-sided bolting, where P is the clamping force (not to scale)

connection performance comparing to the through ordinary bolting which can only apply the clamping force on the external contact surfaces.

The S-SL-AB specimen achieved the highest $K_{e,M}$ (911.0 kN.m/rad), $K_{e,V}$ (1505.8 kN/rad), M_u (19.4 kN.m) and V_u (32.1 kN), which are 342% ($K_{e,M}$ and $K_{e,M}$) and 73% (M_e and V_e) higher than those of the S-AG-OB specimen. This is because S-SL-AB did not experience friction-slip mechanism as S-AG-OB. This ensures the continuous increase of the moment with the rotation angle at the initial linear stage of the moment-rotation curve, leading to a higher elastic stiffness (see Figs. 9.8 and 9.9). In addition, S-SL-AB avoided stress concentration due to the bolts and showed no accumulative damages to the stud. The strain-hardening behaviour of the steel end plate ensures the continuous moment increase of S-SL-AB, leading to higher M_u and V_u values.

Another two interesting observations were further noticed from the shear specimens. All bolted connections (AG and SL) have similar M_u values around 11.2 kN.m to 11.5 kN.m, and similar V_u values around 18.5 kN to 19.0 kN. This is because all bolted connections ultimately failed due to the web/flange separations of the GFRP studs (see Figs. 9.5, 9.6, 9.7). All SL connections (bolted or bonded) have similar M_e and V_e values (i.e. M_e from 7.8 kN.m to 8.3 kN.m, and V_e from 12.9 kN to 13.7 kN). This is because for all SL connections, SL yielded at the end plate before the ultimate failure of the GFRP stud (see Figs. 9.10 and 9.11).

For moment specimens, M-AG-OB showed the lowest $K_{e,M}$ (189.1 kN.m/rad), $K_{e,V}$ (110.6 kN/rad), M_e (6.3 kN.m) and V_e (3.7 kN). For M-SL-OB, these properties are higher by 47% ($K_{e,M}$ and $K_{e,V}$) and 30% (M_e and V_e), respectively. The M_e and V_e of M-SL-BB did not change much (only 4% difference) comparing to M-SL-OB. However, both $K_{e,M}$ and $K_{e,V}$ increased by 15% from ($K_{e,M}$ from 278.0 kN.m/rad to 320.8 kN.m/rad, and $K_{e,V}$ from 162.6 kN to 187.6 kN). The reason can be referred to the fastening mechanism as shown in Fig. 9.12.

Again, M-SL-AB achieves the best results in $K_{e,M}$ (1019.9 kN.m/rad), $K_{e,V}$ (596.4 kN/rad), M_u (≥ 15.5 kN.m) and V_u (≥ 9.1 kN), which are 439% ($K_{e,M}$ and $K_{e,V}$) and $\geq 60\%$ (M_u and V_u) higher than those of M-AG-OB. It should be noted that ' $\geq$ ' is used when describing M_u of M-SL-AB specimen, as the test was stopped because the stretch limit of the load cell was reached. Therefore, higher M_u and V_u should have been achieved if the specimen were loaded further. In addition, the two observations are still valid for moment specimens, i.e. bolted connections (AG and SL) have similar M_u and V_u values, and all SL connections (bolted or bonded) show close M_e and V_e values.

The effects of the shear dominant and the moment dominant loading conditions on the connection behaviours can be hardly identified by the moment properties of the connections, i.e. $K_{e,M}$, M_e and M_u. For example, M_e of bolted shear connections ranges from 6.0 kN.m to 7.8 kN.m which is very similar to the M_e of corresponding bolted moment connections (ranges from 6.3 kN.m to 8.2 kN.m). However, the effects of loading conditions can be obviously observed based on the shear force related properties, i.e. $K_{e,V}$, V_e and V_u. For example, under the shear force dominant condition, bolted AG and SL connections have V_e ranging from 10.0 kN to 12.9 kN, which is on average 173.3% higher than V_e of the bolted connections under the

Table 9.4 Equations for the connection classifications

		Full strength	Partial strength	Nominally pinned
Capacity	Classification	Full strength	Partial strength	Nominally pinned
	Equations	$M_u \geq M_s$	$M_s > M_u \geq 0.25 M_s$	$M_u < 0.25 M_s$
Rigidity	Classification	Rigid	Semi-rigid	Nominally pinned
	Equations	$K_{e,M} \geq 25 EI_S/L_S$	$25 EI_S/L_S \geq K_{e,M} \geq 0.5 EI_S/L_S$	$K_{e,M} < 0.5 EI_S/L_S$

moment dominant condition (ranging from 3.7 kN to 4.8 kN). For the V_u values of the same shear dominant specimens (ranging from 18.5 kN to 19.0 kN), they are on average 207.4% higher than those of the corresponding moment dominant specimens (ranges from 5.7 kN to 6.5 kN). This comparison suggests that a connection may fail at a similar moment capacity but with a largely varied shear force.

9.3.5 Classification of the Connections

With the results obtained from this study, it is able to present the classification of these connections [27]. Considering the lack of the design guidelines for connecting the pultruded GFRP profiles to the steel members, the standard for the classification of the steel connections, Eurocode 3 [57], is adopted. The classifications can be expressed using the equations in Table 9.4.

In Table 9.4, E, I_s and M_s stand for the tensile modulus, the second moment of area and the moment capacity of the stud, respectively. L_s corresponds to the length of the actual structural member (i.e. the stud in this study). Therefore a value of 3.4 m is used considering that the moment dominant connection (1.7 m, see Sect. 9.2.3) is half the length of an actual stud. The values of M_s and EI_s/L_s can be calculated based on the material and geometry properties of the GFRP stud given in Table 9.2 and in Sect. 9.2.2. The boundaries of the capacity classifications are (7.6 kN.m, 30.4 kN.m). The boundaries of the rigidity classifications are (22.5 kN.m/rad, 1127.6 kN.m/rad). As can be seen in Table 9.3, the SL connection using the adhesive bonding is almost four times higher in terms of $K_{e,M}$ and two times higher in terms of M_u than the corresponding bolted connections. The $K_{e,M}$ of SL-AB is very close to the upper bound of the rigidity classification, while the other bolted connections all fall in the semi-rigid range. Therefore, the proposed bonded SL connection can be classified as a rigid, partial strength connection according to the Eurocode 3 [57].

9.4 Conclusions

A steel sleeve connector is proposed in this chapter for connecting the pultruded GFRP SHS to the steel member. The sleeve connector was fabricated by a steel tube welded to a steel end plate. The end plate was connected to both flanges of a steel beam using all threaded rods. The steel tube was connected to the GFRP tubular section by one of the three methods, i.e. ordinary bolts, one-sided bolts and adhesive bond. The mechanical behaviours of these sleeve connections are evaluated through a series of moment-rotation experiments. Specimens using conventional connecting method with steel angles were also prepared to compare with the sleeve connections. Two stud lengths were selected to investigate the effects of the shear force dominant and the moment dominant loading conditions on the behaviours of the connections. Experimental results, including failure modes, moment-rotation curves, shear-rotation curves, strain gauge readings, rotational stiffness, moment and shear capacities are obtained and compared between the sleeve and the angle connections. Based on the experimental results presented in this chapter, the following conclusions can be drawn:

1. A web/flange separation failure of the GFRP section was observed for all the bolted connections (AG with angle connector and SL with sleeve connector). The shear force dominant and moment dominant loading conditions have noticeable effects on this web/flange separation failure mode. For the shear force dominant specimens, the web/flange separation was due to the bolt compression on the bottom flange and steel tube pushing upward on the top flange. While for the moment dominant specimens, the web/flange separation was because of the bolt compression on the top flange and steel tube pushing downward on the bottom flange. It was indicated that all the bolted connectors showed yielding initiation at the corner of the AG or SL connectors. Bonded SL connections showed no damages to the GFRP sections when extensive yielding deformation was observed at the end plates.

2. All bolted connections (AG and SL) showed similar two-stage moment-rotation curves before the ultimate moment capacity M_u was reached, because of the friction-slip mechanisms. After M_u, the moment started to decrease due to the web/flange separation of the GFRP section. For the bonded SL connections, the moment linearly increased with the rotation angle until the yielding of the steel end plate. Since no damages were observed to the GFRP section, the moment continued to increase steadily with the rotation angle, due to the steel strain-hardening behaviour. The 'strain-hardening' feature enables the bonded SL connection to carry the load without sudden failure as observed in the other bolted connections.

3. For either the shear or the moment connections, the bonded SL connection achieved the best performance with the elastic moment-rotation stiffness $K_{e,M}$ and the elastic shear-rotation stiffness $K_{e,V}$ higher by 342%, and M_u and V_u higher by $\geq$60% than those of the conventional AG specimen. The SL-OB specimen (i.e. with sleeve connector and ordinary bolts) was higher by 16% in $K_{e,M}$ and $K_{e,V}$, and by 30% in M_e and V_e than those of AG-OB connection. When replaced with the one-sided bolts, $K_{e,M}$ and $K_{e,V}$ of the SL-BB connection can be improved by 14% in comparison to the AG-OB specimen (i.e. with angle connector and ordinary bolts).

4. The effects of the shear force dominant and the moment dominant loading conditions on the connection behaviour can be obviously identified through the connection shear properties, i.e. $K_{e,V}$, V_e and V_u. It showed that, under the shear force dominant condition, the bolted AG and SL specimens have V_e higher by an average of 173.3% than that of the connections under the moment dominant condition. $K_{e,V}$ and V_u of the shear dominant bolted specimens are higher by 163.1% and 212.8% (on average), respectively, than those of the specimens under the moment dominant loading condition. For the bonded sleeve connection, $K_{e,V}$, V_e and V_u values increase by 152.5%, 174.0% and 252.7% when the moment dominant condition is changed to the shear force dominant condition.
5. It is found the bonded sleeve connection outperformed all the other connection methods and can be classified as a rigid, partial strength connection, according to the connection design standard of Eurocode 3. This classification is only based on the moment properties, i.e. (i.e. $K_{e,M}$ and M_u) of the connection.

References

1. Manalo A (2013) Structural behaviour of a prefabricated composite wall system made from rigid polyurethane foam and magnesium oxide board. Constr Build Mater 41:642–653
2. Schafer BW (2011) Cold-formed steel structures around the world. Steel Constr 4(3):141–149
3. Habashi HR, Alinia MM (2010) Characteristics of the wall-frame interaction in steel plate shear walls. J Constr Steel Res 66(2):150–158
4. Frenette CD, Bulle C, Beauregard R, Salenikovich A, Dominique D (2010) Using life cycle assessment to derive an environmental index for light-frame wood wall assemblies. Build Environ 45(10):2111–2122
5. Sotiropoulos S, GangaRao H, Mongi A (1994) Theoretical and experimental evaluation of FRP components and systems. J Struct Eng 120(2):464–485
6. Dawood M, Taylor E, Ballew W, Rizkalla S (2010) Static and fatigue bending behavior of pultruded GFRP sandwich panels with through-thickness fiber insertions. Compos B Eng 41(5):363–374
7. Evernden MC, Mottram JT (2012) A case for houses to be constructed of fibre reinforced polymer components. Proc ICE-Constr Mater 165(1):3–13
8. Fang H, Sun H, Liu W, Wang L, Bai Y, Hui D (2015) Mechanical performance of innovative GFRP-bamboo-wood sandwich beams: experimental and modelling investigation. Compos B Eng 79:182–196
9. Hutchinson JA, Singleton MJ (2007) Startlink composite housing. Advanced composites for construction (ACIC 2007). University of Bath
10. Halliwell S (2010) Technical papers: FRPs-the environmental agenda. Adv Struct Eng 13(5):783–791
11. Hollaway LC (2010) A review of the present and future utilisation of FRP composites in the civil infrastructure with reference to their important in-service properties. Constr Build Mater 24(12):2419–2445
12. Bakis CE, Bank LC, Brown V, Cosenza E, Davalos JF, Lesko JJ, Machida A, Rizkalla SH, Triantafillou TC (2002) Fiber-reinforced polymer composites for construction-state-of-the-art review. J Compos Constr 6(2):73–87
13. Dhand V, Mittal G, Rhee KY, Park SJ, Hui D (2015) A short review on basalt fiber reinforced polymer composites. Compos B Eng 73:166–180

14. Wu HC, Yan A (2013) Durability simulation of FRP bridge decks subject to weathering. Compos B Eng 51:162–168
15. Bai Y, Keller T, Correia JR, Branco FA, Ferreira JG (2010) Fire protection systems for building floors made of pultruded GFRP profiles–Part 2: modeling of thermomechanical responses. Compos B Eng 41(8):630–636
16. Wu C, Bai Y (2014) Web crippling behaviour of pultruded glass fibre reinforced polymer sections. Compos Struct 108:789–800
17. Wu C, Bai Y, Zhao XL (2015) Improved bearing capacities of pultruded glass fibre reinforced polymer square hollow sections strengthened by thin-walled steel or CFRP. Thin-Walled Structures 89:67–75
18. Keller T (2003) Use of fiber reinforced polymers in bridge construction, Structural engineering documents 7. International Association for Bridge and Structural Engineering (IABSE), Zurich, Switzerland
19. Bank LC (2006) Composites for construction: structural design with FRP materials. John Wiley & Sons, Hoboken
20. Bai JP (2013) Advanced fibre-reinforced polymer (FRP) composites for structural applications. Elsevier
21. Hollaway LC (1993) Polymer composites for civil and structural engineering. Chapman & Hall, London
22. Keller T, de Castro J (2005) System ductility and redundancy of FRP beam structures with ductile adhesive joints. Compos B Eng 36(8):586–596
23. Wu C, Bai Y, Mottram JT (2015) Effect of elevated temperatures on the mechanical performance of pultruded FRP joints with a single ordinary or blind bolt. ASCE J Compos Constr 20(2)
24. Hejll A, Täljsten B, Motavalli M (2005) Large scale hybrid FRP composite girders for use in bridge structures—theory, test and field application. Compos B Eng 36(8):573–585
25. Satasivam S, Bai Y (2014) Mechanical performance of bolted modular GFRP composite sandwich structures using standard and blind bolts. Compos Struct 117:59–70
26. Satasivam S, Bai Y, Zhao XL (2014) Adhesively bonded modular GFRP web-flange sandwich for building floor construction. Compos Struct 111:381–392
27. Mottram JT, Zheng Y (1996) State-of-the-art review on the design of beam-to-column connections for pultruded frames. Compos Struct 35(4):387–401
28. Bank LC, Mosallam AS, McCoy GT (1994) Design and performance of connections for pultruded frame structures. J Reinf Plast Compos 13(3):199–212
29. Carrion JE, Hjelmstad KD, LaFave JM (2005) Finite element study of composite cuff connections for pultruded box sections. Compos Struct 70(2):153–169
30. Mosallam AS, Abdelhamid MK, Conway JH (1994) Performance of pultruded FRP connections under static and dynamic loads. J Reinf Plast Compos 13(5):386–407
31. Smith SJ, Parsons ID, Hjelmstad KD (1998) An experimental study of the behavior of connections for pultruded GFRP I-beams and rectangular tubes. Compos Struct 42(3):281–290
32. Wu C, Feng P, Bai Y (2015) Comparative study on static and fatigue performances of pultruded gfrp joints using ordinary and blind bolts. J Compos Constr 19(4):04014065
33. Bank LC, Mosallam AS, Gonsior HE (1990) Beam-to-column connections for pultruded FRP structures. In: Proceedings of the 1st Materials Engineering Congress, Materials Engineering Division. Denver, Colorado, United States
34. Bank LC, Yin J, Moore L, Evans DJ, Allison RW (1996) Experimental and numerical evaluation of beam-to-column connections for pultruded structures. J Reinf Plast Compos 15(10):1052–1067
35. Mosallam AS, Bank LC (1992) Short term behavior of pultruded fiber reinforced plastic frame. J Struct Eng 118(7):1937–1954
36. Mosallam AS (1995) Connection and reinforcement design details for pultruded fiber reinforced plastic (PFRP) composite structures. J Reinf Plast Compos 14(7):752–784
37. Smith S, Parsons I, Hjelmstad K (1999) Experimental comparisons of connections for GFRP pultruded frames. J Compos Constr 3(1):20–26

38. Mottram JT, Zheng Y (1999) Further tests on beam-to-column connections for pultruded frames: web-cleated. J Compos Constr 3(1):3–11
39. Mottram JT, Zheng Y (1999) Further tests of beam-to-column connections for pultruded frames: flange-cleated. J Compos Constr 3(3):108–116
40. Smith S, Parsons I, Hjelmstad K (1999) Finite-element and simplified models of GFRP connections. J Struct Eng 125(7):749–756
41. Singamsethi S, LaFave J, Hjelmstad K (2005) Fabrication and testing of cuff connections for GFRP box sections. J Compos Constr 9(6):536–544
42. Bai Y, Yang X (2013) Novel joint for assembly of all-composite space truss structures: conceptual design and preliminary study. J Compos Constr 17(1):130–138
43. Luo FJ, Bai Y, Yang X, Lu Y (2015) Bolted sleeve joints for connecting pultruded FRP tubular components. J Compos Constr 20(1)
44. AS 4100-1998 (1998) Steel structures. Standards Association of Australia, Sydney, Australia
45. Eurocomp Designe Code and Handbook (1996) Structural design of polymer composites. The European Structural Polymeric Composites Group. ISBN 0419194509
46. Technical Document CNR-DT 205/2007 (2008) Guide for the Design and Construction of Structures made of FRP Pultruded Elements. Italian National Research Council (CNR), Rome
47. Chamers RE (1997) ASCE design standard for pultruded fiber-reinforced-plastic (FRP) structures. J Compos Constr 1(1):26–38
48. ASTM D3039/D3039M-08 (2008) Standard test method for tensile properties of polymer matrix composite materials. West Conshohocken, United States
49. ASTM D2344/D2344M-00 (2006) Standard test method for short-beam strength of polymer matrix composite materials and their laminates. West Conshohocken, United States
50. ASTM D3171-15 (2015) Standard test methods for constituent content of composite materials, ASTM International. West Conshohocken, United States
51. ASTM E8/E8M-15a (2015) Standard test methods for tension testing of metallic materials. West Conshohocken, United States
52. ASTM F436-11 (2011) Standard specification for hardened steel washers. West Conshohocken, United States
53. AS 1112.1-2000 (2000) ISO metric hexagon nuts. Part 1: Style 1-Product grades A and B. Standards Association of Australia, Sydney, Australia
54. Wu C, Zhao XL, Duan WH, Al-Mahaidi R (2012) Bond characteristics between ultra high modulus CFRP laminates and steel. Thin-Walled Struct 51:147–157
55. Turvey GJ, Cooper C (2004) Review of tests on bolted joints between pultruded GRP profiles. Proc ICE-Struct Build 157(3):211–233
56. Turvey GJ, Cooper C (2000) Semi-rigid column base connections in pultruded GRP frame structures. Comput Struct 76(1–3):77–88
57. Eurocode 3 (2005) Design of steel structures. Part 1.8: Design of joints. EN-1993-1-8. European Committee for Standardization, Brussels

Chapter 10
Connections of Fibre Reinforced Polymer to Steel Members: Numerical Modelling

Zhujing Zhang, Chao Wu, Xin Nie, Yu Bai, and Lei Zhu

Abstract This chapter numerically investigates the proposed bonded sleeve connection for joining tubular glass fibre reinforced polymer (GFRP) composites and steel members. Experimental results focused on mechanical responses of such specimens using bonded sleeve connections and conventional steel angle connections were introduced in previous chapter. These results are used to set the benchmark for detailed finite element (FE) modelling in this chapter. In the detailed FE analysis, bolt geometry including head, shank and washer were accurately modelled. Paired contact elements were used for simulating the contact and slip behaviour between bolt shanks and holes, washers and steel or GFRP. The pretension force in the bolts was also taken into account by implementing pretension elements. The FE models developed were first validated against the experimental results in terms of failure mode, moment-rotation curves and strain responses. Parametric studies were then undertaken to investigate the structural behaviour of the bonded sleeve connections considering the effects of major design parameters such as endplate thickness, bonding length, number of bolts, etc. It was found that the endplate thickness dominates the initial stiffness and the elastic moment capacity of the bonded sleeve connection and the presence of central one-sided bolts may improve the elastic moment capacity of the bonded sleeve connection.

Reprinted from Composite Structures, 157, Zhujing Zhang, Chao Wu, Xin Nie, Yu Bai, Lei Zhu, Bonded sleeve connections for joining tubular GFRP beam to steel member: Numerical investigation with experimental validation, 51–61, Copyright 2016, with permission from Elsevier.

Z. Zhang · C. Wu · Y. Bai (✉)
Department of Civil Engineering, Monash University, Clayton, Australia
e-mail: yu.bai@monash.edu

X. Nie
Department of Civil Engineering, Tsinghua University, Beijing, China

L. Zhu
School of Civil and Transportation Engineering, Beijing University of Civil Engineering and Architecture, Beijing, China

10.1 Introduction

Glass fibre reinforced polymer (GFRP) composites have been increasingly used in civil applications due to their high-strength to low-weight ratio, low maintenance costs, excellent corrosion resistance, and low energy consumption during manufacturing process [1]. They have not been well employed in frame and building applications as there may be a lack of connection approaches between GFRP beams and columns. Pioneering research has revealed the problems of simply mimicking steel beam-to-column connection details on GFRP profile [2], such as the delamination of GFRP connection elements, and separation between web and flange of GFRP profiles. These challenges were all unique to GFRP materials. Therefore studies have been done to develop connection solutions for GFRP structures. A comparative investigation on the behaviour of connections using GFRP I-shaped sections and GFRP tubular sections as beams and columns is conducted in [3]. With the similar strong axis bending properties and same connecting elements, the merits of using pultruded closed sections (e.g. tubular sections, or square hollow sections and rectangular hollow sections) over pultruded open sections (e.g. I-shaped sections) were demonstrated, with considerable improvement in connection stiffness and strength. This is because tubular sections generally have improved torsional rigidity and improved weak axis strength and stiffness over open sections. The effects of GFRP connecting elements and steel connecting elements used for pultruded I-shaped sections has been also studied in [4]. The results showed that both connection strength and stiffness were increased when using steel connecting elements. An innovative cuff connection concept was further proposed in [4], which integrated connection design into a single monolithic connection element and utilized the entire column section as a part of connection system to avoid bolts in the GFRP beam. Such a cuff connection was fabricated later on using vacuum assisted resin transfer moulding in [5]. It was found that the cuff connection using purely adhesive bonding was stiffer and strong. All above studies highlighted the benefits of using tubular members, steel connecting components and cuff geometry when joining GFRP profiles.

For modelling on the connection performance of GFRP structures, many studies have focused on simple versions of plate-to-plate connections, e.g. double-lap or single-lap connections, and limited results have been reported for member connections such as beam-to-column connections. Early work was done by Bank et al. [6] where FE analysis of several connection configurations for I-shaped sections were conducted including angle connection (flange or web cleats), gusset plate connection and wrapped angle connection. Modelling results for the initial moment-rotation behaviour of those tested connection configurations were obtained in comparison to the experimental results. Numerical analyses of angle connections for I-shaped GFRP sections were also performed in [7], where all members were simulated via two-dimensional (2D) solid plane stress elements using Ansys software. Contacts between different components were modelled via applying a spring element with assigned stiffness between nodes along the contact surface. Two methods were considered for modelling bolts. The first approach was to use 2-node link elements to represent the positions of the bolt head and nut. The second approach was to use quadrilateral plane stress elements to represent the bolt shank. The pretension of the bolts was modelled via reducing their temperature. Reasonable results were obtained for the initial stiffness of the flange cleat connections (when the plane elements were used for simulating bolts), but not for combined flange and web cleat connections.

Three-dimensional (3D) FE modelling on a series of beam-to-column connections was developed in [8] using Abaqus. 3D 8-node shell elements were used to represent GFRP beams and columns which were connected via node coupling along the interface. Similar 3D FE modelling approaches were used in [5] to predict the connection behaviour of proposed composite cuff connections. In that work, the composite cuff and GFRP beams and columns were considered to be perfectly bonded using the adhesive (modelled via node coupling method again). Such modelling approaches commonly presented linear elastic responses of the connection and thus were unable to track material damage initiations of GFRP profiles. Another FE study was performed in [9] on the composite cuff connections similar to those in [5], but taking into account the material failure of GFRP profiles. In their modelling approach, to cope with the difficulty in selection of the reference surfaces (i.e. contact surfaces) when shell elements were used, 20-node quadratic solid elements were employed to model tubular GFRP profiles. Material damage of GFRP members was indicated by employing Tsai-Wu failure criterion [10]. The resulting failure region matched well with the observed material damage in the tests. However, no bolts were used in the composite cuff connection investigated, as the composite cuff and GFRP beams and columns were connected via adhesive bonding. Therefore, application of this node coupling approach may not characterise the interaction of bolted joints at the connected locations.

On the other hand, as bolts are commonly used in steel connections, published results from the modelling of steel bolted endplate connections may provide valuable information. Among the considerable FE modelling work on bolted endplate connections, the study by Shi et al. [11] using Ansys demonstrated good comparisons and convenient implementation of predicting moment-rotation behaviour, the mode of failure and the load capacity of connection systems. In their modelling approach, 3D 10-node solid elements were used to represent all components including bolts and nuts; pretension of bolts was considered in the model via applying pretension elements; the contacts between different components were modelled through contact pairs. This modelling approach was successfully applied for predicting the structural responses of the connection systems in GFRP lattice frames [12] and in GFRP space frames [13].

A sleeve connection, where a steel tube is inserted as a sleeve into GFRP profiles, was proposed for connections between tubular GFRP and steel members. Previous experimental study on the proposed sleeve connection for joining tubular GFRP and steel showed promising performance [14], with significant improvement in both connection stiffness and moment capacity in comparison to the use of steel seated angle connections. However, only limited parameters such as different connection type (angles versus sleeve connector) with predetermined geometries were examined. Experimental studies may be costly and time-consuming if parametric study is required. Modelling techniques, on the other hand, may develop more comprehensive understanding of the effects of major design parameters on overall structural responses. Furthermore, from the above review of modelling techniques, it can be concluded that 3D detailed modelling should be developed in order to take into account elements such as bolts and nuts. It also becomes possible to model in detail the bolt interaction with other components, contact and slip between different components (such as shanks and holes), and pretension forces in bolts. Tsai-Wu failure criterion is a useful option to indicate possible damage of GFRP members.

This chapter therefore presents a detailed numerical investigation of the proposed bonded sleeve connection for joining tubular GFRP and steel members, with consideration of the aforementioned merits. The modelling approach developed was further validated by experimental results introduced in the previous chapter, in terms of moment-rotation responses, failure modes and local strain responses. With the validated modelling approach, a parametric study was then performed to investigate design parameters including endplate thickness, bonding length, number of bolts, and the presence of central bolts for connecting the centre of the steel endplate to the flange of the steel column. The effects on the overall performance of the proposed bonded sleeve connection were thus clarified.

10.2 Experimental Summary

The mechanical behaviour of bonded sleeve connections for joining tubular GFRP and steel members has been experimentally examined through four specimens with comparison to conventional steel angle connections. Figure 10.1 shows the details for the bonded sleeve specimens and the steel angle specimens. The GFRP beam was a square hollow section with dimensions of 102 × 102 × 9.5 mm (width × depth × thickness), and the steel column had an I-shaped section with dimensions of 158 × 153 × 9.4 × 6.6 mm (width × depth × flange thickness × web thickness). The sleeve connector was made by welding a 175 mm-long cold formed steel tube of 80 × 80 × 10 mm (width × depth × thickness) with a 10 mm-thick steel endplate.

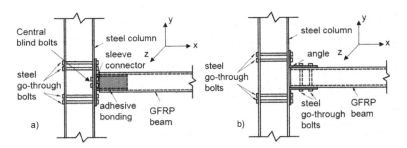

Fig. 10.1 Experimental specimens with **a** bonded sleeve connection (L of 600 mm for SB, of 1700 mm for BB); **b** steel angle connection (L of 600 mm for SA, of 1700 mm for BA)

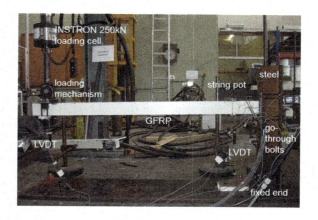

Fig. 10.2 Set-up for experimental specimens

The steel tube was inserted into the GFRP beam and they were bonded using epoxy adhesive, thereby achieving a bond length of 160 mm and allowing a 15 mm gap for the welding. The steel endplate was connected to the steel column using M12 steel high-tensile (Class 8.8) go-through bolts and four M8 one-sided bolts (BB) as shown in Fig. 10.1a. In the angle specimens, the GFRP beam and the steel column were connected by two steel angles with dimensions of 150 × 90 × 10 mm (width × depth × thickness). The two angles were connected via the same M12 go-through bolts as in the bonded sleeve connections (Fig. 10.1b).

Two beam lengths (600 and 1700 mm) were adopted to enforce the shear and bending dominant loading conditions. Therefore a total of four specimens have been tested. The specimens are referred to as "S" for shear dominant loading or "B" for bending dominant loading; the second part of the specimen name describes the connection method, either "A" for steel angle connections or "B" for bonded sleeve connections.

All specimens were tested in a cantilever set-up (see Fig. 10.2) where the steel column was fixed on the ground while a displacement-controlled loading was applied at the end of the GFRP beam. To accommodate the large rotations expected in the connections, a loading mechanism was developed to ensure free rotation at the end of the GFRP beam.

10.3 Detailed FE Modelling

Ansys was used to model the bonded sleeve specimens and steel angle specimens in detail. Ansys parametric design language (APDL) tools were adopted [15] for the subsequent parametric study.

10.3.1 Material Properties

Three different materials were used, GFRP, steel and adhesive. Table 10.1 summarises the material properties used in the modelling. To represent the elastic–plastic behaviour of the steel components a bilinear model was applied, considering a linear isotropic hardening stage after the yield stress was reached. The tangent slope (E_t) of the linear hardening stage was assumed to be 2% of the corresponding Young's modulus [11]. The Young's modulus (E_s), Poisson's ratio (v_s), and yielding strength (f_y) of the steel tube, steel endplate and steel column were determined through coupon tests according to ASTM E-8/E-8M [16]. The properties of the steel go-through bolts and one-sided bolts were adopted from references [17]. Young's modulus (E_a) of Araldite 420 epoxy is 1.9 GPa and Poisson ratio (v_a) is 0.3 according to [18]. Its tensile strength was found to be 28.6 MPa and shear strength was 25 MPa according to the test results reported by Fawzia et al. [19]. In the FE model, Araldite 420 epoxy was considered as an elastic isotropic material where its E_a and v_a were the inputs. The adhesive layer was defined with a thickness of 1.5 mm between GFRP beam and steel tube.

GFRP was assumed to be transversely isotropic in the study. For such materials, only five independent constants are required as input [20], i.e. Young's modulus (E_1, E_2), shear modulus (G_{12}) and Poisson's ratio (v_{12}, v_{23}) where '1' represents the pultrusion direction, '2' represents the in-plane transverse direction and '3' represents the through-thickness direction. In this study, E_1 was determined through tensile coupon tests according to [21]. The other four required constants were estimated based on the properties of constituents and the fibre volume fraction (V_f). V_f was tested and determined as 46.7% according to [22] as the density of E-glass fibres is 2.57 g/cm^3. The constituent properties, matrix (polyester) and fibre (E-glass fibre)

Table 10.1 Material properties used in FE modelling

Components	Young's modulus (GPa)	Poisson's ratio	Yielding strength (MPa)	Tangent slope (MPa)	Shear modulus (GPa)
Steel column	216.4	0.33	361.8	4328	–
Steel endplate	218.8	0.3	332.4	4376	–
Steel tube	216.9	0.3	326.2	4338	–
Steel angle	220.7	0.28	349.7	4413.6	–
Steel go-through bolt	235	0.42	1043	4700	–
Steel one-sided bolt	200	0.42	640	4000	–
Epoxy adhesive	1.9	0.3	–	–	–
GFRP beam	30.2 (E_1) 11.9 (E_2) 11.9 (E_3)	0.29 (v_{12}) 0.41 (v_{23}) 0.29 (v_{13})	–	–	3.65 (G_{12}) 4.22 (G_{23}) 3.65 (G_{13})

were taken as typical values according to [23, 24]. As such, v_{12} could be determined using micromechanics relations [9], E_2, G_{12} were determined through the Halpin–Tsai method [25] and v_{23} was estimated according to [26]. The resulting values for the GFRP properties are presented in Table 10.1.

10.3.2 Failure Criteria

The yielding of steel is governed by von Mises yield criterion [27], i.e. the steel is deemed to yield when the von Mises stress reaches its yield strength. In the FE modelling, the stress state ratio (R_Y) was used as the ratio of von Mises stress to the yield strength. For the assessment of the material failure of GFRP profiles, Tsai-Wu failure criteria was adopted in the form of the inverse of Tsai-Wu strength ratio index (I_F) and failure occurred for the GFRP material if the value was greater than 1. Since the GFRP material was considered to be transversely isotropic, I_F was then dependent on non-zero strengths including its tensile strength in pultrusion direction, its transverse tensile and compressive strength as well as its interlamimar shear strength. The tensile strength in the pultrusion direction was 306.5 MPa as per the coupon test results according to [21]. The compressive strength in the pultrusion direction was determined as 277.2 MPa through short column compression tests in [28]. The transverse tensile strength was taken as 44 MPa according to [25] based on materials with similar constituent and fibre architecture. Its transverse compressive strength was adopted as 110 MPa as a conservative estimation, considering the transverse compressive strength of commercially produced pultruded E-glass/polyester profiles ranging from 110 to 140 MPa suggested in [24]. Interlaminar shear strength was determined through short-beam tests in [14] as 26.7 MPa according to [29].

10.3.3 Model Set-Up

The components of the specimens, including GFRP and steel members, sleeve connector, steel angle (with the root radius), steel go-through bolts, steel one-sided bolts and adhesive layer, were established first according to their geometries and then meshed using solid elements Solid45 (see Fig. 10.3a for bonded sleeve specimens and Fig. 10.3b for angle specimens). Solid45 is a 3D 8-node solid element with three degrees of freedom (DoFs) per node and allows for stress stiffening, large deflection and large strain capabilities [30]. The bolt geometry including bolt head, shank and washer was also meshed using Solid45. To achieve accuracy with less computational processing time, partition and mapped mesh techniques were used (see Fig. 10.3c). The meshing size in the connection zone was controlled to be no more than 5 mm, while the meshing size away from the connection zone was around 10 to 20 mm.

Previous studies [5, 9] have evidenced that the node coupling method can effectively simulate the adhesive bonding between different components. As illustrated

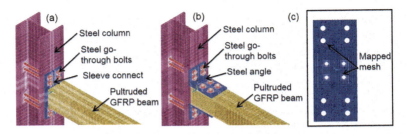

Fig. 10.3 Detailed FE modelling after meshing for **a** angle specimens (SA and BA); **b** bonded sleeve specimens (SB and BB); **c** illustration of mapped meshes of a steel endplate

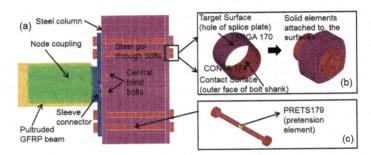

Fig. 10.4 Illustrations of **a** node coupling method for bonding; **b** creating contact pairs (Targa170 and Conta174) between bolt shank and bolt hole; **c** creating pretension element in a bolt

in Fig. 10.4a, nodes at the outer face of the steel tube and those at the inner face of the adhesive layer, and nodes at the outer face of the adhesive layer and those at the inner face of the GFRP Beam shared the same location. The DoFs of these coincident nodes were coupled in the x, y and z directions. The contact/slip behaviour between shanks and holes, washers and endplates/angles, endplates/angles and column were modelled via surface-to-surface contact elements with predetermined friction coefficients, i.e. Conta174 and Targa170. The former was a 4-node surface element and the latter was an 8-node surface element with extra midnodes. The friction coefficient between GFRP and steel was 0.25 and the coefficient between steel components was 0.44 [11, 31]. Conta174 elements were located as the surface of solid elements and had the same geometric characteristics as the solid element with which they were in contact. They overlay the solid elements describing the boundary of a deformable body and were potentially in contact with Targa170. Such contact occurred when the surface of the Conta174 element penetrated one of the target elements Targa170. Figure 10.4b illustrates how contact pairs were created between a hole in the steel column and the bolt shank, which was done by selecting the inner face of the hole to be target surface and selecting the outer face of the bolt shank to be the contact surface. In addition, the pretension force of a bolt was taken into account in the detailed modelling approach. This function technically defined a meshed bolt into two parts and generated the pretension elements (Psmesh) between these two parts (see Fig. 10.4c). The magnitude of the pretension force could be specified according to the torque applied in this study [14], i.e. 100 N.m for steel-to-steel components and 30 N.m when GFRP was involved.

All the nodes at the steel base were restricted in all directions to represent the fixed boundary condition. There were two load steps in the FE analysis. The first load step was to apply the pretension forces to the bolts. The second load step was to apply the vertical displacement at the beam end to start the static loading. The Newton–Raphson iterative method with a force convergence criterion of 0.01 was used in the solution process.

10.4 Verification of Modelling Results

To validate the modelling approach developed, comparisons were made against experimental results in terms of the failure modes, the moment-rotation behaviour and the local strain responses.

In the experiments, both bonded sleeve specimens (SB and BB) failed by extensive yielding of the steel endplates, which caused further bending of the steel go-through bolts. This is well captured in the FE results of deformed shapes as shown in Fig. 10.5. The location on the endplate of the initiation of yielding can be identified by checking the stress state ratio (R_Y). As indicated by Fig. 10.6a,b, yielding of endplate first occurred near the junction at the top flange of the steel tube with a load of 8.82 kN.m for SB and 8.98 kN.m for BB. Those load levels were therefore considered as the elastic moment capacity of the bonded sleeve specimens, indicating the moment at the yielding initiation. The modelling results of elastic moment capacity compared well (about 5% difference) to the measured value in [14] (8.3 kN.m for SB and 8.5 kN.m for BB). In addition, no damage of the GFRP beam was observed in either of the bonded sleeve specimens, indicated by Tsai-Wu failure criterion (as $I_F < 1$ for all regions on the GFRP member in SB and BB at the experimental ultimate loads). However, certain regions near the bolt holes of the steel angle specimens SA

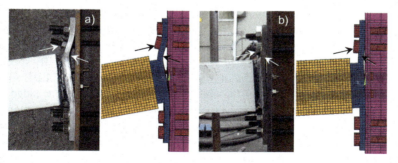

Fig. 10.5 Comparison of failure modes between experimental observations and FE modelling for **a** SB; **b** BB (arrows indicate yielding of steel endplate and bending of steel go-through bolts)

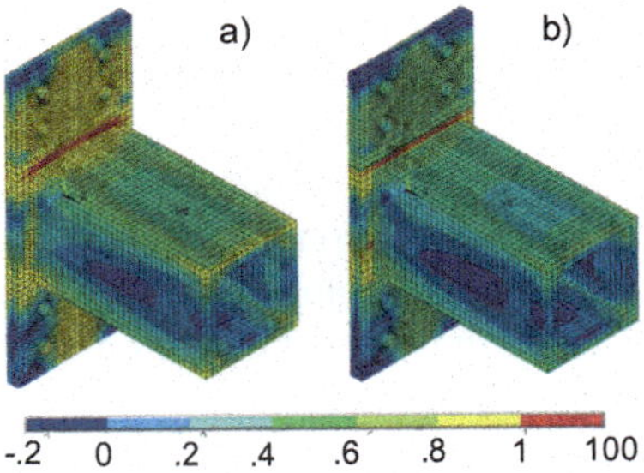

Fig. 10.6 Contour plot of stress state ratio (R_Y) on sleeve connectors for **a** SB at load level 8.82 kN.m; **b** BB at load level of 8.98 kN.m

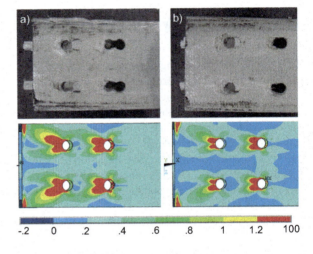

Fig. 10.7 Contour plots of inverse of Tsai-Wu strength ratio index (IF) for bolt hole regions indicating material failure of GFRP in **a** SA; **b** BA, with comparison to observed GFRP material failure from experiments

and BA were highlighted in yellow and red, corresponding to I_F values greater than 1. This result indicates that damage may have occurred to GFRP materials there. These damage regions compared well with those observed from experiments (see Fig. 10.7).

The experimental moment-rotation curves are compared with modelling results in Fig. 10.8, where overall satisfactory agreement is found for all specimens. For SB as shown in Fig. 10.8a, the moment linearly increased with the rotation angle until nonlinearity occurred at about 8.3 kN.m due to yielding of the steel endplate. Although the top rows of one-sided bolts failed in testing for the BB specimen (marked on the experimental curve in Fig. 10.8b), the load capacity of BB showed only a slight decrease (2.14% of the applied moment). Failure of one-sided bolts was not simulated in the modelling as such a slight drop in the applied moment was not present in the modelling results. The experimental results of the initial rotational stiffness for SB and BB were calculated using the initial slopes of the experimental curves as 911 kN.m/rad and 1020 kN.m/rad respectively. The values from FE analysis were 996 kN.m/rad for SB specimen (i.e. 8.53% discrepancy) and 1122.3 kN.m/rad for BB specimen (i.e. 9.12% discrepancy).

10 Connections of Fibre Reinforced Polymer …

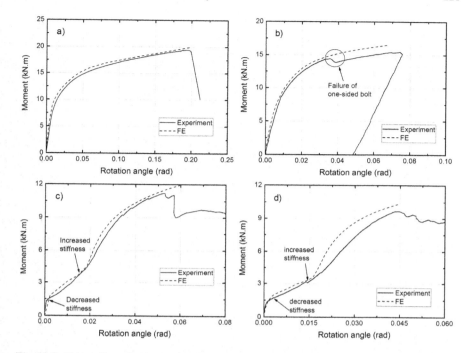

Fig. 10.8 Comparison of moment–rotation curves between experimental and FE results for **a** SB; **b** BB; **c** SA; **d** BA

In the angle specimens SA and BA, the development of connection rotation with applied moment showed three stages during the loading process. After the initial linear stage, the tangent of the moment-rotation curves, i.e. the rotational stiffness, decreased at around 1.5 kN.m for both angle specimens. This was due to the contact and slip between the bolts and bolt holes. After the bolts came into full contact with the bolt holes, the rotational stiffness increased again. It is clear in Fig. 10.8c, d that such changes in the rotational stiffness of the experimental moment–rotation curves were captured by the FE modelling curves. The drop of load capacity in the experimental curves was due to failure of the GFRP beam in both specimens, and such failure initiation could be indicated by Tsai-Wu failure criterion as discussed previously.

Strain responses measured at key locations from experiments were used to further validate the modelling approach. The locations of interest (considering an indication of steel yielding) were the junction locations between the steel tube and the endplate for the bonded sleeve specimens, and the locations between the wall plate and the bottom plate of the top steel angle for the angle specimens. Reasonable agreement was found for the moment-strain curves obtained from FE analysis and experiments as evidenced in Fig. 10.9.

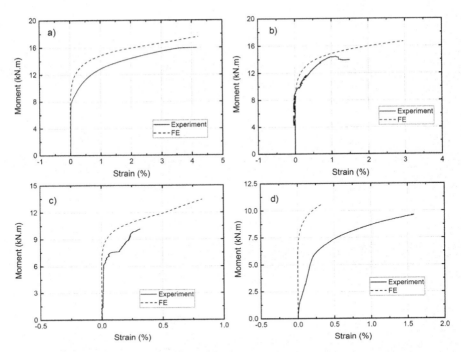

Fig. 10.9 Comparison of moment–strain curves between experimental and FE results for **a** SB; **b** BB; **c** SA; **d** BA

10.5 Parametric Study of Bonded Sleeve Connections

The validated modelling approach is used in this Section to evaluate the bonded sleeve connection with various design parameters, including i) the thickness of the steel endplate; ii) the bonding length between steel tube and GFRP beam; iii) the number of go-through bolts; and iv) the presence of one-sided bolts for connecting the centre of the steel endplate to the flange of the steel column. Such connection details were only applied to specimen BB with the length of 1700 mm, for more realistic bending moment dominant scenarios. Also, the parametric study below was applicable only to the yielding of steel or the material failure of GFRP according to the aforementioned failure criteria. As the steel tube and the GFRP beam were considered to be perfectly bonded (modelled via node coupling method), failure of the adhesive bonding was not considered in this study as also no such failure occurred in either SB or BB specimens during experiments.

10.5.1 Effect of Endplate Thickness

The steel endplate is a part of the sleeve connector that welds to the steel tube (sleeve) and connects to the steel column with bolts. To understand the effect of changing endplate thickness on the overall structural behaviour of the bonded sleeve connection, the other design parameters remained the same as those in the FE model of specimen BB. The thickness values selected in the parametric analysis were 6 mm (Model T6), 8 mm (Model T8), 10 mm (Model T10, the same model as in the experimental validation in Section 10.4) and 12 mm (Model T12).

The resulting moment–rotation curves of the bonded sleeve connection models with different endplate thicknesses are shown in Fig. 10.10a. The initial stiffness of the connection (taken as the linear slopes of the moment-rotation curves) was plotted against the endplate thickness in Fig. 10.10, showing the increase of rotational stiffness with endplate thickness. This trend is reasonable because the endplate is under prying effects and a large thickness provides an increased second moment of area for the bonded sleeve connection, which in turn improves its overall stiffness when subjected to such prying effects. More specifically, when compared with Model T6, a 60% increase in rotational stiffness was found when the thickness of the endplate was doubled (Model T12), compared to a 29% or 48% increase in initial stiffness for Model T8 or Model T10 respectively.

All models showed extensive yielding of endplates and bending of bolts, as evidenced in Fig. 10.11, whereas no material damage on GFRP was observed according to Tsai-Wu failure criterion. At the same applied displacement at the beam end (200 mm, for example), more gaps were observed between the endplate and the column with a thinner endplate, as shown in Fig. 10.11. The yielding initiation in all models were found at the location of the steel endplate near the top flange of the steel tube. The corresponding elastic moment capacities at the yielding initiation were 10.86 kN.m for T12, 8.98 kN.m for T10, 7.42 kN.m for T8 and 6.95 kN.m for T6. An increase of 56% in elastic moment capacity was found in Model T12 compared with Model T6, but only a 7% increase from Model T6 to Model T8.

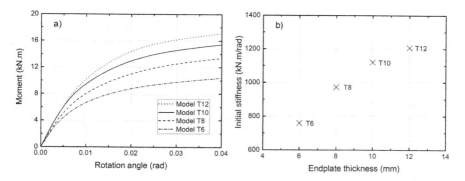

Fig. 10.10 a Moment–rotation curves; b initial rotational stiffness for bonded sleeve connection models with different endplate thicknesses

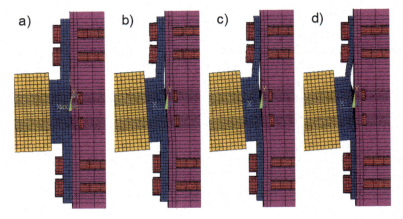

Fig. 10.11 Deformed shapes of bonded sleeve connections with 160 mm bonding length and eight go-through bolts and different endplate thicknesses for **a** Model T12; **b** Model T10; **c** Model T8; **d** Model T6

10.5.2 Effect of Bonded Sleeve Length

In a bonded sleeve connection, adhesive bonding between the GFRP section and steel tube provides full composite action between these two components. The length of the steel tube thus determines the bonded sleeve length. The bond length was 160 mm in the FE model for specimen BB (Model L160). In order to investigate the effect of the bonded sleeve length on the overall performance of the connection, this length was selected as 80 mm (Model L80, equal to the outer width of the steel tube) and 240 mm (Model L240, three times the outer width of the steel tube) in the parametric study. The maximum shear stress at the adhesive layer was about 17 MPa when the elastic moment capacity of Model L80 was achieved. It is understandable, therefore, that a bond length too short might cause premature debonding failure between the steel tube and the GFRP, while a length too long would certainly be a waste of material and create difficulties in assembly.

As shown in Fig. 10.12a, the moment–rotation response of Model L240 is similar to that of Model L160; both presented initial stiffness and elastic moment capacity higher than those of Model L80. It is clear from Fig. 10.12b that the initial rotational stiffness increased with longer bond length, with a 58% increase in initial stiffness from Model L80 to L160 but only 14% increase from Model L160 to L240. The increased initial stiffness was due to the improved second moment area of the composite section of GFRP and steel sections with a longer length, resulting in a stiffer beam end engaged in the connection. The elastic moment capacity of the connection system was found to be similar for bonded sleeve connections with different sleeve lengths (7.77 kN.m for Model L240; 7.73 kN.m for Model L160 and 7.63 kN.m for Model L80), with a 1.31% increase from Model L80 to L160 and a 1.84% increase from Model L160 to Model L240. The reason is that the elastic

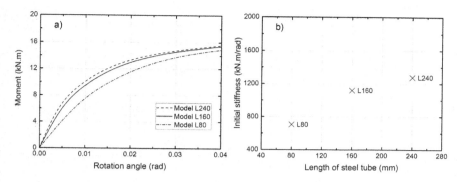

Fig. 10.12 Effect of bonded sleeve length on **a** moment–rotation curves; **b** initial rotational stiffness for bonded sleeve connections with 10 mm endplate thickness and eight go-through bolts

moment capacity was determined mainly by the yielding of the endplate and its thickness was 10 mm for all the models in this study.

10.5.3 Effect of Number of Go-Through Bolts

The number of go-through bolts connecting the steel endplate to the steel I-shaped beam is another design parameter. Eight steel high-strength go-through bolts were used in four parallel rows (Model N8) in specimen BB (see Fig. 10.1b). In this parametric analysis, four go-through bolts in two rows (Model N4) and 12 go-through bolts in six rows (Model N12) were comparatively studied. It should be noted that the depth of the steel endplate had to be adjusted according to the steel design requirements to accommodate the corresponding number of bolts. This resulted in depth of the steel endplate of 183 mm for Model N4 and of 303 mm for Model N12.

It appears in Fig. 10.13a that moment-rotation responses were scarcely affected by the number of bolt rows for the models with 10 mm endplate and 160 mm bonding length. The initial stiffness of these three models was found to be almost the same (1122 kN.m/rad). The elastic moment capacity of these three models was also found to be similar, at 8.73 kN.m for Model N4 and 8.98 kN.m for both Model N8 and Model N12. This result can be supported by the stress state ratio in each go-through bolt in Fig. 10.13b. Most loads were carried by the two rows of bolts close to centre of the connection, while the other rows of bolts did not contribute much in Model N8 and Model N12. Thus, adding more rows of bolts with a diameter of 12 mm in this connection detail did not contribute much to the overall stiffness of the connection system. The modelling results showed that the endplates yielded prior to the bending of the first row of bolts (which was close to the top surface of the steel tube) in all three models. Again, the elastic moment capacity was dominated mainly by the thickness of the steel endplates.

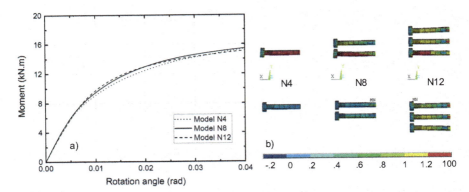

Fig. 10.13 Effect of different number of bolts on **a** moment–rotation curves; **b** stress state ratio of four bolts at rotation of 0.04 rad, for bonded sleeve connections with 10 mm endplate and 160 mm bonding length

10.5.4 Effect of Central One-Sided Bolts for Connecting Endplate

Four one-sided bolts were applied for connecting the centre of the steel endplate to the flange of steel I section in specimen BB (see Fig. 10.1a). In order to understand the effects of the one-sided bolts on the connection performance, a FE model for the bonded sleeve connection without the one-sided bolts was developed and analysed.

The resulting moment-rotation responses are shown in Fig. 10.14a, indicating that the presence of one-sided bolts introduced minor effects on the initial stiffness of the bonded sleeve connection (1123.2 kN.m/rad with one-sided bolts and 1109.4 kN.m/rad without one-sided bolts). However, the effect of one-sided bolts was significant on the elastic moment capacity of the bonded sleeve connection.

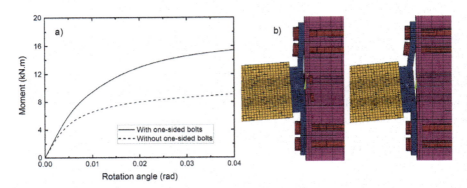

Fig. 10.14 Effect of presence of one-sided bolts on **a** moment–rotation curve; **b** deformed shape for bonded sleeve connection with 10 mm endplate thickness and 160 mm bonding length and eight go-through bolts

The elastic moment capacity (when the endplate started to yield) dropped significantly (from 8.98 kN.m to 3.72 kN.m) if no one-sided bolts were applied. This is because the one-sided bolts could provide the endplate with additional constraint to the column section and such constraint could effectively reduce the prying effects of the endplate. This mechanism is clearly illustrated in Fig. 10.14b, where a much smaller gap is evident in the model with one-sided bolts in comparison to the model without one-sided bolts at the same applied displacement at the beam end.

10.6 Conclusions

This chapter presented a numerical analysis with experimental validation for the bonded sleeve connection of GFRP beam to steel. A parametric study was conducted to investigate the effects on initial stiffness and elastic moment capacity of major design parameters, including the steel endplate thickness, the bonding length between steel tube and GFRP beam, the number of go-through bolts, and the presence of one-sided bolts at the centre of the endplate. The following conclusions can be drawn:

1. The modelling approach proposed was found to be able to describe the failure mechanism of the yielding of steel endplates and the bending of go-through bolts (taking into account the pretension of bolts) as observed in the experiments. It produces close estimations of the initial stiffness and elastic moment capacity of both bonded sleeve specimens and steel angle specimens. It can describe the contact and slip of bolt shanks and holes, washers and steel or GFRP, and thereby accurately present the change of rotational stiffness due to such effects. It is limited to indicating material failure in the forms of steel yielding and GFRP material damage only.
2. The parametric study clarified the dominant effects of endplate thickness on both initial stiffness and elastic moment capacity of the bonded sleeve connection. A 60% increase in stiffness and a 56% increase in elastic moment capacity was found when the endplate thickness was increased from 6 to 12 mm. This was because the endplate was under prying effects and a larger thickness provided an increased second moment of area for the bonded sleeve connection.
3. Longer bonded sleeve length resulted in an improvement in initial stiffness of the bonded sleeve connection, with a 58% increase if the length of the bonded sleeve was increased from 80 to 160 mm. This improvement became limited (only 14%) when the bonding length was further increased to 240 mm. The increase in the bonded sleeve length showed no effects on the elastic moment capacity of the bonded sleeve connection.
4. An increase of the number of go-through bolts from four to twelve bolts had little effect on either the initial stiffness or the elastic moment capacity for the configuration details of the bonded sleeve connection in this study. This was because the use of four go-through bolts in two parallel rows appeared sufficient

to connect the steel endplate and steel in the bonded sleeve connection, and most load was carried by the two rows of bolts close to centre of the connection.

5. The effect of central one-sided bolts on the initial stiffness of the bonded sleeve connection was minor. However, their effects on the elastic moment capacity of the bonded sleeve connections were considerable. This is because the one-sided bolts were able to provide the endplate with additional constraint to the steel member, which could effectively reduce the prying effects of the endplate.

References

1. Hollaway LC (2010) A review of the present and future utilisation of FRP composites in the civil infrastructure with reference to their important in-service properties. Constr Build Mater 24(12):2419–2445
2. Bank LC, Mosallam AS, McCoy GT (1994) Design and performance of connections for pultruded frame structures. J Reinf Plast Compos 13(3):199–212
3. Smith SJ, Parsons ID, Hjelmstad KD (1998) An experimental study of the behavior of connections for pultruded GFRP I-beams and rectangular tubes. Compos Struct 42(3):281–290
4. Smith SJ, Parsons ID, Hjelmstad KD (1999) Experimental comparisons of connections for GFRP pultruded frames. J Compos Constr 3(1):20–26
5. Singamsethi SK, LaFave JM, Hjelmstad KD (2005) Fabrication and testing of cuff connections for GFRP box sections. J Compos Constr 9(6):536–544
6. Bank LC, Yin J, Moore L, Evans DJ, Allison RW (1996) Experimental and numerical evaluation of beam-to-column connections for pultruded structures. J Reinf Plast Compos 15(10):1052–1067
7. Harte AM, McCann D (2001) Finite element modelling of the semi-rigid behaviour of pultruded FRP connections. J Mater Process Technol 119(1–3):98–103
8. Smith SJ, Parsons ID, Hjelmstad KD (1999) Finite-element and simplified models of GFRP connections. J Struct Eng 125(7):749–756
9. Carrion JE, Hjelmstad KD, LaFave JM (2005) Finite element study of composite cuff connections for pultruded box sections. Compos Struct 70(2):153–169
10. Tsai SW, Wu EM (1971) A General theory of strength for anisotropic materials. J Compos Mater 5(1):58–80
11. Shi G, Shi Y, Wang Y, Mark AB (2008) Numerical simulation of steel pretensioned bolted end-plate connections of different types and details. Eng Struct 30(10):2677–2686
12. Luo F, Bai Y, Yang X, Lu Y (2015) Bolted sleeve joints for connecting pultruded FRP tubular components. J Compos Constr 20(1):04015024
13. Yang X, Bai Y, Ding F (2015) Structural performance of a large-scale space frame assembled using pultruded GFRP composites. Compos Struct 133:986–996
14. Chao W, Zhang Z, Bai Y (2016) Connections of tubular GFRP wall studs to steel beams for building construction. Compos B 95:64–75
15. ANSYS Inc. (2013) ANSYS Mechanical APDL Command Reference Release 15.0
16. ASTM E8/E8M-15a (2008) Standard test methods for tension testing of metallic materials. West Conshohocken, PA
17. Song Q, Heidarpour A, Zhao X, Han L (2015) Performance of unstiffened welded steel I-beam to hollow tubular column connections under seismic loading. Int J Struct Stab Dyn 15(01):1450033
18. Wu C, Zhao X, Duan W, Al-Mahaidi R (2012) Bond characteristics between ultra high modulus CFRP laminates and steel. Thin-Walled Struct 51:147–157

19. Fawzia S, Zhao X, Al-Mahaidi R (2010) Bond-slip models for double strap joints strengthened by CFRP. Compos Struct 92(9):2137–2145
20. ANSYS Inc. (2013) ANSYS Mechanical APDL Material Reference Release 15.0
21. ASTM D3039/D3039M-08 (2008) Standard test method for tensile properties of polymer matrix composite materials. West Conshohocken, PA
22. ASTM D3171-15 (2015) Standard test methods for constitute content of composite materials. West Conshohocken, PA
23. Agarwal BD, Broutman LJ, Bert CW (1981) Analysis and performance of fiber composites. J Appl Mech 48(1):213–213
24. Bank LC (2006) Composites for construction: structural design with FRP material. John Wiley & Sons, Inc., pp 78–127
25. Clarke JL (2003) Structural design of polymer composites: Eurocomp design code and background document. Taylor & Francis, London, UK
26. Chamis CC (1969) Failure Criteria for Filamentary Composites. DTIC Document
27. Mises RV (1913) Mechanics of solid bodies in the plastically-deformable state. Göttin. Nachr. Math. Phys. 1:582–592
28. Xie L, Qi Y, Fang Y, Bai Y, Caprani C, Wang H (2016) Experimental and numerical investigations on pultruded GFRP SHS columns under axial compression. Advanced Composites Innovation Conference, Melbourne, Australia
29. ASTM D2344/D2344M-00 (2006) Standard test method for short-beam strength of polymer matrix composite materials. West Conshohocken, PA
30. ANSYS, Inc. (2013) ANSYS Mechanical APDL Element Reference Release 15.0.
31. Mottram J (2004) Friction and load transfer in bolted joints of pultruded fibre reinforced polymer section. In: Proceedings of the 2rd International Conference on FRP composites in Civil Engineering

Chapter 11
Cyclic Performance of Bonded Sleeve Beam-Column Connections

Zhujing Zhang, Yu Bai, Xuhui He, Li Jin, and Lei Zhu

Abstract This chapter presents the cyclic performance of bonded sleeve connections for joining tubular glass fibre reinforced polymer (GFRP) beams and columns. Specimens with different endplate thickness and number of bolts are examined under cyclic loading. The hysteretic moment-rotation responses of specimens, including rotational stiffness, ultimate moment and rotation capacity, and local strain responses are experimentally obtained and comparatively investigated. The cyclic performance of beam-column specimens is also characterized in terms of their ductility and energy dissipation capacity. Excellent ductility and energy dissipation capacity can be achieved through yielding of the steel endplate prior to the final connection failure. Detailed finite element analysis is also performed to describe the cyclic performance of beam-column specimens with bonded sleeve connections. Numerical and experimental results agree well in terms of hysteretic moment-rotation responses, ductility and energy dissipation capacity. Further parametric study of the endplate thickness provides evidence that reduction in endplate thickness may decrease the moment capacity with satisfactory ductility and energy dissipation capacity through the full development of steel yielding.

Reprinted from Composites Part B: Engineering, 142, Zhujing Zhang, Yu Bai, Xuhui He, Li Jin, Lei Zhu, Cyclic performance of bonded sleeve beam-column connections for FRP tubular sections, 171–182, Copyright 2018, with permission from Elsevier.

Z. Zhang
Department of Civil Engineering, Monash University, Clayton, Australia

Y. Bai (✉)
Department of Civil Engineering, Monash University, Clayton, Australia
e-mail: yu.bai@monash.edu

X. He
School of Civil Engineering, Central South University, Changsha, China

L. Jin
Xi'an Jiaotong University, Xi'an, China

L. Zhu
School of Civil and Transportation Engineering, Beijing University of Civil Engineering and Architecture, Beijing, China

11.1 Introduction

Fibre reinforced polymer (FRP) composites have attracted increasing attention in civil applications due to their light weight, high strength, and excellent corrosion resistance and durability [1–3]. In structural applications, FRP composites have been used to repair and strengthen existing steel and concrete structures [4–10]. Glass fibre reinforced polymers (GFRPs) in particular also show great potential as primary loading members in building applications, although proper connection detailing, as well as their brittle material behaviour, should be well addressed [11–13]. Delaminations of GFRP angles or separation between web-flange junctions of GFRP columns have often been observed when GFRP I-shaped beams and columns are connected by GFRP angles and bolts [14, 15] and no ductility or energy dissipation capacity can be provided. A bonded sleeve connection has been proposed for GFRP stud to steel beam and GFRP beam-column applications [16, 17] as illustrated in Fig. 11.1a. It utilizes an innovative sleeve connector made from a steel tube welded with a steel endplate. The steel tube can be inserted into a tubular GFRP beam using adhesive and the steel endplate forming the sleeve connector can be connected to a tubular GFRP column and another steel endplate using steel through bolts. Static experiments on the tubular GFRP beam-column specimens with bonded sleeve connections indicated that the specimens with sufficient bond length failed predominantly through yielding of the steel endplate [17]. Those specimens presented semi-rigid rotational stiffness of 300 to 390 kN·m/rad with a tubular section size of $102 \times 102 \times 9.5$ mm, corresponding to 10 to 15% of the rotational stiffness of the rigid connection boundary according to Eurocode 3 [18]. This percentage range was much greater than the range found in a few existing connections for I-shaped sections (mostly less than 5%) such as wrapped angle connection [14] and universal connection [19], and was comparable to the percentage range for tubular sections such as steel connection and cuff connection with both bolted and bonded methods [20, 21]. The ductile failure manner and improved stiffness and capacity of the proposed bonded sleeve connections further suggests a promising energy dissipation capacity of such beam-column connections for tubular GFRP members under cyclic loading.

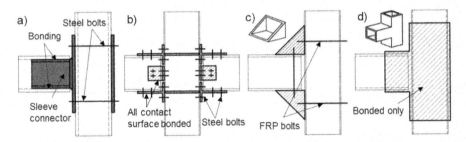

Fig. 11.1 a Bonded sleeve connection for beam-column applications, and connection configuration from the literature, **b** T-stub and web angle connection, **c** universal connection, and **d** cuff connection

Several studies have been conducted to investigate the cyclic performance of plate-to-plate connections for FRP members [22–24]. However, studies of the cyclic performance of beam-column connections for FRP structures are still limited. Early work on the cyclic performance of GFRP beam-column connection was carried out in 1994 [25]. The GFRP beam and column were I-shaped sections with dimensions of 203.2 × 101.6 × 9.5 mm. The connection configuration used T-stub connectors, web angles, and column stiffeners, as shown in Fig. 11.1b. Steel bolts were used to connect T-stub connectors and web angles to the GFRP beam and column. All contact surfaces were bonded. Failure was initiated by the separation of the column stiffener from the column flange at the second load cycle of 3.52 kN. The connection achieved an ultimate capacity of 3.65 kN with no ductility presented. The ultimate failure mode was separation of the column flange from its web.

A "universal connector" system was proposed in [19] to connect GFRP I-shaped members with the dimensions 101.6 × 101.6 × 6.35 mm. The universal connector utilized a GFRP angle stiffened by the side gusset plate, as shown in Fig. 11.1c. Two specimens with universal connectors were tested under cyclic loading [26]; one specimen was bolted-only and the other specimen was both bolted and bonded. Both specimens were loaded with 36 cycle groups with a drift angle up to 0.16 rad. The bolted-only specimen failed initially at the top universal connector with cracking at the 6th cycle group with a drift angle of 0.0145 rad. It achieved a moment capacity of 2.59 kN·m and then lost most load carrying capacity due to the sudden failure of the FRP bolts. The bolted and bonded specimen achieved an improved rotational stiffness 5 times greater than that of the bolted-only specimen, and the failure initially occurred in the adhesive at the 14th cycle group with drift angle of 0.022 rad and a moment capacity of 3.09 kN·m. No ductility was developed in either specimen as evidenced in their linear moment-rotation curves.

A cuff connection was proposed in [20]. The cuff connection was a single monolithic unit made by vacuum-assisted resin molding as shown in Fig. 11.1d, and connected GFRP beam and column using adhesive only. In investigation of the cyclic performance of cuff connections in [21], three beam-column specimens (102 × 51 × 6.35 × 3.18 mm tubular sections) were tested. The first specimen with cuff connections achieved a peak load of 17.8 kN and then lost all load-carrying capacity due to a bond failure at the adhesive interface between the GFRP beam and the cuff. To prevent the bond failure that occurred in the first specimen, pre-treatment was used in the second specimen with the cuff connection (i.e. the inside cuff surface was mechanically roughed prior to the adhesive bonding process). This resulted in a 5% improvement in the rotational stiffness compared to the first cuff specimen, and reached a higher peak load of 23.2 kN. After the peak load, the second specimen failed with crushing at the top of the GFRP beam. Both the first and second specimens showed linear elastic responses under cyclic loading with almost no ductility. The third specimen with cuff connection was equipped with additional GFRP angle stiffeners inside the GFRP beam and showed similar rotational stiffness and moment capacity to the second specimen. Progressive matrix cracking was seen at both bottom and top corners of the cuff during earlier cycles and pseudo-ductility was shown in the moment-rotation curve due to the progressive damage.

From the very limited investigations reported for the cyclic performance of GFRP beam-column connections, the foregoing results showed a brittle failure manner of the connection configurations under cyclic loading, with little or no ductility and energy dissipation capacity. The bonded sleeve beam-column connection proposed may favorably show a ductile failure manner under static loading. As one step further, this chapter presents an investigation on the cyclic performance of the bonded sleeve connection for GFRP beam-column applications including rotational stiffness, moment and rotation capacity, and local strain responses. The specimens are also characterized and compared in terms of their ductility and energy dissipation capacity. Furthermore, a detailed finite element (FE) analysis is performed to describe the cyclic moment-rotation responses of FRP beam-column specimens with the proposed bonded sleeve connections.

11.2 Experimental Program

11.2.1 Materials

Both beams and columns were GFRP square hollow sections (SHS) with measured dimensions of 102 × 102 × 9.5 mm. The tensile properties in the pultrusion direction of the GFRP were measured through tensile coupon tests according to ASTM D-3039 [27]. The longitudinal tensile modulus was measured as 25.2 GPa, tensile strength as 331 MPa, and Poisson's ratio as 0.28. The in-plane shear modulus of 3.02 GPa and strength of 36.6 MPa were measured through 10° off-axis tests [28]. The interlaminar shear strength of the GFRP was measured as 31.2 MPa by short beam tests according to ASTM D-2344 [29]. The tensile modulus of 6.20 GPa and tensile strength of 88.5 MPa in the transverse direction were measured by three-point bending of transverse beams.

The sleeve connector was formed by welding a steel tube with the dimensions of 80 × 80 × 6 mm to a 6 or 8 mm thick steel endplate. The back endplate was 6 mm thick, as shown in Fig. 11.2. The steel tube and endplates were of Grade 250. Their Young's modulus of 201 GPa, Poisson's ratio of 0.28, and yield stress of 314 MPa (corresponding to 0.15% yield strain) were measured according to ASTM E8/E8M [30]. The steel through bolts used (see Fig. 11.1a) had a diameter of 11 mm (M12) and were of Class 8.8 with a total length of 120 mm. The yield stress of the bolts was measured as 1043 MPa in [31]. Elevated and high temperatures may cause the degradation in strength and stiffness of structural adhesives [32]. In the study, the adhesive (Araldite 420) used to bond the steel tube and GFRP beam has its service temperature up to 70 °C according to the manufacturer' recommendations [33]. At room temperature (around 25 °C), Araldite 420 had the tensile modulus of 1.9 GPa, Poisson's ratio of 0.36, tensile strength of 28.6 MPa, and shear strength of 25 MPa [34–37]. In higher temperature scenarios such as fire, passive and/or active

11 Cyclic Performance of Bonded Sleeve Beam-Column Connections 235

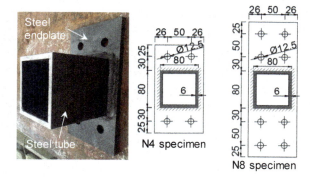

Fig. 11.2 Steel sleeve connector and its geometry for beam-column specimens

fire protection techniques may be used [38–40] to mitigate temperature effects on its strength and stiffness degradation.

11.2.2 Specimens

To study the cyclic behavior of bonded sleeve connections for GFRP tubular beam-column members, three full-scale specimens were tested under cyclic loading. Two parameters were varied, with the number (N) of steel through-bolts (four or eight bolts) connecting the steel endplates and the GFRP column, and the thickness (T, 6 or 8 mm) of the front endplate that was welded to the steel tube. All specimens were designed to have a bond length (B) of 160 mm between the steel tube and the GFRP beam, in accordance with the previous static experimental investigation [17]. These three specimens were therefore named as N4B160T6, N8B160T6, and N8B160T8. The geometry of the sleeve connector for the N4 and N8 specimens is presented in Fig. 11.2, considering both steel and FRP standards [15, 41–43].

To ensure bonding quality, the inner surface of the GFRP beam was sanded and the outer surface of the steel tube was sandblasted prior to the bonding application. A thick layer of adhesive was then hand brushed on both surfaces and any extra adhesive was squeezed out along the interface after the steel tube had been inserted into the GFRP beam to ensure minimal voids in the interface. It was then cured for two weeks before cyclic testing at room temperature. After that, the steel endplates and GFRP column were bolted together using through bolts with a clamping force of 20 N·m corresponding to a snug tight condition, to form the full specimens.

11.2.3 Cyclic Loading Program

The loading procedure suggested by ANSI/AISC 341-16 [44] for steel structures was adopted in this study because no specific loading procedures had been defined for FRP

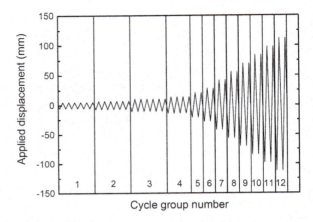

Fig. 11.3 Cyclic loading procedure applied for GFRP beam end

structures. As shown in Fig. 11.3, the applied cyclic displacement was determined as the multiplication of the drift angle suggested by ANSI/AISC 341-16 [44] and the length of the GFRP beam (1400 mm). There was a total of 12 cyclic loading groups (C1 to C12) and the calculated maximum applied cyclic displacement was 112 mm in both directions. Such a traveling distance was within the ultimate stretch of the loading cell used.

11.2.4 Experimental Setup

The experimental setup shown in Fig. 11.4 was used to investigate the cyclic performance of bonded sleeve connections in GFRP beam-column specimens. The cyclic

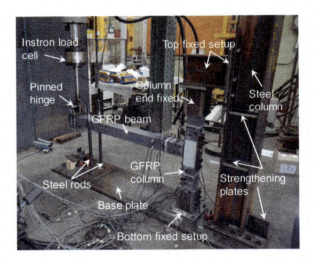

Fig. 11.4 Experimental setup for cycle testing of bonded sleeve specimens

displacement was applied at a distance of 200 mm from the end of the cantilever GFRP beam, through an Instron 250 kN universal testing machine controlled by a dynamic testing software at a loading rate of 0.5 mm/s. A pinned loading configuration was applied at the loading point of the GFRP beam to accommodate possible large rotations at the loading position while maintaining the verticality of the actuator. During the cyclic loading, a solid timber block was inserted into the tubular GFRP beam to avoid any local crushing at the loading point.

In this setup, both top and bottom ends of the GFRP column were fixed. Four steel endplates (10 mm thick) with 13 mm bolt holes were bonded to outer surface of the GFRP column to strengthen the section locally and to prevent typical bolt failures such as bearing failure, net-tension failure, and shear-out failure [45–47]. These strengthened GFRP column ends were then bolted within a π-shaped steel base (formed by two vertical 16 mm thick steel side plates and a 20 mm thick steel end plate as shown in Fig. 11.4). This resulted in six steel through-bolts in three rows for the bolted connection at each column end.

As shown in Fig. 11.4, one steel base was bolted to a 35 mm thick large steel base plate (mounted to the strong floor) to achieve a bottom fixed setup. The other steel base was bolted under a cantilever setup to achieve the top fixed setup. This top end setup was formed by a steel beam welded to a 20 mm thick endplate that was then bolted to a steel column. To minimize the movement of the steel column during the test, one steel plate was welded to the bottom of the steel column for strengthening and four steel plates were also welded to the steel column as column stiffeners (see Fig. 11.4).

11.2.5 Instrumentation

The instrumentation installed is shown in Fig. 11.5. The rotation of the connection was measured for each specimen using two string pots, SP1 and SP2. One end of each string pot was fixed on the GFRP column (at a vertical distance of 150 mm from

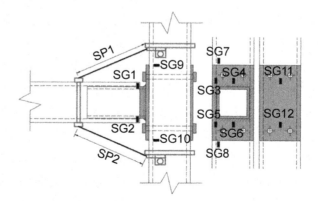

Fig. 11.5 Instrumentation for all specimens

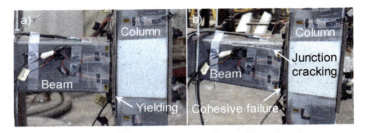

Fig. 11.6 **a** Failure initiation at 3.5 kN in C8 (first cycle), **b** ultimate failure mode at 4.3 kN in C11 in front view of specimen N4B160T6 with zoomed-in connection area

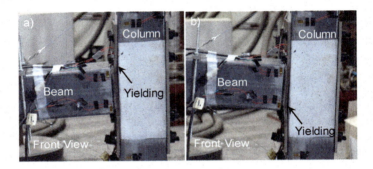

Fig. 11.7 Failure mode **a** at 5.7 kN in C11 in downward loading direction, **b** at 5.6 kN in C11 for upward loading direction of specimen N8B160T6 with zoomed-in connection area

the top/bottom outer surface of the GFRP beam) and the other end was fixed on the GFRP beam (at a horizontal distance of 200 mm from the surface of the steel front endplate). Thus, the top and bottom rotation angles could be calculated based on the change in string length. Six strain gauges (SG1 to SG6) were installed on the front steel endplate and back endplate close to the junction with the steel tube to indicate possible yielding of steel endplates. Six strain gauges (SG7 to SG12) were installed at several critical positions on the GFRP beam and column to monitor potential damage to GFRP material. The critical positions were selected where damages had been observed in previous static experiments [17].

11.3 Experimental Results and Discussion

11.3.1 Failure Modes

Detailed observations were made during the experiments to indicate failure modes, as shown in Figs. 11.6, 11.7, and 11.8 for specimens N4B160T6, N8B160T6, and N8B160T8, respectively.

11 Cyclic Performance of Bonded Sleeve Beam-Column Connections

Specimens N4B160T6 and N8B160T6 (with front endplate thickness of 6 mm) showed elastic responses in the bonded sleeve connections for the first six cycle groups (C1 to C6). For specimen N4B160T6, yielding of the front endplate was visually observed from the 8th cycle group at around 3.5 kN, evidenced by the deformed endplate at the junction with the bottom flange of the steel tube as shown in Fig. 11.6a (photo taken from the 1st cycle of C8). Cohesive failure at the interface between the inner surface of the GFRP tubular section (beam) and the outer surface of the steel tube was noticed at the 10th cycle group (C10 with the maximum applied displacement of 84 mm) as constant cracking sounds were heard. N4B160T6 reached its highest load capacity of 4.83 kN (upward direction) in the 10th cycle group (C10). After that, cohesive failure in N4B160T6 propagated, eventually causing web-flange junction cracking at the GFRP beam end at the start of the 11th cycle group (C11), as shown in Fig. 11.6b. At this point, the specimen lost its load capacity due to this junction separation at the GFRP beam end, with no failure observed on the GFRP column.

For specimen N8B160T6 with 6 mm front endplate thickness and eight bolts connecting endplates and the GFRP column, no failure was observed on either GFRP beam or column during the experiment. Yielding of the front steel endplate was observed as evidenced in Fig. 11.7 where the front endplate deformed at the junction between the bottom flange of the steel tube in both downward (refer to Fig. 11.7a) and upward (refer to Fig. 11.7b) directions. A peak load was achieved at 5.74 kN (downward direction) in the last cycle (C12). The loading was then stopped due to the stretch capacity of the load cell and the excessive yielding deformation of the front steel endplate. Visual inspection indicated that no further failure occurred in specimen N8B160T6 except yielding of the steel endplate.

For specimen N8B160T8 with the front endplate thickness of 8 mm, no obvious yielding (i.e. large deformation) of the front steel endplate was visually observed during the experiment, although yielding initiation was evidenced at the junction with the bottom flange of the steel tube by the local strain results greater than 0.15% (refer to Section 11.3.4). Cohesive failure was visually observed in specimen N8B160T8 at 3.7 kN in the 7th cycle group (C7), as shown in Fig. 11.8a. The specimen reached its maximum load capacity of 5.41 kN (in the downwards direction) in the first cycle of 8th cycle group (C8). The cohesive failure then propagated, eventually causing

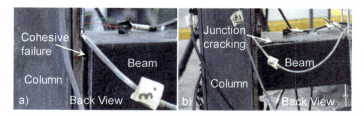

Fig. 11.8 a Failure initiation at 3.7 kN in C7, b ultimate failure mode at 4.5 kN in C9 in back view of specimen N8B160T8 with zoomed-in connection area

web-flange junction separation at the GFRP end in the second cycle of 9th cycle group (C9), as shown in Fig. 11.8b.

11.3.2 Moment-Rotation Responses

The moment-rotation relationship is the principal characterization of the mechanical performance of beam-column connections. In this study, rotation of the connection system was measured using two string pots; the moment was simply the multiplication of the recorded load and the beam length (1400 mm). The resulting moment-rotation hysteretic curves for all specimens are shown in Fig. 11.9, where elastic responses were found in the first six cycle groups (C1 to C6) for all specimens, reflected by their linear moment rotation relationship in these ranges. Starting from 7th cycle group (at the load level of 2.6 kN·m for specimens N4B160T6 and N8B160T6, load level of 3.4 kN·m for specimen N8B160T6), the stiffness of all specimens gradually degraded and the connections gradually entered the elasto-plastic stage. After reaching the peak moment values in the moment-rotation hysteretic curves for specimens N4B160T6 and N8B160T8, a large drop in load capacity was recorded due to the junction cracking of the GFRP beam, as previously mentioned. That was not the case for specimen N8B160T6, in which only yielding of the steel endplate was observed.

For better understanding of the stiffness degradation in the testing specimens, the measured rotational stiffness of each cycle group is plotted in Fig. 11.10. It is evident that the rotational stiffness of specimen N8B160T8 with the 8 mm endplate thickness is greater than that in specimens N4B160T6 and N8B160T6 with the 6 mm endplate thickness. Figure 11.10 also suggests that the stiffness of all specimens shows minor change in the first six cycle groups (C1 to C6). Starting from the 7th circle group (C7), degradation of stiffness in the connection with an increase in beam displacement becomes prominent in all specimens, as illustrated in Fig. 11.10. For specimens N4B160T6 and N8B160T6, the main reason for the stiffness degradation was the yielding of the front endplate. The stiffness degradation of N4B160T6 in the last two cycles (C10 and C11) may be due to the combination of yielding of the front

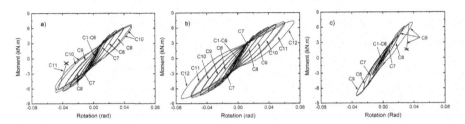

Fig. 11.9 Moment-rotation hysteretic curves of **a** specimen N4B160T6, **b** specimen N8B160T6, and **c** specimen N8B160T8

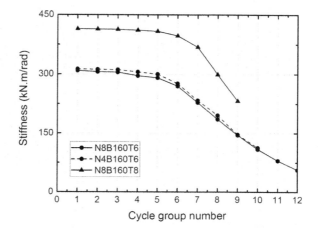

Fig. 11.10 Calculated stiffness of each cycle group for all testing specimens

endplate and progressive cohesive failure at the interface between the GFRP beam and the steel tube (see this failure mode shown in Fig. 11.6b). This was also the main reason for the stiffness degradation in specimen N8B160T8 in the last two cycles.

The envelop curves for all specimens are presented in Fig. 11.11a–c, where the effects of the endplate thickness and number of bolts on the overall moment-rotation responses of such beam-column connections can be identified. Except for the last two cycle groups (C11 and C12), the envelop curves for specimens N4B160T6 and N8B160T6 are almost identical, indicating that the number of through-bolts (i.e., either four or eight bolts) in the connection of the GFRP column and steel front endplate has a minor effect on the moment-rotation hysteretic responses. On the other hand, comparison of specimens N8B160T6 and N8B160T8 shows that an increase in the front endplate thickness from 6 to 8 mm results in an improvement of around 30% in the rotational stiffness of the bonded sleeve connections for the same rotation level. This result can be explained by the improved second moment of area due to the increased front endplate thickness from 6 to 8 mm. Similar conclusions were observed from the static experiments on specimens with different endplate thickness and number of bolts.

11.3.3 Local Strain Responses

Yielding of the front endplate was monitored using the strain gauges (SG1 to SG4) during the tests. The enveloped strain results of SG1 are plotted in Fig. 11.12, where the yield strains (0.15%) are also shown. It is evident that the front endplate of all specimens has yielded during the tests. The yielding of front endplate occurred at around 3 kN·m in specimens N4B160T6 and N8B160T6. In specimen N8B160T8, the yielding occurred at the higher load level of 5 kN·m due to the increased stiffness of the specimen.

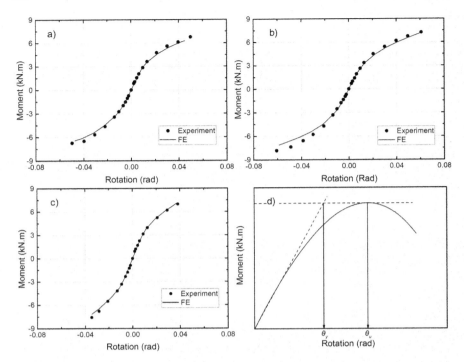

Fig. 11.11 Moment rotation envelop curves for **a** specimen N4B160T6, **b** specimen N8B160T6, **c** specimen N8B160T8, and **d** illustration of defining yield rotation, θ_y in a typical moment-rotation curve

Fig. 11.12 Strain results of SG1 for all specimens

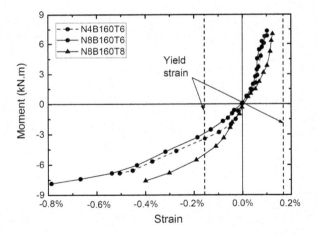

Table 11.1 Measured rotation and moment capacity, and ductility of all beam-column specimens with bonded sleeve connection under cyclic loading

	Ultimate moment capacity (kN·m)		Ultimate rotation capacity (rad)		Ductility	
	Upward direction	Downward direction	Upward direction	Downward direction	Upward direction	Downward direction
N4B160T6	6.76	6.74	0.050	0.050	4.18	4.20
N8B160T6	>7.37*	>8.04*	>0.073*	>0.073*	4.95*	4.05*
N8B160T8	6.99	7.57	0.038	0.035	3.50	2.54

Note *The test was stopped due to the exceeding the stretch capacity of loading cell and excessive deformation of steel endplate

11.3.4 Ultimate Rotation, Moment Capacity and Ductility of Connections

The measured ultimate moment and rotation capacities for all specimens are presented in Table 11.1. It appears that although specimens N4B160T6 and N8B160T8 show similar moment-rotation responses at the same cycle groups as discussed earlier, specimen N8B160T6 presents the ultimate moment capacity greater than 7.37 kN·m for the upwards direction and 8.04 kN·m for the downwards direction, corresponding to 8.3 to 16.2% higher than the moment capacity of specimen N4B160T6. Specimen N8B160T6 achieved the ultimate rotation capacity of 0.073 rad in both directions, that is also higher than the rotation capacity of 0.05 rad in both directions achieved in specimen N4B160T6. The difference in ultimate rotation and moment capacity between the two specimens may not be due to the number of bolts (see Sect. 11.3.2), but most likely due to the imperfect bonding quality associated with specimen N4B160T6, as evidenced by its cohesive failure. Specimen N8B160T8, with a thicker front endplate, achieved the ultimate moment capacity comparable to the specimen with a thinner front endplate (5.1 to 5.8% less than specimen N8B160T6) and a reduced rotation capacity of 0.035 rad (48% of that of N8B160T6) in the downwards direction and 0.038 rad in the upwards direction.

Ductility is the ability of a structure to undergo large plastic deformation without significant loss of strength. It can be quantified through rotation measurements, as the ratio between the ultimate rotation θ_u and the yield rotation, θ_y. In this study, θ_y was defined as the intersection between the initial tangent slope and the horizontal line at the ultimate moment of a moment-rotation curve [48, 49] as illustrated in Fig. 11.11d. The calculated ductility is also listed in Table 11.1 for all specimens. It is found that specimens N4B160T6 and N8B160T6 can achieve ductility ranging from 4.05 to 4.95 and specimen N8B160T8 can achieve ductility of 2.54 in the upwards direction and 3.5 in the downwards direction. The ductility of beam-column specimens N4B160T6 and N8B160T6 with 6 mm endplate thickness is comparable to that of bolted moment endplate connections reported in steel applications (ranging from 3.35 to 10.87) [50].

11.3.5 Energy Dissipation Capacity

The energy dissipation capacity of a connection system can be calculated as the summation of the enclosed area under hysteresis loops in the moment-rotation curve. The energy dissipation capacity of all specimens in each cycle was computed and the accumulative energy capacity in each cycle group is presented in Fig. 11.13.

In the first six cycle groups, all three specimens exhibited almost no energy dissipation capacity because of their elastic responses. In the subsequent cycles, specimen N8B160T8 showed less accumulative energy dissipation capacity than N4B160T6 and N8B160T6. This was explained by the consideration that, under the same loading level, a smaller area of the front endplate of specimen N8B160T8 undergone yielding, which also supported by the strain results shown in Fig. 11.12. It seemed that the accumulative energy dissipation of specimen N8B160T8 approached that of the other two specimens at the 9th cycle group, but N8B160T8 lost all energy dissipation capacity subsequently due to the sudden junction failure in the GFRP beam. On the other hand, both specimens N4B160T6 and N8B160T6 showed excellent energy dissipation capacity resulting from the yielding of front endplate. These energy dissipation capacities of specimens contributed by the yielding area of steel endplate.

11.4 Finite Element Modelling

11.4.1 Modelling Approach

The FE modelling was completed through Ansys [51]. As shown in Fig. 11.13a, GFRP material was assumed to be transverse isotropic in this study. The elastic–plastic behavior of the steel components under cyclic loading was modelled via a

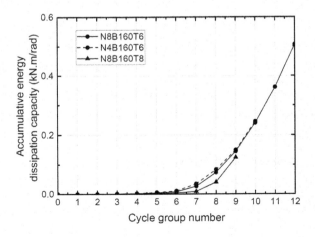

Fig. 11.13 Accumulated energy dissipation capacity of each cycle group for all beam-column specimens

bilinear kinematic hardening model, considering that 2% of the Young's modulus to be the tangent slope of the hardening stage after reaching its yield strength. The Young's modulus, Poisson's ratio and the yield strength of steel components required in the model were determined via coupon tests as mentioned in Sect. 11.2.1. The adhesive layer was modeled as a linear elastic material [52]. The material properties of the GFRP, steel, and adhesive layer used in the model were defined as the values introduced in Sect. 11.2.1. All components including the GFRP beam and column, steel endplates, steel tube, adhesive layer, and steel through bolts were modelled using 3D solid elements, Solid45, as shown in Fig. 11.14a, b. The bonding between the GFRP beam and steel tube was achieved by the node coupling method, considering full composite action between them. Previous studies [17, 21, 31, 53] have evidenced that the node coupling method can effectively simulate the shear stiffness of the bonding between difference components. The bolt-hole contact behavior was simulated via surface contact elements Targa170 and Conta174 with predetermined friction coefficients (0.25 between GFRP and steel; 0.44 between steel components) [54–56]. The pretension of bolts was also taken into account by implementation pretension elements, Prets179, through which the pretension was applied to the middle of bolts before the loading process. All degrees of freedom for the nodes at both end of the GFRP column were restricted to represent a fixed boundary condition as in the experimental setup.

The cyclic loading was implemented through vertical downward and upward displacements at the end of the beam. Three failure criterions were considered in the FE study, Tsai-Wu failure criteria was adopted for the assessment of the material failure of GFRP profiles; the yielding of steel is governed by von Mises yield criterion; and the cohesive failure initiation observed in the tests was checked against the transverse tensile strength of 28.6 MPa of the adhesive layer (see Fig. 11.14b), as the critical stress component was found to be its tensile normal stress in the xy plane [17]. It should be noted that material damages such as cracking of web-flange junction of GFRP beam are not modelled in this study, as such damages were not observed prior to the yielding of steel endplates and/or cohesive failure between the steel tube and GFRP beam.

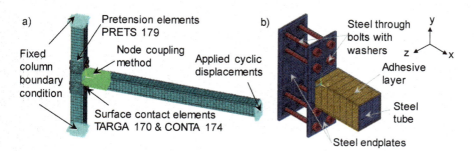

Fig. 11.14 **a** FE modelling and approaches for specimen N8B160T6 as an example, **b** detailed modelling for sleeve connector welded with steel tube and steel endplate

11.4.2 Modelling Validation

To validate the numerical results, the moment-rotation hysteretic curves from experimental and numerical results are compared for selected cycle groups (C4, C8, and the last two cycle groups for example) as shown in Fig. 11.15. The cohesive failure initiations predicted from the FE analysis according to the transverse tensile strength of the adhesive are also marked in Fig. 11.15, where the FE results of the moment-rotation responses indicated that the tensile normal stress reached the transverse tensile strength at the 11th cycle (C11) for specimen N4B160T6 and the 9th cycle group (C9) for specimen N8B160T8. From the FE predictions, it was also found that the cohesive failures were first seen at the load level of 7 kN·m for specimens N4B160T6 and N8B160T8, a result that tended to overestimate (by less than 15%) the load level when cohesive failures were noticed in the tests. This resulted in slight discrepancies of the moment-rotation hysteretic curves at the 11th cycle group (C11) for specimen N4B160T6 and at the 9th cycle group (C9) for N8B160T8. Moreover, the FE results predicted that cohesive failure initiation would occur at the load level of 7.3 kN·m in specimen N8B160T6. Such cohesive failure initiation was not visually observed during testing up to 8.04 kN·m (see Fig. 11.9b).

The hysteresis moment-rotation behavior and accumulative dissipated energy at the 8th cycle group (C8) for all specimens obtained from the FE modeling and experiments are compared as an example, with the values summarized in Table 11.2. The results suggest that the developed model is able to accurately simulate the cyclic performance of all specimens within 10% difference from the experimental results in terms of the peak moment (M), peak rotation (θ), rotational stiffness (k), and dissipated energy (E). Moreover, comparisons between the moment-rotation envelop curves from the FE and the test results up to the predicted cohesive failure point for all three specimens are plotted in Fig. 11.11a–c (C1 to C10 for specimen N4B160T6; C1 to C11 for specimen N8B160T6; C1 to C9 for specimen N8B160T8). The comparisons suggest an agreement between the FE results and the experimental results up to the cohesive failure point, for all three specimens in terms of the peak moment and rotation capacity in each cycle. The discrepancies (less than 15%)

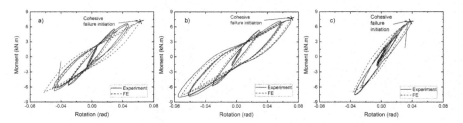

Fig. 11.15 Comparison of moment-rotation hysteretic curves between FE and test results for **a** specimen N4B160T6 at C4, C8, C10, and C11; **b** specimen N8B160T6 at C4, C8, C11, and C12; **c** specimen N8B160T8 at C4, C8, and C9

11 Cyclic Performance of Bonded Sleeve Beam-Column Connections 247

Table 11.2 Comparison of moment and rotation capacity, rotational stiffness, and accumulative dissipated energy for all testing specimens at the 8th cycle group (C8) from FE analysis and experimental results

		Peak M (kN·m)		Peak θ (rad)		k (kN·m/rad)		E (kN·m/rad)	
		Exp	FE	Exp	FE	Exp	FE	Exp	FE
N4B160T6	C8(+)	0.030	0.027	5.61	5.25	195	192	0.050	0.047
	C8(−)	0.031	0.027	5.68	5.26				
N8B160T6	C8(+)	0.030	0.030	5.42	5.27	187	181	0.048	0.042
	C8(−)	0.030	0.030	5.79	5.27				
N8B160T8	C8(+)	0.029	0.027	6.20	6.08	299	303	0.031	0.036
	C8(−)	0.028	0.027	6.79	6.09				

Note "+" means upward direction and "−" means downward direction

observed in specimen N8B160T6 in the downward direction may result from the possible asymmetrical loading applied to the specimen.

11.4.3 Effects of Endplate Thickness

To gain a full understanding of the effect of front endplate thickness on the failure mode, hysteretic moment-rotation behavior, ductility, and energy dissipation capacity of the bonded sleeve connections for the beam-column specimens, a parametric study on the front endplate thickness (4, 6, and 8 mm) was then performed through detailed FE modelling with validation and reference to the experimental specimen N4B160T6. This resulted in three FE models of N4B160T4, N4B160T6, and N4B160T8 with the same bond length of 160 mm and four-bolts configuration (see details in Fig. 11.2). The modeling results of the moment-rotation hysteretic curves of these bonded sleeve connection models with marked cohesive failure initiation are shown in Fig. 11.16a and the corresponding moment-rotation envelop curves are shown in Fig. 11.16b.

The results suggested that the rotational stiffness of the bonded sleeve connections improved with increased front endplate thickness (about 30% increase from 4 to 6 mm, and about 20% when the endplate thickness increased from 6 to 8 mm). This finding is reasonable and consistent with that from previous static experiments [11]. The cohesive failure initiations for N4B160T6 and N4B160T8 were marked at the moment capacity of 7 kN·m as confirmed by the experimental observations for N4B160T6. On the other hand, no cohesive failure was suggested in model N4B160T4 because the maximum tensile stress achieved was lower than the tensile capacity of the adhesive. Based on the constructed moment-rotation envelop curves as shown in Fig. 11.16b, the calculated ductility for models N4B160T4, N4B160T6, and N4B160T8 was 6.1, 4.7, and 2.8, respectively. Model N4B160T4 had the highest ductility of the three models, more than 1.5 times the value in model N4B160T6 and more than three times the value in model N4B160T8.

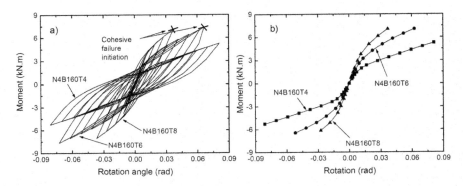

Fig. 11.16 Modelling results of **a** moment-rotation hysteretic curves, **b** moment-rotation envelop curves for beam column specimens using bonded sleeve connections with different endplate thicknesses

For better understanding on the energy dissipation capacity of bonded sleeve beam-column connections with different endplate thicknesses, the comparison of moment-rotation hysteretic curves of all models at the 5th cycle group (C5) and the 10th cycle group (C10) are plotted in Fig. 11.17a, b. The computed accumulative dissipated energy in all models at the selected cycle groups is also presented in the figures. At C5 as shown in Fig. 11.17a, both models N4B160T4 and N4B160T6 showed energy dissipation capacity resulting from the yielding of the steel endplate in bonded sleeve connections. Model N4B160T4 dissipated 60% more energy than model N4B160T6. In contrast, model N4B160T8 showed almost no energy dissipation capacity, due to its elastic response observed under the curve (with the maximum load of 3 kN·m). At C10, both models N4B160T4 and N4B160T6 showed considerable energy dissipated capacity, as evident in Fig. 11.17b, resulting from the yielding of the steel endplate. However, model N4B160T8 might lose its energy dissipation capacity, as cohesive failure was predicted. The results also suggested that model N4B160T6 dissipated 27% more energy than the energy dissipated in model

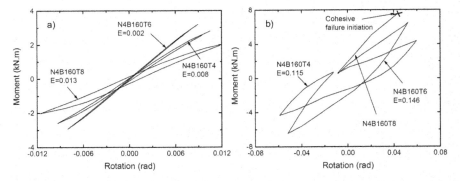

Fig. 11.17 Comparison of dissipated energy (unit: kN·m/rad) in all FE models **a** at C5, **b** at C10

N4B160T4. This difference is due to the increase in endplate thickness (from 6 to 8 mm) resulting in the increased area in the steel endplate that undergone yielding in model N4B160T6.

11.5 Conclusions

The cyclic performance of bonded sleeve connections for tubular GFRP beam-column applications was investigated through experimental and numerical studies. Specimens with different endplate thicknesses and number of bolts were examined under cyclic loading applied at the beam end. From the experimental investigation, the moment-rotation hysteretic behaviors were characterized through rotational stiffness, ultimate moment and rotation capacity, and local strain responses. The cyclic performance of the beam-column specimens was compared in terms of ductility and energy dissipation capacity. Detailed FE modeling was developed to describe the moment-rotation responses under cyclic loading and to understand the ductility and energy dissipation capacity of the proposed beam column connections. With validation from the experimental results, a parametric study on the effects of front endplate thickness as a critical design parameter was carried out. The following conclusions can be drawn from this work:

1. From the failure modes observed experimentally, it was found that the beam-column specimens with bonded sleeve connection failed initially through yielding of the steel endplate. After that yielding, cohesive failure developed at the interface between steel tube and GFRP column, especially in specimens with a thicker endplate such as N8B160T8 (with eight bolts, 160 mm bond length and 8 mm endplate thickness). These cohesive failures were then propagated, eventually caused web-flange junction cracking at the GFRP beam end. On the other hand, yielding of the steel endplate was the dominant mode in specimens with smaller endplate thickness, such as N8B160T6 (with eight bolts, 160 mm bond length and 6 mm endplate thickness), in which case the experiment was stopped due to the large deflection at the beam end caused by the yielding deformation of the endplate in the connection.
2. From the measured moment-rotation hysteretic curves and the corresponding envelop curves, it was found that specimens N4B160T6 and N8B160T6 with the same endplate thickness (6 mm) showed similar moment-rotation responses at the same cycle groups, as the number of bolts connecting the GFRP column and the steel front endplate showed only minor effect on the ultimate rotation and moment capacity of the bonded sleeve connections. Specimen N8B160T8, with a thicker steel endplate (8 mm), had about 5% less ultimate moment capacity and about 50% less ultimate rotation capacity (due to the cohesive failure observed) but 30% more rotational stiffness, compared to the specimens with thinner steel endplates.

3. Specimens N4B160T6 and N8B160T6 both showed a favorable ductile failure process through well-developed yielding of the steel endplate in comparison to the brittle failure observed in other forms of connections in GFRP beam-column applications. They also exhibited excellent ductility in the range from 4.05 to 4.95, comparable to the ductility of similar bolted moment endplate connections in steel applications. Energy dissipation capacity was evidenced by specimens N8B160T6 and N4B160T6 as characterized by the area under the hysteretic moment-rotation curves, again contributed from the yielded steel endplate. On the other hand, N8B160T8 showed less ductility and energy dissipation capacity than specimens N4B160T6 and N8B160T6, because of the higher load and the induced cohesive failure with less development of steel yielding of the endplate.
4. The proposed modelling approach using solid elements, contact modeling method, with consideration of the pretension of bolts, was found to produce accurate descriptions of the moment-rotation responses of under cyclic loading, with the maximum discrepancy of 10% for the experimental results of the peak moment and rotation, initial rotational stiffness, and energy dissipation capacity. Although the proposed modelling approach gave an overestimate of up to 15% for the cohesive failure initiation of specimens N4B160T6 and N8B160T8, the predicted envelop curves matched the experimental results well in terms of the peak moment and rotation capacity in each cycle. Therefore, the models were used for the subsequent parametric study on the effect of endplate thickness.
5. Results from the parametric study suggested that the model with the thinnest endplate (4 mm for N4B160T4) achieved the highest ductility of three models, because the thinner steel endplate was associated with more comprehensive development of yielding. The modeling results also suggested that the model N4B160T6 dissipated 27% more energy than model N4B160T4, because the increase in endplate thickness (from 6 to 8 mm) resulted in a larger area in the steel endplate that undergone yielding in model N4B160T6. Although the bonded sleeve connection model with 4 mm thickness (model N4B160T4) had the highest ductility and might mitigate the initiation of cohesive failure, it was also accompanied by the decrease in the initial rotation stiffness, moment capacity, and energy dissipation capacity.

This work presented successful development of achieving ductility and energy dissipation capacity using bonded sleeve connections in GFRP beam-column applications. This may then enable the assembly of GFRP frames using bonded sleeve connections with expected cyclic performance with ductility and energy dissipation capacity.

References

1. Hollaway LC (1994) Handbook of polymer composites for engineers. Woodhead Publishing Ltd.
2. Ku H, Wang H, Pattarachaiyakoop N, Trada M (2011) A review on the tensile properties of natural fiber reinforced polymer composites. Compos Part B: Eng 42(4):856–873
3. Bakis CE, Bank LC, Brown VL, Cosenza E, Davalos JF, Lesko JJ, Machida A, Rizkalla SH, Triantafillou TC (2002) Fiber-reinforced polymer composites for construction - state-of-the-art review. J Compos Constr 6(2):73–87
4. Fang H, Zou F, Liu W, Wu C, Bai Y, Hui D (2017) Mechanical performance of concrete pavement reinforced by CFRP grids for bridge deck applications. Compos Part B: Eng 110:315–335
5. Marouani S, Curtil L, Hamelin P (2012) Ageing of carbon/epoxy and carbon/vinylester composites used in the reinforcement and/or the repair of civil engineering structures. Compos Part B: Eng 43(4):2020–2030
6. Shi H, Liu W, Fang H, Bai Y, Hui D (2017) Flexural responses and pseudo-ductile performance of lattice-web reinforced GFRP-wood sandwich beams. Compos Part B: Eng 108:364–376
7. Napoli A, Bank LC, Brown VL, Martinelli E, Matta F, Realfonzo R (2013) Analysis and design of RC structures strengthened with mechanically fastened FRP laminates: a review. Compos Part B: Eng 55:386–399
8. Zoppo MD, Ludovico MD, Balsamo MD, Prota A, Manfredi G (2017) FRP for seismic strengthening of shear controlled RC columns: experience from earthquakes and experimental analysis. Compos Part B: Eng 129:47–57
9. Feng P, Zhang Y, Bai Y, Ye L (2013) Combination of bamboo filling and FRP wrapping to strengthen steel members in compression. J Compos Constr 17(3):347–356
10. Feng P, Zhang Y, Bai Y, Ye L (2013) Strengthening of steel members in compression by mortar-filled FRP tubes. Thin-Walled Struct 64:1–12
11. Turvey GJ, Cooper C (2004) Review of tests on bolted joints between pultruded GRP profiles. Struct Build 157:211–233
12. Coelho AMG, Mottram JT (2014) A review of the behaviour and analyse of mechanically fastened joints in pultruded fibre reinforced polymers. Mater Des. ISSN: 0261-3069
13. Keller T, Vallée T (2005) Adhesively bonded lap joints from pultruded GFRP profiles. Part I: stress-strain analysis and failure modes. Compos Part B: Eng 36:331–340
14. Bank LC, Yin J, Moore L, Evans DJ, Allison RW (1996) Experimental and numerical evaluation of beam-to-column connections for pultruded structures. J Reinf Plast Compos 15(10):1052–1067
15. Clarke JL (1996) Structural design of polymer composites: EUROCOMP design code and handbook. E&FN Spon, pp 707–718
16. Wu C, Zhang Z, Bai Y (2016) Connections of tubular GFRP wall studs to steel beams for building construction. Compos Part B: Eng 95:64–75
17. Zhang Z, Bai Y, Xiao X (2018) Bonded sleeve connections for joining tubular GFRP beams and columns: an experimental and numerical study. J Compos Constr 22(4):04018019
18. Eurocode 3 (2015) Design of steel structures. Part 1.8: design of joints. European Committee for Standardization
19. Mosallam AS, Abdelhamid MK, Conway JH (1994) Performance of pultruded FRP connections under static and dynamic loads. J Reinf Plast Compos 13(5):386–407
20. Smith SJ, Parsons ID, Hjelmstad KD (1999) Experimental comparisons of connections for pultruded frames. J Compos Constr 3(1):20–26
21. Carrion JE, LaFave JM, Hjelmstad KD (2005) Experimental behavior of monolithic composite cuff connections for fiber reinforced plastics box sections. Compos Struct 67:333–345
22. Giannopoulos IK, Dawes DD, Kourousis KI, Yasaee M (2017) Effects of bolt torque tightening on the strength and fatigue life of airframe FRP laminate bolted joints. Compos Part B: Eng 125:19–26

23. Heshmati M, Haghani R, Emrani AE (2015) Environmental durability of adhesively bonded FRP/steel joints in civil engineering applications: state of the art. Compos Part B: Eng 81:259–275
24. Luccio GD, Michel L, Ferrier E, Martinelli E (2017) Seismic retrofitting of RC walls externally strengthened by flax–FRP strips. Compos Part B: Eng 127:133–149
25. Bruneau M, Walker D (1994) Cyclic testing of pultruded fiber-reinforced plastic beam-column rigid connection. J Struct Eng 20(9):2637–2652
26. Mosallam AS (1997) Design considerations for pultruded composite beam-to-column connections subjected to cyclic and sustained loading conditions. In: SPI/CI 52rd annual conference and exposition, California, Session 14-B, pp 1–18
27. ASTM D-3039 (2000) Standard test method for tensile properties of polymer matric composite materials. West Conshohocken, United States
28. Chamis CC, Sinclair JH (1977) Ten-deg off-axis test for shear properties in fiber composites. Exp Mech 17(9):339–346
29. ASTM D-2344 (2000) Standard test method for short-beam strength of polymer matrix composite materials and their laminates. West Conshohocken, United States
30. ASTM E8/E8M (2008) Standard test methods for tension testing of metallic materials. West Conshohocken, United States
31. Zhang Z, Wu C, Nie X, Bai Y, Zhu L (2016) Bonded sleeve connections for joining tubular GFRP beam to steel member: numerical investigation with experimental validation. Compos Struct 157:51–61
32. Bai Y, Keller T (2011) Effects of thermal loading history on structural adhesive modulus across glass transition. Constr Build Mater 25:2162–2168
33. Huntsman Advanced Materials (2004) Structural adhesive - aerospace adhesives araldite 420 A/B two component epoxy adhesive. Huntsman Advanced Materials No. A161gGB
34. Satasivam S, Bai Y, Zhao XL (2014) Adhesively bonded modular GFRP web–flange sandwich for building floor construction. Compos Struct 111:381–392
35. Satasivam S, Bai Y (2016) Mechanical performance of modular FRP-steel composite beams for building construction. Mater Struct 49(10):4113–4129
36. Yang X, Bai Y, Luo FJ, Zhao XL (2017) Fibre-reinforced polymer composite members with adhesive bonded sleeve joints for space frames structures. J Mater Civ Eng 29(2):04016208
37. Qiu C, Feng P, Yang Y, Zhu L, Bai Y (2017) Joint capacity of bonded sleeve connections for tubular fibre reinforced polymer members. Compos Struct 163:267–279
38. Zhang L, Bai Y, Chen W, Ding F, Fang H (2017) Thermal performance of modular GFRP multicellular structures assembled with fire resistant panels. Compos Struct 172:22–33
39. Bai Y, Hugi E, Ludwig C, Keller T (2011) Fire performance of water-cooled GFRP columns part I: fire endurance investigation. ASCE J Compos Constr 15(3):404–412
40. Bai Y, Keller T (2011) Fire performance of water-cooled GFRP columns part II: post-fire investigation. ASCE J Compos Constr 15(3):413–421
41. Chambers RE (1997) ASCE design standard for pultruded fiber-reinforced-plastic (FRP) 537 structures. J Compos Constr 1:26–38
42. AS 4100 (1998) Steel structures. Australian Standard, Sydney, Australia
43. CNR DT 205/2007 (2008) Guide for the design and construction of 556 structures made of thin FRP pultruded elements. National Research Council of Italy (CNR), Rome, Italy
44. ANSI/AISC 341-16 (2002) Seismic provisions for structural steel buildings. American Institute of Steel Construction, Chicago
45. Lee YG, Choi E, Yoon SJ (2015) Effect of geometric parameters on the mechanical behavior of PFRP single bolted connection. Compos Part B: Eng 75:1–10
46. Thoppul SD, Finegan J, Gibson RF (2009) Mechanics of mechanically fastened joints in polymer-matrix composite structures - a review. Compos Sci Technol 69(3–4):301–329
47. Zhou A, Zhao L (2013) Connection design for FRP structural members. In: Zoghi M (ed) The international handbook of FRP composites in civil engineering. CRC Press, Hoboken
48. Babum SS, Sreekumar S (2012) A study on the ductility of bolted beam-column connections. Int J Mod Eng Res 2(5):3517–3521

49. Tizani W, Wang ZY, Hajirasouliha I (2013) Hysteretic performance of a new blind bolted connection to concrete filled columns under cyclic loading: an experimental investigation. Eng Struct 46:535–546
50. Ceolho AMG, Bijlaard FSK, Silva LS (2004) Experimental assessment of ductility of extended end plate connections. Eng Struct 26:1185–1206
51. ANSYS Inc (2013) ANSYS mechanical APDL command reference release 15.0
52. Shi G, Wang M, Bai Y, Wang F, Shi Y, Wang Y (2012) Experimental and modeling study of high-strength structural steel under cyclic loading. Eng Struct 37:1–13
53. Singamsethi S, Lafave J, Hjelmstad K (2005) Fabrication and testing of cuff connections for GFRP box sections. J Compos Constr 9(6):536–544
54. Shi G, Shi Y, Wang Y, Mark AB (2008) Numerical simulation of steel pretensioned bolted endplate connections of different types and details. Eng Struct 30(10):2677–2686
55. Luo FJ, Yang X, Bai Y (2015) Member capacity of pultruded GFRP tubular profile with bolted sleeve joints for assembly of latticed structures. J Compos Constr 20(3):04015080
56. Luo FJ, Bai Y, Yang X, Lu Y (2016) Bolted sleeve joints for connecting pultruded FRP tubular components. J Compos Constr 20(1):04015024

Chapter 12
Joint Capacity of Bonded Sleeve Connections for Tubular Fibre Reinforced Polymer Members

Chengyu Qiu, Peng Feng, Yue Yang, Lei Zhu, and Yu Bai

Abstract Bonded sleeve joints formed by telescoping a steel tube connector for bolt-fastening have been shown as effective means for assembling tubular fibre reinforced polymer (FRP) members into more complex structures such as planar or space frames. A theoretical formulation is developed in this chapter to estimate the capacity of such joints in axial loading for circular tube sections and the predictions are validated by experimental results covering various section sizes and bond lengths. The formulation is based on the bilinear bond-slip constitutive relationship considering elastic, softening and debonding behaviour at the adhesive bonding region. Finite element (FE) analysis is also conducted to estimate the joint capacity and to describe shear stress distribution in the adhesive layer, validating the theoretic results. The theoretical formulation is therefore further used to study the effects of design parameters including bond length and adherend stiffness ratio, again validated by FE results. An effective bond length can be calculated by the theoretical formulation for the joint capacity at both the elastic limit and the ultimate state. Given a bond length, an optimal adherend stiffness ratio can also be identified to achieve the maximum joint capacity at the elastic limit or the ultimate state.

Reprinted from Composite Structures, 163, Chengyu Qiu, Peng Feng, Yue Yang, Lei Zhu, Yu Bai, Joint capacity of bonded sleeve connections for tubular fibre reinforced polymer members, 267–279, Copyright 2017, with permission from Elsevier.

C. Qiu
Department of Civil Engineering, Monash University, Clayton, Australia

P. Feng · Y. Yang
Department of Civil Engineering, Tsinghua University, Beijing, China

L. Zhu
School of Civil and Transportation Engineering, Beijing University of Civil Engineering and Architecture, Beijing, China

Y. Bai (✉)
Department of Civil Engineering, Monash University, Clayton, Australia
e-mail: yu.bai@monash.edu

12.1 Introduction

With the properties such as high strength-to-weight ratio, corrosion resistance and low maintenance requirements, fibre reinforced polymer (FRP) structural members have gained broad recognition in the construction industry [1, 2]. In particular, glass FRP (GFRP) composites, with sufficient strength and stiffness at moderate cost, have received considerations as construction materials for structural members [3–5]. Aided by the pultrusion manufacturing technique [6, 7], mass production of GFRP structural components with constant cross-section has been facilitated at reduced cost and with satisfactory quality control. Such components have been employed as bridge decks [8–11], reinforcements [12], roof structures [13], trusses [14], floor systems [15–17], and components in hybrid members [18–20], in shapes of open or closed sections. However, their application in structural construction still requires the development of reliable and convenient connection approaches, especially for closed tubular sections.

Early efforts to develop connections for GFRP tubular sections imitated practices in steel structures, for example, using bolted-through web-gusset plates or flange-angle cleats for beam-column connections [21]. Later, a cuff connection for pultruded GFRP tubular sections was proposed and demonstrated enhanced strength and stiffness [22, 23]. The benefits of a sleeve connection formed by telescoping steel and GFRP tubular members have been underlined for applications in space frame structures [24]. The steel sleeve connector not only enables versatile connection forms but also provides ductile failure mode through steel yielding. In such connections, the GFRP-steel telescoping portion or the sleeve portion acts compositely through adhesive bonding [25] or mechanical fastening [26, 27]. Although mechanical fastening facilitates convenient in-situ fabrication, its employment in fixing sleeve joints exposes the relatively weak shear strength of anisotropic GFRP materials and such uses become inapplicable when circular profiles are used. If adhesive bonding is used, a more uniform stress transfer to GFRP materials is achieved. Bonded sleeve connections have also been developed into beam-to-column scenarios [28], where pultruded GFRP and steel tubular square members were used. Such bonded sleeve connections have exhibited significant improvement in both stiffness and strength in comparison with traditional steel angle connections and bolted sleeve connections.

Bonded sleeve connections transfer axial loading through shear between the GFRP and steel adherends. As a special type of bonded lap joints, the joint capacity may be determined through the shear mechanism of the adhesive bonding when the adherends (steel and GFRP in this case) are sufficiently strong. Pioneering theoretical works on single and double lap adhesive-bonded joints have been conducted to understand such shear mechanisms where only elastic [29, 30] or elastic–plastic [31, 32] bonding behaviour was considered. Joint capacity was therefore evaluated against allowable material stress or strain at critical regions. These approaches are limited to identifying possible failure location or initiation.

Fracture mechanics-based approaches were further introduced to understand the full range behaviour of several types of single and double lap joints. This required understanding of the nonlinear bond-slip relationship at the bonded region. Simplification of the nonlinear bond-slip relationship into a bilinear form enabled the development of closed form analytical solutions for joint capacity and shear stress distribution in the adhesive layer. Such a bilinear relationship is characterised by a linearly ascending stage of shear stress with slip, followed by a linear decrease to zero shear stress at the debonding slip (and slip may further increase while shear stress remains zero). This bilinear relationship was adopted by Ranisch [33] to study the post-cracking behaviour of bonded steel-to-concrete single lap joints subjected to axial loading. By incorporating Volkersen's classical stress analysis and the bilinear bond-slip model, analytical solutions for the capacities of a pull–push plate-to-block bonded joint at elastic and ultimate limit states were developed by Brosens and Van Gemert [34]. This work was further expanded in [35], where analytical solutions were developed to identify several stages of the debonding process. Also utilising the bilinear model, Wu et al. [36] analytically derived the expressions for the joint capacity, adhesive shear stress distribution and the debonding processes of pull–pull and pull–push single lap joints. Experimental investigations were also conducted to understand the debonding behaviour of single and double lap joints, such as steel plate-to-concrete block double lap joints in [37], FRP plate-to-steel pull–push joints in [38], single and double lap FRP plate joints for structural rehabilitation in [39], and single and double lap aluminium plate joints in [40]. Such experimental results were able to validate analytical and numerical models that employed the aforementioned bilinear bond-slip relationship. Furthermore, experimental results of shear stress and slip in the adhesive layer measured from FRP-to-concrete [41, 42] and FRP-to-steel bonded lap joints [38, 43, 44] provided evidence that, when a brittle adhesive was used, the bilinear bond-slip relationship provided a satisfactory approximation to the experimental measurements.

In addition to the bilinear form, several representations of the bond-slip relationships were also proposed to study the behaviour of single and double lap bonded joints. A nonlinear bond-slip curve including softening behaviour was employed by Täljsten [45] for numerical calculation of the capacity of CFRP plate-to-concrete bonded joints; however, it may be sophisticated for analytical derivation. A bond-slip relationship characterised by abrupt failure after the linearly elastic region was adopted by Yuan et al. [35, 46]; and a bond-slip relationship with exponential softening behaviour was also considered in the former. For ductile adhesives, bond-slip relationships with trapezoidal shape were considered as more appropriate [44, 47]. The load–displacement behaviour of a double lap joint connecting aluminium plates under pull-out test was numerically studied in [48], with consideration of four shapes of bond-slip relationship, i.e. bilinear, linear-parabolic, exponential and trapezoidal. It was concluded that, given identical initial stiffness, peak shear stress and fracture energy (as the area under the bond-slip curve) of the bonded interface, the load–displacement behaviour was practically independent of the shape of the bond-slip relationship. It appears that previous analytical and numerical studies have focused mainly on the adhesive shear stress distribution and capacity of adhesive-bonded joints in single or double lap forms. Such responses of a bonded sleeve connection require theoretic understanding and experimental validation to enhance confidence in their application.

This chapter therefore formulates theoretical modelling for the joint capacity of a bonded sleeve connection for GFRP members with circular tube sections (i.e. circular hollow sections, or CHS). This theoretic formulation employs the aforementioned well-accepted bilinear bond-slip relationship. Experimental results of joint capacities were obtained from tension tests conducted on steel/GFRP bonded sleeve connection specimens with various cross-section sizes and bond lengths. Finite element (FE) models were also developed for comparison of the joint capacities and shear stress distribution along the bond length. After validation by experimental and FE results, the theoretical formulation was further utilised to investigate the effects of major design parameters including bond length and adherend stiffness ratio, i.e. $E_s A_s / E_g A_g$ where E_s or E_g are the Young's modulus of steel or GFRP, and A_s or A_g are cross-section area of the steel or GFRP component.

12.2 Summary of Experimental Results

A series of bonded sleeve joint specimens consisting of a steel tube connector and a GFRP component were prepared and examined, as shown in Fig. 12.1a–c [24, 25]. Their joint capacities were tested for assembly into planar trusses [49] or space frames with assistance of the Octatube nodal joints [25] (Fig. 12.1d). The tested specimens were categorized into five groups (G1 to G5) based on different cross-sections and geometrical characteristics of steel and GFRP components. Such information is illustrated in Fig. 12.2 with relevant dimensions provided in Table 12.1. The material properties of the steel and GFRP components in each group had been measured previously [24, 25] and are summarised in Table 12.1. The adhesive used was a two-component epoxy based structural adhesive. Its tensile strength and Young's modulus were tested to be 28.6 MPa and 1.9 GPa; Poisson ratio of 0.36 was obtained [24].

The bonded sleeve connection specimens in G1 and G2 were formed by GFRP rods inserted into steel tubes; those in G3 to G5 were formed by steel tubes inserted

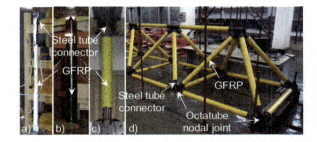

Fig. 12.1 Bonded sleeve connections for **a** specimens G1&G2, **b** specimens G3&G4, **c** specimen G5, **d** an assembled large-scale space frame

Fig. 12.2 Geometries of specimens in different groups (magnitudes of the symbols are given in Table 12.1 for each group)

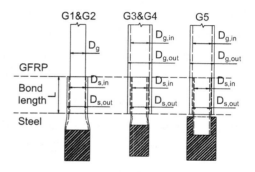

Table 12.1 Geometries and material properties of specimens

Group	Description	GFRP diameter $D_{g,out}/D_{g,in}$ (mm)*	Steel diameter $D_{s,out}/D_{s,in}$ (mm)*	Bond length for pull-out failure L (mm)	Young's modulus E (GPa)*	Strength (MPa)*
G1	GFRP rod into steel tube	15.5/–	21.3/17.3	30, 40, 50	51.5/198.3	207.0/487.5
G2		25.3/–	33.7/27.5	30, 40, 50, 60, 70, 80	47.4/209.7	207.0/397.4
G3	Steel tube into GFRP tube	38.0/30.0	26.9/22.3	30, 40	21.6/182.4	240.0/294.7
G4		31.0/24.6	21.3/17.3	30, 40	31.4/198.3	240.0/487.5
G5		92.0/76.0	73.0/65.0	80	39.3/196.8	300.0/451.3

*For GFRP or steel component: outer diameter/inner diameter; for Young's modulus: longitudinal direction of GFRP/steel; for strength: longitudinal tensile strength of GFRP/yield strength of steel

into GFRP tubes (Figs. 12.1 and 12.2). These groups of specimens cover possible scenarios in practice where the GFRP component may be either the outer [25] or inner adherend [49]. The adoption of GFRP rods instead of tubes in G1 and G2 was to increase their cross-section areas thus axial capacities, preventing potential failure of the GFRP component. In G1 to G4, the free ends of the steel tubes were flattened for clamping or bolt-fastening to nodal joints in structural assembly. With a similar purpose, those in G5 were slot-welded with a steel gusset plate. The specimens were loaded in tension to failure under displacement control. Among all specimens, two types of failure modes were observed, namely pull-out failure (Fig. 12.3a) and excessive yielding or fracture of the steel tube connector (Fig. 12.3b). The former failure mode was characterised by adhesive failure close to the inner adherend or interface failure between the adhesive and the inner adherend. Also listed in Table 12.1 are the bond lengths that incurred pull-out failure within each group of specimens. In Table 12.2, these specimens are named following the convention of group index and bond length, together with the experimental results of their ultimate joint capacity $P_{u,E}$.

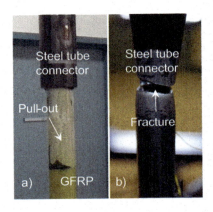

Fig. 12.3 Typical failure modes of bonded sleeve connection specimens: **a** pull-out failure, **b** fracture of steel tube connector

Table 12.2 Ultimate joint capacity (P_u) of specimens with pull-out failure mode

Specimen	Experiment $P_{u,E}$ (kN)	Theoretical $P_{u,TH}$ (kN)	$P_{u,E}/P_{u,TH}$	Finite element $P_{u,FE}$ (kN)	$P_{u,E}/P_{u,FE}$
G1-30	40.0	33.8	1.183	34.1	1.172
G1-40	48.0	44.2	1.085	43.3	1.109
G1-50	53.0	53.9	0.983	51.6	1.028
G1-60, 70 and 80 failed by fracture of steel					
G2-30	52.0	54.9	0.947	52.5	0.990
G2-40	72.0	72.3	0.996	69.6	1.035
G2-50	82.0	89.0	0.922	85.7	0.957
G2-60	97.0	104.6	0.927	102.7	0.944
G2-70	111.0	119.1	0.932	114.7	0.968
G2-80	116.0	132.1	0.878	125.5	0.924
G2-90 and 100 failed by excessive yielding of steel					
G3-30	41.0	57.4	0.714	52.9	0.776
G3-40	57.0	73.7	0.773	64.9	0.878
G3-50, 60 and 70 failed by fracture of steel					
G4-30	42.0	46.8	0.897	42.5	0.988
G4-40	56.0	60.7	0.922	55.4	1.012
G4-50, 60 and 70 failed by fracture of steel					
G5-80	362.0	387.6	0.934	378.0	0.958
G5-100 failed by fracture of steel					
Mean			0.935		0.981
Standard deviation			0.114		0.095

12.3 Theoretical Formulation of Joint Capacity

12.3.1 Governing Differential Equations

Differential governing equations are formulated in this Section for adhesive shear stress and adherend normal stress based on Volkersen's stress analysis. When a joint is under axial loading (P), the adhesive layer is predominantly subjected to shear deformation, as illustrated in Fig. 12.4. Here the outer adherend is presumed to be a GFRP tube and the inner adherend a steel tube (Fig. 12.4a), though the deduced equilibrium is equally applicable to other materials or to the case in which the inner adherend is a circular rod. Moreover, despite the derivation being based on axial tension, a compression scenario would produce identical force equilibrium. In this study, L, A and b are the bond length, cross-section area and circular circumference of the bonded face respectively; σ, u and ε are axial stress, displacement and strain of adherend; E is Young's modulus and τ is shear stress in the adhesive layer; subscripts 'g' or 's' denotes the material of GFRP or steel.

A few assumptions or simplifications are made in the derivation. Adherends are linear elastic thus the formulation is valid before yielding of the steel tube. Normal stress is assumed as uniformly distributed over the cross-section of adherends. The adhesive layer is thin compared to the thickness of the adherend; therefore, shear stress is considered uniform through the thickness of adhesive. Cross-sections of adherends and adhesive are constant along the bonded region.

As a result, the slip between adherends δ is defined as the relative displacement between the GFRP and steel components (Fig. 12.4b):

$$\delta = u_s - u_g \qquad (12.1)$$

where u_s and u_g are the axial displacement of the steel and GFRP adherend respectively.

Taking derivatives of δ with respect to x gives:

$$\frac{d\delta}{dx} = \frac{du_s}{dx} - \frac{du_g}{dx} = \frac{\sigma_s}{E_s} - \frac{\sigma_g}{E_g} \qquad (12.2)$$

$$\frac{d^2\delta}{dx^2} = \frac{d^2u_s}{dx^2} - \frac{d^2u_g}{dx^2} \qquad (12.3)$$

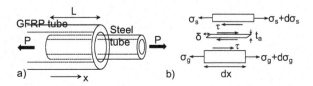

Fig. 12.4 A circular bonded sleeve connection under axial tension **a** overall geometry, **b** typical differential element

For a differential length of the steel tube (Fig. 12.4b), the shear stress in the adhesive layer τ and the normal stress in the steel σ_s can be expressed according to force equilibrium in the x direction:

$$\tau b_s dx = A_s d\sigma_s \rightarrow \frac{d\sigma_s}{dx} = \frac{\tau b_s}{A_s} \tag{12.4}$$

$$\sigma_s = E_s \frac{du_s}{dx} \rightarrow \frac{d\sigma_s}{dx} = E_s \frac{d^2 u_s}{dx^2} \tag{12.5}$$

Likewise, for a differential length of GFRP tube (Fig. 12.4b), shear stress in the adhesive layer τ and normal stress in the GFRP σ_g can be expressed accordingly:

$$\tau b_g dx = -A_g d\sigma_g \rightarrow \frac{d\sigma_g}{dx} = -\frac{\tau b_g}{A_g} \tag{12.6}$$

$$\sigma_g = E_g \frac{du_g}{dx} \rightarrow \frac{d\sigma_g}{dx} = E_g \frac{d^2 u_g}{dx^2} \tag{12.7}$$

For a full cross-section within the bond length, the force equilibrium in the axial direction gives:

$$\sigma_s A_s + \sigma_g A_g = P \rightarrow \sigma_s = \frac{P - \sigma_g A_g}{A_s} \tag{12.8}$$

Combining Eqs. 12.3–12.7, the governing equation for the shear stress in the adhesive layer τ can be formed as:

$$\frac{d^2 \delta}{dx^2} - \left(\frac{b_s}{E_s A_s} + \frac{b_g}{E_g A_{gs}} \right) \tau = 0 \tag{12.9}$$

Because of the thin adhesive layer, in Eq. 12.9 the bonded circumference around the steel b_s is approximately equal to that of the GFRP b_g, and both these values can be considered as the centre line perimeter of the adhesive layer. The governing equation for the normal stress in the GFRP component σ_g can be obtained by substituting Eq. 12.8 into Eq. 12.2:

$$\sigma_g = \frac{1}{\frac{A_g}{A_s}\frac{1}{E_s} + \frac{1}{E_g}} \left(\frac{P}{E_s A_s} - \frac{d\delta}{dx} \right) \tag{12.10}$$

Knowing the normal stress in the GFRP component σ_g from Eq. 12.10, the normal stress in the steel component σ_s can be easily obtained from Eq. 12.8.

Equations 12.9 and 12.10 are the governing differential equations for the adhesive shear stress and adherend normal stress of a bonded sleeve joint under axial loading. They are valid regardless of the exact shape of the bond-slip relationship.

In the following sections, a bilinear bond-slip relationship is implemented into the governing equations. The stress distributions and corresponding joint capacities are solved under possible cases of boundary conditions.

12.3.2 Bond-Slip Relationship of Adhesive Layer

Figure 12.5 shows the bilinear bond-slip relationship for the adhesive-bonded region. Defined by the origin (0, 0), the peak shear stress point (δ_1, τ_f) and the debonding point (δ_f, 0), such a bilinear curve is commonly described with three stages: elastic, softening and debonding stages, as indicated in Fig. 12.5. The area bounded by the curve (i.e. elastic and softening stage) is termed the interface fracture energy (G_f). Equations 12.11 and 12.12 were used to determine the peak shear stress τ_f and the debonding slip δ_f, according to the correlations formed in [38]. This is because similar adherend materials (steel and pultruded FRP), adhesive thickness (1 to 2 mm), adhesive tensile strength (20 to 30 MPa) and surface preparation of adherends (sandblasted and cleaned with Acetone) were adopted. With the assumption of uniform shear stress through the thickness of adhesive layer, the slip at peak shear stress δ_1 can be obtained from Eq. 12.13 based on the corresponding shear strain (τ_f/G_a).

$$\tau_f = 0.8 f_{t,a} \tag{12.11}$$

$$\frac{1}{2} \tau_f \delta_f = 31 \left(\frac{f_{t,a}}{G_a} \right)^{0.56} t_a^{0.27} \tag{12.12}$$

$$\delta_1 = \tau_f t_a / G_a \tag{12.13}$$

where $f_{t,a}$ is the tensile strength of the adhesive; G_a is the shear modulus of the adhesive; and t_a is the thickness of the adhesive layer. In Eq. 12.12, τ_f, $f_{t,a}$ and G_a are in MPa, δ_f and t_a in mm. As a result, the bond-slip relationship is represented in

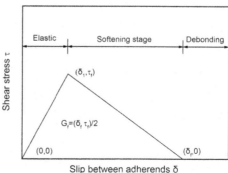

Fig. 12.5 Typical bilinear bond-slip relationship for adhesive bonding between steel and GFRP

mathematical form as Eq. 12.14

$$\tau = f(\delta) = \begin{cases} \frac{\tau_f}{\delta_1} \delta & 0 \leq \delta \leq \delta_1 \\ \frac{\tau_f}{\delta_f - \delta_1}(\delta_f - \delta) & \delta_1 \leq \delta \leq \delta_f \\ 0 & \delta_1 \leq \delta \end{cases} \quad (12.14)$$

where τ and δ are the shear stress in the adhesive layer and slip between the adherends.

As the axial load (P) increases, the joint capacity at the elastic limit (P_e) is attained when any location within the bonded area is loaded to completion of the elastic stage, i.e. when τ reaches τ_f. Further, the joint capacity at the ultimate state is attained when the debonding slip δ_f is reached at either end of the bonded area [35] (i.e. $\delta(x = 0$ or $L) = \delta_f$) with softening of only one end ($P_{u,s1}$) or both ends ($P_{u,s2}$), or when the full bond length is loaded to the softening stage ($P_{u,sf}$, i.e. for $0 \leq x \leq L$ and $\delta_1 \leq \delta < \delta_f$). In what follows, the joint capacities are solved for the elastic limit (P_e) and three different scenarios of the ultimate state ($P_{u,s1}$, $P_{u,s2}$ or $P_{u,sf}$). Finally, the joint capacities can be compared to the capacities of the adherends (steel tube connector or GFRP component) before the critical capacity can be determined accordingly.

12.3.3 Joint Capacity P_e at Elastic Limit

When the applied load P is smaller than P_e, i.e. the slip δ along the bond length is no greater than δ_1, the entire bond length L remains in the elastic region, i.e. $L = L_e$, where L_e denotes the length of the bonded region at the elastic stage. Figure 12.6 presents a typical shear stress distribution along the bond length when δ ($x = 0$) reaches δ_1, i.e. the applied load P reaches P_e.

Since $0 \leq \delta < \delta_1$, the corresponding shear stress expression in Eq. 12.14 can be substituted into Eq. 12.9 as:

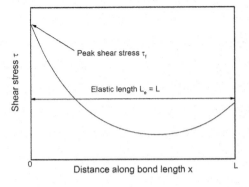

Fig. 12.6 Typical shear stress distribution when $P = P_e$

$$\frac{d^2\delta}{dx^2} - \left(\frac{b_s}{E_s A_s} + \frac{b_g}{E_g A_g}\right)\frac{\tau_f}{\delta_1}\delta = 0 \tag{12.15}$$

The general solution form of Eq. 12.15 for slip at the elastic stage δ_{el} can be given as

$$\delta_{el}(x) = A_1 \cosh(\lambda_1 x) + B_1 \sinh(\lambda_1 x) \tag{12.16}$$

where the subscript 'el' refers to the region of bond length at the elastic stage.

Substituting Eq. 12.16 into Eq. 12.10, the normal stress $\sigma_{g,el}$ in the GFRP at the elastic stage of adhesive bonding can be obtained as

$$\sigma_{g,el}(x) = \eta\left(\frac{P}{E_s A_s} - \lambda_1 A \sinh(\lambda_1 x) - \lambda_1 B \cosh(\lambda_1 x)\right) \tag{12.17}$$

where the constants λ_1 and η are

$$\lambda_1^2 = \left(\frac{b_s}{E_s A_s} + \frac{b_g}{E_g A_g}\right)\frac{\tau_f}{\delta_1} \tag{12.18}$$

$$\eta = \frac{1}{\left(\frac{A_g}{E_s A_s} + \frac{1}{E_g}\right)} \tag{12.19}$$

By substituting the boundary conditions (Eqs. 12.20 and 12.21) at the two free ends of the bond length

$$\sigma_{g,el}(x=0) = \frac{P}{A_g} \tag{12.20}$$

$$\sigma_{g,el}(x=L) = 0 \tag{12.21}$$

The unknown constants A_1 and B_1 can be solved as:

$$A_1 = \frac{P}{\lambda_1}\left(\frac{1}{E_s A_s \sinh(\lambda_1 L)} - \frac{1}{\tanh(\lambda_1 L)}\left(\frac{1}{E_s A_s} - \frac{1}{\eta A_g}\right)\right) \tag{12.22}$$

$$B_1 = \frac{P}{\lambda_1}\left(\frac{1}{E_s A_s} - \frac{1}{\eta A_g}\right) \tag{12.23}$$

P_e is then obtained when δ_{el} at either end ($x = 0$ or L) reaches δ_1 and the shear stress along the bond length $\tau(x)$ when $0 \leq P \leq P_e$ can be obtained as

$$\tau(x) = \frac{\tau_f}{\delta_1}\delta_{el}(x) \tag{12.24}$$

12.3.4 Joint Capacity $P_{u,s1}$ at Ultimate State for Softening of Only One End

A bonded sleeve joint may reach its ultimate capacity $P_{u,s1}$ (the subscript 's1' denotes softening of one end) when the debonding slip δ_f is attained at one end while the other remains in the elastic stage (i.e. δ at this end is no greater than δ_1). The full bond length L in this case can be divided into two portions: the softening length L_s where the τ-δ relationship has entered the softening stage, and the elastic length L_e where the τ-δ relationship remains at the elastic stage (Fig. 12.7). If the steel component is stiffer than the GFRP component (i.e. $E_s A_s > E_g A_g$), the GFRP end of the bonded area may be subjected to a higher level of slip between adherends [29, 32]. Figure 12.7 depicts a typical shear stress distribution along the bond length when δ reaches δ_f at the GFRP end and the corresponding applied load P therefore reaches the ultimate load $P_{u,s1}$.

Within the elastic length, i.e. $L_s < x \leq L$, and based on Eqs. 12.15 and 12.16, the slip between adherends δ_{el} and normal stress in the GFRP component $\sigma_{g,el}$ can be expressed as:

$$\delta_{el}(x) = A_2 \cosh(\lambda_1(x - L_s)) + B_2 \sinh(\lambda_1(x - L_s)) \tag{12.25}$$

$$\sigma_{g,el}(x) = \eta \left(\frac{P_{u,s1}}{E_s A_s} - \lambda_1 A_2 \sinh(\lambda_1(x - L_s)) - \lambda_1 B_2 \cosh(\lambda_1(x - L_s)) \right) \tag{12.26}$$

Within the softening length, i.e. $0 < x \leq L_s$, the corresponding shear stress expression from Eq. 12.14 can be substituted into Eq. 12.9, and Eq. 12.27 can be obtained as:

$$\frac{d^2\delta}{dx^2} - \left(\frac{b_s}{E_s A_s} + \frac{b_g}{E_g A_g} \right) \left(\frac{\tau_f}{\delta_f - \delta_1} \right) \delta = \left(\frac{b_s}{E_s A_s} + \frac{b_g}{E_g A_g} \right) \left(\frac{\tau_f}{\delta_f - \delta_1} \right) \delta_f \tag{12.27}$$

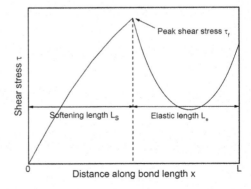

Fig. 12.7 Typical shear stress distribution when $P = P_{u,s1}$ (softening of only one end)

The general solution for δ in Eq. 12.27 can be formed as:

$$\delta_{sl}(x) = C_1 sinh(\lambda_2 x) + D_1 sinh(\lambda_2 x) + \delta_f \qquad (12.28)$$

where the subscript 'sl' refers to the region of bond length at the softening stage, and the constant λ_2 is

$$\lambda_2^2 = \left(\frac{b_s}{E_s A_s} + \frac{b_g}{E_g A_g}\right)\left(\frac{\tau_f}{\delta_f - \delta_1}\right) \qquad (12.29)$$

Substituting Eq. 12.28 into Eq. 12.10 gives

$$\sigma_{g,sl}(x) = \eta\left(\frac{P_{u,sl}}{E_s A_s} - \lambda_2 C_1 cos(\lambda_2 x) - \lambda_2 D_1 sin(\lambda_2 x)\right) \qquad (12.30)$$

where the values of A_2, B_2, C_1, D_1, $P_{u,sl}$, L_s can be determined by applying the following six boundary conditions:

$$\delta_{sl}(x=0) = \delta_f \qquad (12.31)$$

$$\delta_{el}(x=L_s) = \delta_1 \qquad (12.32)$$

$$\delta_{sl}(x=L_s) = \delta_1 \qquad (12.33)$$

$$\sigma_{g,sl}(x=0) = \frac{P_{u,sl}}{A_g} \qquad (12.34)$$

$$\sigma_{g,sl}(x=L) = 0 \qquad (12.35)$$

$$\sigma_{g,el}(x=L_s) = \sigma_{g,sl}(x=L_s) \qquad (12.36)$$

When $P = P_{u,sl}$, the shear stress $\tau(x)$ at a given location x along the bond length can be obtained as:

$$\tau(x) = \begin{cases} \frac{\tau_f}{\delta_f - \delta_1}(\delta_f - \delta_{sl}(x)) & 0 \leq x \leq L_s \\ \frac{\tau_f}{\delta_1}\delta_{el}(x) & L_s \leq x \leq L \end{cases} \qquad (12.37)$$

When the applied load P is less than $P_{u,sl}$ (while larger than P_e), the values of A_2, B_2, C_1, D_1 and L_s can be obtained by applying Eqs. 12.32–12.36, and the shear stress $\tau(x)$ and normal stress $\sigma_{g,el}(x)$ can be solved accordingly using Eqs. 12.25, 12.26, 12.28, 12.30, and 12.37. It is also necessary to check that the steel end is within the elastic stage (i.e. $\delta_{el}(x=L) < \delta_1$) and the softening length is less than the bond length (i.e. $L_s < L$), for the validity of this ultimate state (when $P = P_{u,s1}$), as softening of the bond length occurs only at one end.

12.3.5 Joint Capacity $P_{u,s2}$ at Ultimate State for Softening of Both Ends

This case of the ultimate state represents the scenario when the debonding slip δ_f is attained at one end while the other end is loaded into the softening stage of the bond-slip curve (i.e. $\delta_1 \leq \delta < \delta_f$). Therefore, the full bond length L can be divided into three portions: with two softening lengths at two ends (L_{s1} at the GFRP end and L_{s2} at the steel end), a portion of the bond length in between remains in the elastic stage (L_e). Figure 12.8 shows a typical shear stress distribution for this ultimate state and the divisions of bond length, when the applied load P reaches the ultimate load $P_{u,s2}$ (the subscript 's2' denotes softening of two ends) for this case.

Within the elastic length, i.e. $L_{s1} < x \leq L_{s1} + L_e$ (Fig. 12.8), and based on Eqs. 12.25 and 12.26, the slip δ_{el} between adherends and normal stress $\sigma_{g,el}$ of the GFRP component can be expressed as:

$$\delta_{el}(x) = A_3 cosh(\lambda_1(x - L_{s1})) + B_3 sinh(\lambda_1(x - L_{s1})) \tag{12.38}$$

$$\sigma_{g,el}(x) = \eta(\frac{P_{u,s2}}{E_s A_s} - \lambda_1 A_3 sinh(\lambda_1(x - L_{s1})) - \lambda_1 B_3 cosh(\lambda_1(x - L_{s1}))) \tag{12.39}$$

Within the left softening length, i.e. $0 \leq x < L_{s1}$, and based on Eqs. 12.28 and 12.30, the slip δ_{sl1} and normal stress $\sigma_{g,sl1}$ of the GFRP component can be expressed as:

$$\delta_{sl1}(x) = C_2 sin(\lambda_2 x) + D_2 sinh(\lambda_2 x) + \delta_f \tag{12.40}$$

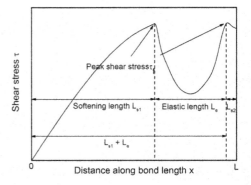

Fig. 12.8 Typical shear stress distribution when $P = P_{u,s2}$ (softening of both ends)

$$\sigma_{g,sl1}(x) = \eta\left(\frac{P_{u,s2}}{E_s A_s} - \lambda_2 C_2 \cos(\lambda_2 x) + \lambda_2 D_2 \sin(\lambda_2 x)\right) \quad (12.41)$$

where the subscript 'sl1' refers to the left softening length.

Within the right softening length, i.e. $L_{s1} + L_e < x \leq L$ (Fig. 12.8), and also based on Eqs. 12.28 and 12.30, slip δ_{sl2} and normal stress $\sigma_{g,sl2}$ of the GFRP component can be expressed as:

$$\delta_{sl2}(x) = C_3 \sin(\lambda_2(x - L_{s1} - L_e)) + D_3 \sinh(\lambda_2(x - L_{s1} - L_e)) + \delta_f \quad (12.42)$$

$$\sigma_{g,sl2}(x) = \eta\left(\frac{P_{u,s2}}{E_s A_s} - \lambda_2 C_3 \cos(\lambda_2(x - L_{s1} - L_e)) + \lambda_2 D_3 \sin(\lambda_2(x - L_{s1} - L_e))\right) \quad (12.43)$$

where the subscript 'sl2' refers to the right softening length.

The values of $A_3, B_3, C_2, D_2, C_3, D_3, P_{u,s2}, L_{s1}, L_e$ can be determined by applying the following nine boundary conditions:

$$\delta_{sl1}(x = 0) = \delta_f \quad (12.44)$$

$$\delta_{sl1}(x = L_{s1}) = \delta_1 \quad (12.45)$$

$$\delta_{el}(x = L_{s1}) = \delta_1 \quad (12.46)$$

$$\delta_{el}(x = L_{s1} + L_e) = \delta_1 \quad (12.47)$$

$$\delta_{sl2}(x = L_{s1} + L_e) = \delta_1 \quad (12.48)$$

$$\sigma_{g,sl1}(x = 0) = \frac{P_{u,s2}}{A_g} \quad (12.49)$$

$$\sigma_{g,sl2}(x = L) = 0 \quad (12.50)$$

$$\sigma_{g,sl1}(x = L_{s1}) = \sigma_{g,el}(x = L_{s1}) \quad (12.51)$$

$$\sigma_{g,el}(x = L_{s1} + L_e) = \sigma_{g,sl2}(x = L_{s1} + L_e) \quad (12.52)$$

When $P = P_{u,s2}$, the resulting shear stress at a given location x along the bond length $\tau(x)$ can be calculated from Eq. 12.53:

$$\tau(x) = \begin{cases} \frac{\tau_f}{\delta_f - \delta_1}(\delta_f - \delta_{sl1}(x)) & 0 \leq x \leq L_{s1} \\ \frac{\tau_f}{\delta_1}\delta_{el}(x) & L_{s1} \leq x \leq L_{s1} + L_e \\ \frac{\tau_f}{\delta_f - \delta_1}(\delta_f - \delta_{sl2}(x)) & L_s + L_e \leq x \leq L + L_e \end{cases} \quad (12.53)$$

If the applied load P continuously increases from P_e to the ultimate load $P_{u,s2}$, one end of the bond length (i.e. the GFRP end) starts to soften first, followed by the softening of both ends and then pull-out failure occurs when the debonding slip δ_f is attained at the GFRP end. Thus the form of shear stress distribution changes accordingly, depending on the applied load level. When $P_e < P < P_{u,s2}$, shear stress distribution $\tau(x)$ thus may involve the softening of one or both ends, depending on the load level. It is also necessary to check that at this ultimate state (when $P = P_{u,s2}$), the slip at the steel end is within the softening region (i.e. $\delta_1 < \delta_{sl2}(x = L) < \delta_f$), and the softening length at the steel end is positive (i.e. $L_{s2} > 0$).

12.3.6 Joint Capacity $P_{u,sf}$ at Ultimate State for Softening of Full Bond Length

This ultimate state corresponds to the scenario in which the full bond length is loaded into the softening stage of the bilinear bond-slip curve (i.e. $L = L_s$) and the peak shear stress τ_f is achieved within the bond length at a location $x = L_p$, as illustrated by Fig. 12.9.

As $L = L_s$ and based on Eqs. 12.28 and 12.30, the slip δ_{sl} and normal stress of the GFRP component $\sigma_{g,sl}$ can be expressed as:

$$\delta_{sl}(x) = C_2 \sin(\lambda_2 x) + D_2 \sin(\lambda_2 x) + \delta_f \quad (12.54)$$

$$\sigma_{g,sl}(x) = \eta \left(\frac{P_{u,sf}}{E_s A_s} - \lambda_2 C_2 \cos(\lambda_2 x) + \lambda_2 D_2 \sin(\lambda_2 x) \right) \quad (12.55)$$

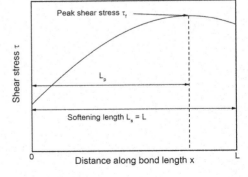

Fig. 12.9 Typical shear stress distribution when $P = P_{u,sf}$ (softening of full bond length)

When this ultimate state is attained, i.e. $P = P_{u,sf}$ (the subscript 'sf' denotes softening of the full bond length), the peak shear stress τ_f appears at the location $x = L_p$, and the values of $C_2, D_2, P_{u,sf}, L_p$ can be determined by applying the following four boundary conditions:

$$\delta_{sl}(x = L_p) = \delta_1 \tag{12.56}$$

$$\frac{d\delta_{sl}}{dx}(x = L_p) = 0 \tag{12.57}$$

$$\sigma_{g,sl}(x = 0) = \frac{P_{u,sf}}{A_g} \tag{12.58}$$

$$\sigma_{g,sl}(x = L) = 0 \tag{12.59}$$

The corresponding shear stress distribution $\tau(x)$ can be formed by Eq. 12.60 for this ultimate state.

$$\tau(x) = \frac{\tau_f}{\delta_f - \delta_1}(\delta_f - \delta_{sl}(x)) \; 0 \leq x \leq L \tag{12.60}$$

When the applied load increases from P_e to $P_{u,sf}$, softening of one end is initiated first (i.e. the GFRP end), followed by softening of the other end, and finally the two softening regions merge into one (i.e. $L_s = L$) and pull-out failure occurs. The forms of shear stress distribution may change accordingly, depending on the applied load level. When $P_e < P < P_{u,s2}$, shear stress distribution $\tau(x)$ thus may involve softening of one or two ends, depending on the load level. Again, it is necessary to validate this ultimate state (when $P = P_{u,sf}$) by checking that the slip at the GFRP end is within the softening stage (i.e. $\delta_1 < \delta_{sl}(x = 0) < \delta_f$) and the same condition at the steel end (i.e. $\delta_1 < \delta_{sl}(x = L) < \delta_f$).

It should be noted that failure may also occur in the steel or GFRP component if the joint capacity is greater than the axial capacity of the adherend components. In this case, the joint capacity P_s is determined by the steel yielding strength as $A_s f_{sy}$ or by the GFRP tensile strength as $A_g f_{gu}$, where f_{sy} and f_{gu} are the yield strength and ultimate strength of the steel and GFRP component respectively.

12.4 FE Analysis

12.4.1 Geometries and Materials

FE analysis was conducted using Ansys to describe the mechanical response of bonded sleeve joints under axial tension. Figure 12.10a presents an example of the geometry of a bonded sleeve joint connecting steel and GFRP tubes. Focusing on

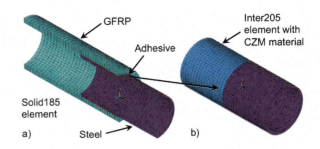

Fig. 12.10 FE modelling **a** overall geometry, **b** interface elements

verifying the theoretical results and estimating the experimental pull-out failure load, the FE models were constructed only concerning the bonded regions of the specimens. The material properties of the GFRP, steel and adhesive are provided in Table 12.1. The GFRP was defined as a linearly elastic and transversely isotropic material with its longitudinal direction in alignment with the x-axis (Fig. 12.10a). An isotropic bilinear work hardening material model was employed for the steel component, with von Mises's yield criterion and flow rule after yielding. The post-yield hardening modulus of the steel was taken 2% of its Young's modulus ($2\% E_s$), an upper limit of practical steel material as suggested in [50]. The material definition for the adhesive was isotropic and linearly elastic.

12.4.2 Model Establishment

All elements of the GFRP, steel and adhesive components were meshed in Ansys with Solid185, a 3D 8-node element. Three or seven layers of elements were used across the thickness of each GFRP tube or rod, while two layers were used for the steel and one for the adhesive. Success in simulating bonded zone behaviour hinges on correct implementation of the cohesive zone modelling (CZM) technique in Ansys. In general, the pull-out failure observed from experiments took place in the adhesive layer (near the inner adherend) or at the adhesive-inner adherend interface. This is because of the slightly smaller bonded area associated to the inner adherend, thus larger shear stress on the inner surface of the adhesive layer. Accordingly, Inter205, a planar 3D 8-node cohesive element, was applied between the adhesive and the inner adherend (Fig. 12.10b). This 8-node element has four pairs of overlapping nodes; each pair is attached to adjacent elements of steel and adhesive or GFRP and adhesive. The overlapping nodes were separated upon loading through a 'Mode II' separation, parallel to the bonded interface, as observed from the experiments (Fig. 12.3a). The shear traction–separation relationship was defined with the CZM material property option in Ansys, and was established in accordance to the bilinear bond-slip relationship presented in Fig. 12.5. Mesh studies were performed for specimens in G2 and G5 with respect to the number of elements along the bond length, ensuring that further refinement resulted in less than 1% change in the ultimate joint capacities.

As a result, elements sized at approximately 2 mm in the bond length direction were adopted.

12.4.3 Boundary Condition and Loading

In application of the axial loading, the GFRP end was restrained in all directions before the steel end was displaced in a tensile manner until the ultimate load was attained. Effects of material nonlinearity were accounted for. The displacement was applied in two load steps and solved with the sparse direct equation solver. The first load step was a ramped-load where the displacement was increased uniformly. As the ultimate load was approached, a second load step was introduced for better convergence, in which the automatic time stepping option was activated. This enabled the program to determine the size of displacement increment based on the structural response, and upon convergence failure after a certain number of iterations, to restart with a halved increment size.

12.5 Results and Discussion

12.5.1 Comparison of Ultimate Joint Capacity

Summarised in Table 12.2 are the experimental results of the ultimate joint capacities ($P_{u,E}$), along with theoretical estimates $P_{u,TH}$ (either $P_{u,s1}$, $P_{u,s2}$ or $P_{u,sf}$) and numerical estimates ($P_{u,FE}$) for the specimens observed with the pull-out failure mode. Among these 14 specimens, the average ratio between experimental and theoretical results of joint capacity ($P_{u,E}/P_{u,TH}$) was 0.935 and that between experimental and FE results ($P_{u,E}/P_{u,FE}$) was 0.981, with standard deviations of 0.114 and 0.095 respectively. These indicated good prediction and slight overestimation by the theoretical and FE modelling, which may be attributed to fabrication defects in the bonding zones of the specimens examined in the experiments.

According to the theoretical and FE modelling of all the experimental specimens, their ultimate states were achieved when the full bond length was loaded into the softening stage, i.e. $P = P_{u,sf}$, therefore the resulting shear stress distribution was formulated in Sect. 12.3.6. Because the same bond-slip relationship was used, it was anticipated that close values of ultimate joint capacity (i.e. $P_{u,TH}$ versus $P_{u,FE}$) would be obtained from the theoretical and FE modelling. However, the FE modelling produced slightly lower estimates than the theoretical ones (the average ratio $P_{u,FE}/P_{u,TH}$ was 0.952). This may be attributed to a smaller bonded circumference outside the inner adherend in the FE modelling compared to that defined at the adhesive centreline in the theoretical formulation, i.e. due to the assumption of a thin adhesive layer. Still, a discrepancy to notice is that, besides the yielding of steel, the FE modelling further considered shear deformation of adherends whereas the theoretical formulation did not. The effects of shear deformation of adherends may be taken into account by adopting the improved Volkersen method developed in [51]. Such effects were revealed by trial FE analysis to be insignificant and were thus neglected in the theoretical formulation. It should be noted that FE modelling was also conducted for the same specimens under compression loading. Identical joint capacity and adhesive shear stress distribution were produced as those under tension, which was also in accordance to the theoretical formulation.

12.5.2 Shear Stress Distribution

Verified against the experimental ultimate joint capacities, the theoretical and FE modelling are capable of providing insight into the shear stress distribution, which is difficult to measure from experiments. Figure 12.11 shows examples of three forms of shear stress distribution corresponding to different states of joint capacity through theoretical and FE modelling. Using specimen G1-50 as an example, Figs. 12.11a or 12.10b present the stress distribution when the applied load P reached P_e or $P_{u,sf}$. In Fig. 12.11a, the full bond length was in the elastic stage (i.e. $\delta \leq \delta_1$ and $L = L_e$) with the peak shear stress τ_f attained at the GFRP end. The FE modelling gave higher estimates of slip δ (or shear stress τ) in the middle portion, resulting in 12.4% overestimation of P_e. For the same specimen, in Fig. 12.11b, the full bond length was in the softening stage (i.e. $\delta_1 \leq \delta < \delta_f$ and $L = L_s$) corresponding to the joint capacity $P_{u,sf}$ of 53.9 kN by theoretical formulation or 51.6 kN by FE modelling. The peak shear stress τ_f appeared closer to the steel end, indicating that the GFRP end was subjected to higher level of slip. The shear stress distributions provided by theoretical and FE modelling were practically identical.

The section configuration of specimens G2, with a longer bond length of 140 mm, can be used for theoretical and numerical investigation. Figure 12.11c presents the shear stress distribution when $P = P_{u,s2}$, i.e. softening occurred at both ends and the debonding slip δ_f was attained at the GFRP end. In Fig. 12.11c, the two peaks, representing the peak shear stress τ_f, divide the full bond length into three portions: L_{s1}, L_e and L_{s2} from left to right. This was in accordance with the form of shear stress distribution presented in Fig. 12.8. Compared to the theoretical distribution, the FE result exhibited shorter L_{s1} (84.6% of the theoretical result) and longer L_e (120.6% of the theoretical result). In general, however, a satisfactory agreement of joint capacity $P_{u,s2}$ can be found from Fig. 12.11c. It should be noted that experimentally such a specimen would fail in the steel component because of the joint capacity being greater than the 118.4 kN of the steel adherend.

Figure 12.12 shows, through FE analysis, the adhesive shear stress distribution at different load levels. For both the specimen G1-50 and the numerical example of G2-140, when the applied load (85% P_e in Fig. 12.12a, b) was less than P_e, the full bond length remained in the elastic stage and the maximum shear stress appeared at the GFRP end due to its lower stiffness ($E_g A_g$). As the axial load (85% $P_{u,sf}$ in

Fig. 12.11 Shear stress distribution along bond length **a** specimen G1-50 when $P = P_e$ (20.9 kN or 23.5 for theoretical or FE modelling), **b** specimen G1-50 when $P = P_{u,sf}$ (53.9 kN or 51.6 kN for theoretical or FE modelling), **c** theoretical and FE example G2-140 when $P = P_{u,s2}$ (167.8 kN or 160.4 kN for theoretical or FE modelling)

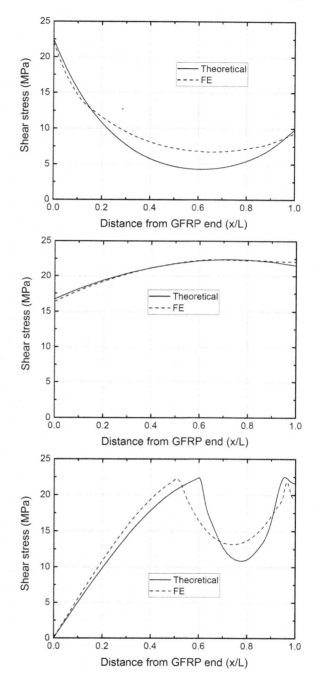

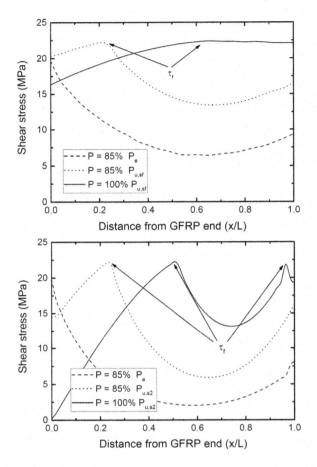

Fig. 12.12 Shear stress distribution along bond length from FE modelling at different load level **a** specimen G1-50, **b** FE example G2-140

Fig. 12.12a and 85% $P_{u,s2}$ in Fig. 12.12b) increased beyond the elastic limit while still below the ultimate state (i.e. $P_e < P < P_{u,sf}$ or $P_{u,s2}$), softening was introduced at the GFRP end, resulting in the peak shear stress (τ_f) appearing and shifting towards the steel end while the rest of bond length remained in the elastic stage. Approaching the ultimate state, specimen G1-50 (see Fig. 12.12a) had softening at both ends, and these two regions merged into one when the load increased to 100% $P_{u,sf}$, resulting in softening of the entire bond length. The peak shear stress (τ_f) appeared at the confluence point of the two softening regions, i.e. at the location of L_p in Fig. 12.9. In contrast, for the numerical example G2-140 at 100% $P_{u,s2}$ (see Fig. 12.12b), where the ultimate state corresponds to softening of both ends, the softening length at the GFRP end extended towards the steel end such that the slip $\delta(x/L = 0)$ reached δ_f. Meanwhile, softening was also introduced at the steel end (i.e. $\delta_1 < \delta$ ($x/L = 1$) < δ_f). The peak shear stress (τ_f), appearing at two locations corresponding to those of L_{s1} and $L_{s1} + L_{se}$ (also see Fig. 12.8), defines the boundaries of the bond length remaining in the elastic stage.

12.5.3 Effect of Bond Length on Joint Capacity

Variation of bond length (*L*), resulting in different forms of shear stress distribution as illustrated in the theoretical formulation, is believed to have a significant effect on the joint capacity. Figure 12.13 presents the results of a parametric study with respect to bond length (*L*), using section configuration of specimens G2. Both the theoretical and the FE results in Fig. 12.13 indicated that the joint capacity at the elastic limit P_e ceased to improve when the bond length reached a certain level, demonstrating the existence of an effective bond length within the elastic stage, which in this case was around 40 mm. For bond lengths longer than the effective value, the theoretical and FE modelling presented joint capacity P_e around 46 kN and 55 kN respectively.

Clearly, the applied load may be increased beyond the elastic stage. Experimental, theoretical and FE results of the relationship between ultimate joint capacity and bond length are also presented in Fig. 12.13. Represented by the black triangles, the experimental results increased almost linearly until a bond length of around 70 mm, after which little improvement in joint capacity was observed, due to yielding of the steel tube connector as observed from experiments. The theoretical and FE results, showing similar patterns, produced close agreement to experimental data. It is worth clarifying that steel yielding was considered in both theoretical and FE calculations, resulting in the joint capacity at 119 and 126 kN formed as a horizontal line as the representation of the capacity of the steel tube connector.

Furthermore, in order to investigate the effect of bond length on ultimate joint capacity that is dominated not by adherend capacity (e.g. steel yielding) but by adhesive bonding, theoretical and FE modelling were conducted on a series of specimens with the same cross-section configuration as those in G2 but with the different steel yield strength (f_{sy}) of 580 MPa. Shown in Fig. 12.13, the theoretical and FE ultimate joint capacities of such specimens still reached a plateau that was not defined by the capacity of the adherends (steel yielding in this case), corresponding to an effective bond length of around 125 mm at the ultimate state. At and beyond this

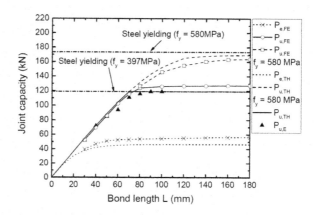

Fig. 12.13 Effect of bond length on joint capacity (specimens based on cross-section configuration in G2)

effective value, the ultimate joint capacity was calculated as 169 kN from theoretical modelling and 163 kN from FE modelling.

It is also worthwhile to examine the change in the form of shear stress distribution at the ultimate state of adhesive bonding (rather than adherend) when the bond length approached the effective value. For the section configuration G2 with steel yield strength (f_{sy}) of 580 MPa, Fig. 12.14a shows that, at the bond length of 100 mm (slightly below the effective value of 125 mm), the type of shear stress distribution at the ultimate state (i.e. $P = P_{u,sf}$) was characterised by softening of the full bond length. In contrast, with a bond length of 120 mm, which was close to the effective value, Fig. 12.14b shows the type of distribution when softening occurred at both ends, with the debonding slip δ_f attained at the GFRP end, corresponding to the ultimate state illustrated in Fig. 12.8, i.e. $P = P_{u,s2}$. This outcome occurred because, as the bond length became sufficiently long, the debonding slip δ_f was attained at one end before the two softening ends merged. It should also be noted that as the bond length passed the effective value, other than the change of ultimate state from $P_{u,sf}$ to

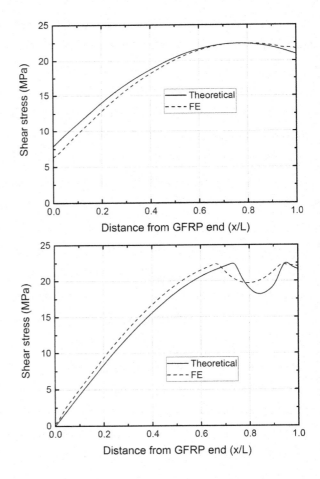

Fig. 12.14 Change of shear stress distribution form as bond length approached effective value **a** G2-100 with elevated adherend strength ($P_{u,sf} = 152.4$ kN or 145.9 kN for theoretical or FE modelling), **b** G2-120 with elevated adherend strength ($P_{u,s2} = 165.1$ kN or 155.8 kN for theoretical or FE modelling)

$P_{u,s2}$ as illustrated above, a change from $P_{u,sf}$ to $P_{u,s1}$ could also occur depending on the section geometries and bond-slip parameters. Similarly, this change was due to the redistribution of shear stress. With a sufficiently long bond length, the debonding slip δ_f might be attained at one end before the other was loaded into the softening stage.

12.5.4 Effect of Stiffness Ratio on Joint Capacities

The stiffness ratio between adherends (i.e. E_sA_s/E_gA_g) is another important parameter to be considered in the design of such bonded sleeve connections. In previous elastic analysis of single or double lap joints, the inequality between E_sA_s and E_gA_g, known as the adherend stiffness imbalance, could cause incapacity of a joint to develop full strength at the elastic stage. This is because the bonded zone at the end with the less stiff adherend could be loaded to its stress limit prior to the other end [32, 52, 53]. Using the developed theoretical formulation, the effect of this stiffness imbalance on both the elastic and the ultimate state joint capacity of a bonded sleeve connection is illustrated in Fig. 12.15, based on the cross-section configurations and bond length of specimen G5-80. The variation of stiffness ratios was achieved by changing the outer diameter of the GFRP tube or the inner diameter of the steel tube, while maintaining the section geometry of the adhesive layer. This resulted in stiffness ratios E_sA_s/E_gA_g from 0.2 to 4 as shown in Fig. 12.15.

Figure 12.15 shows that, consistent with the cases of single or double lap joints, the greatest joint capacity at the elastic limit (P_e) of a bonded sleeve connection was achieved when the of the adherends had equal stiffness i.e. $E_sA_s/E_gA_g = 1$. At the ultimate state (in this case $P_{u,sf}$), the greatest joint capacity was achieved at a stiffness ratio of around 0.8, as shown in Fig. 12.15. The variation of joint capacity at the elastic limit (P_e) with the stiffness ratio E_sA_s/E_gA_g was much more significant

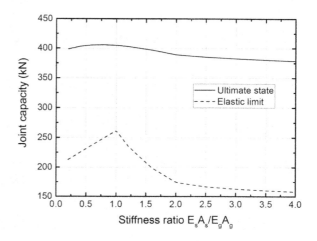

Fig. 12.15 Effect of stiffness ratio (E_sA_s/E_gA_g) on joint capacity at elastic limit (P_e) and at ultimate state ($P_{u,sf}$) based on specimen G5-80

than that at the ultimate state ($P_{u,sf}$). This finding is attributed to the steeper slope of the bilinear bond-slip relationship at the elastic stage than at the softening stage (see Fig. 12.5). Similar results were also observed from such investigations on specimens with other section configurations and bond lengths.

Figure 12.16 shows the shear stress distribution calculated by the theoretical formulation for the configuration of specimen G5-80 with three stiffness ratios. For the elastic limit, i.e. $P = P_e$, Fig. 12.16a indicates that the shear stress distribution was symmetrical when $E_s A_s / E_g A_g = 1$, and the peak shear stress τ_f was attained simultaneously at both ends. Otherwise, for the unbalanced stiffness ratios (i.e. $E_s A_s / E_g A_g = 0.4$ or 2.5), the peak shear stress τ_f only appeared at the end of the less stiff adherend. Figure 12.16b presents the shear stress distribution at the ultimate state, i.e. $P = P_{u,sf}$. Likewise, the balanced adherend stiffness (i.e. $E_s A_s / E_g A_g = 1$) produced a symmetrical shear stress distribution which had the peak shear stress τ_i appear at the middle of the bond length. The unbalanced adherend stiffness resulted in the peak shear

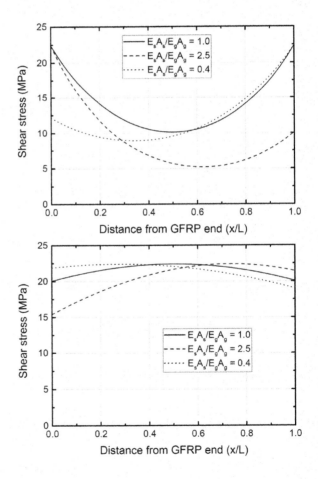

Fig. 12.16 Effect of stiffness ratio ($E_s A_s / E_g A_g$) on shear stress distribution from theoretical modelling based on specimen G5-80 **a** when $P = P_e$, **b** when $P = P_{u,sf}$

stress τ_f being located closer to the end of the stiffer adherend, indicating a higher level of slip δ at the other end.

12.6 Conclusions

This chapter presents a theoretic formulation to describe the adhesive shear stress distribution and to predict the joint capacity of bonded sleeve connections. A bilinear bond-slip relationship was implemented in the governing equations to consider the elastic, softening and debonding behaviour of the adhesive bonding. The results of ultimate joint capacity calculated by the theoretic formation were validated by results from FE modelling and from experimental specimens subjected to axial tension, where various cross-sections and bond lengths were covered. Utilising the validated theoretical and FE modelling, the effects of bond length and stiffness ratio on the adhesive shear stress distribution and joint capacity were discussed and clarified for such bonded sleeve connections. The work presented in this chapter allows the following conclusions to be drawn:

1. Employing the bilinear bond-slip relationship, when the bonded sleeve connection is loaded to elastic limit, the full bond length is within the elastic stage and the peak shear stress (τ_f) appears at the end of the less stiff adherend. The joint capacity at the ultimate state shows three different scenarios, each corresponding to shear stress distribution that features softening of one end or both ends or along the full length of the bonding region. The theoretic formulation, developed based on linear elastic adherends, gives satisfactory prediction of the pull-out joint capacities for the experimental specimens with different cross-sections and bond lengths; the average test/prediction ratio was obtained as 0.935.

2. FE modelling can simulate adhesive bonding using the cohesive interface elements associated with Mode II separation and the bilinear bond-slip relationship. The FE modelling produced slightly lower joint capacities compared to the theoretical modelling, which could be explained by the more conservative representation of the actual bond area in the FE modelling and the consideration of material nonlinearity in the adherend. The FE modelling yielded accurate prediction of the experimental joint capacities for the pull-out failure mode, with an average test/prediction ratio of 0.981.

3. For cases with no adherend failure, effective bond lengths do exist for the joint capacities at both the elastic limit and the ultimate state. Beyond the effective lengths, no further increase in joint capacity was observed. Such effective bond lengths can be well identified by the theoretic formulation and further well validated by the FE modelling results. Attainment of effective bond length may further cause transformation of the shear stress distribution at the ultimate state. For example, the ultimate state of $P_{u,sf}$ may change to $P_{u,s2}$ if the bond length increases from 100 to 120 mm for the G2 cross-section. Such transformation

of shear stress distribution may also occur from $P_{u,sf}$ to $P_{u,sl}$ depending on the cross-section configuration or bond-slip parameters. Finally, taking into account the capacity of adherends, the relationship between bond length and ultimate joint capacity can be developed via theoretical and FE modelling for a specific bonded sleeve connection.

4. The maximum joint capacity at the elastic limit (P_e) occurs when the stiffnesses of adherends are balanced (i.e. $E_sA_s/E_gA_g = 1$). This finding is consistent with those from single or double lap joints. Whereas the joint capacity at the ultimate state ($P_{u,s}$) is optimal at the stiffness ratio (E_sA_s/E_gA_g) of around 0.8, the effect of this stiffness ratio on the ultimate joint capacity is much less obvious in comparison to the effect on the elastic joint capacity. For both the elastic limit and ultimate state, a balanced adherend stiffness (i.e. $E_sA_s/E_gA_g = 1$) generates symmetrical shear stress distribution, while a difference in adherend stiffness (i.e. $E_sA_s/E_gA_g \neq 1$) leads to more slip (δ) at the end of the less stiff adherend.

References

1. Bakis C, Bank LC, Brown V, Cosenza E, Davalos J, Lesko J et al (2002) Fiber-reinforced polymer composites for construction-state-of-the-art review. J Compos Constr 6(2):73–87
2. Hollaway L (2010) A review of the present and future utilisation of FRP composites in the civil infrastructure with reference to their important in-service properties. Constr Build Mater 24(12):2419–2445
3. Nagaraj V, GangaRao HV (1997) Static behavior of pultruded GFRP beams. J Compos Constr 1(3):120–129
4. Di Tommaso A, Russo S (2003) Shape influence in buckling of GFRP pultruded columns. Mech Compos Mater 39(4):329–340
5. Keller T, Schollmayer M (2004) Plate bending behavior of a pultruded GFRP bridge deck system. Compos Struct 64(3):285–295
6. Bank LC (2006) Composites for construction: structural design with FRP materials. Wiley, New York
7. Meyer R (2012) Handbook of pultrusion technology. Springer Science & Business Media
8. Bank LC, Gentry TR, Nuss KH, Hurd SH, Lamanna AJ, Duich SJ et al (2000) Construction of a pultruded composite structure: case study. J Compos Constr 4(3):112–119
9. Keller T (2003) Use of fibre reinforced polymers in bridge construction. International Association for Bridge and Structural Engineering (IABSE), Zurich
10. Turner MK, Harries KA, Petrou MF, Rizos D (2004) In situ structural evaluation of a GFRP bridge deck system. Compos Struct 65(2):157–165
11. Zou X, Feng P, Wang J (2016) Perforated FRP ribs for shear connecting of FRP-concrete hybrid beams/decks. Compos Struct 152:267–276
12. Benmokrane B, El-Salakawy E, El-Gamal S, Goulet S (2007) Construction and testing of an innovative concrete bridge deck totally reinforced with glass FRP bars: Val-Alain Bridge on Highway 20 East. J Bridg Eng 12(5):632–645
13. Keller T, Haas C, Vallée T (2008) Structural concept, design, and experimental verification of a glass fiber-reinforced polymer sandwich roof structure. J Compos Constr 12(4):454–468
14. Hagio H, Utsumi Y, Kimura K, Takahashi K, Itohiya G, Tazawa H (2003) Development of space truss structure using glass fiber reinforced plastics. In: Proceedings: advanced materials for construction of bridges, buildings, and other structures III, Davos, Switzerland. ECI Digital Archives

15. Awad ZK, Aravinthan T, Zhuge Y (2012) Experimental and numerical analysis of an innovative GFRP sandwich floor panel under point load. Eng Struct 41:126–135
16. Satasivam S, Bai Y, Zhao X-L (2014) Adhesively bonded modular GFRP web–flange sandwich for building floor construction. Compos Struct 111:381–392
17. Satasivam S, Bai Y (2014) Mechanical performance of bolted modular GFRP composite sandwich structures using standard and blind bolts. Compos Struct 117:59–70
18. Feng P, Cheng S, Bai Y et al (2015) Mechanical behavior of concrete-filled square steel tube with FRP-confined concrete core subjected to axial compression. Compos Struct 123:312–324
19. Feng P, Zhang Y, Bai Y, Ye L (2013) Combination of bamboo filling and FRP wrapping to strengthen steel members in compression. J Compos Constr 17(3):347–356
20. Feng P, Zhang Y, Bai Y, Ye L (2013) Strengthening of steel members in compression by mortar-filled FRP tubes. Thin-Walled Struct 64(64):1–12
21. Smith S, Parsons I, Hjelmstad K (1998) An experimental study of the behavior of connections for pultruded GFRP I-beams and rectangular tubes. Compos Struct 42(3):281–290
22. Smith S, Parsons I, Hjelmstad K (1999) Experimental comparisons of connections for GFRP pultruded frames. J Compos Constr 3(1):20–26
23. Singamsethi S, LaFave J, Hjelmstad K (2005) Fabrication and testing of cuff connections for GFRP box sections. J Compos Constr 9(6):536–544
24. Yang X, Bai Y, Luo FJ, Zhao X-L, He X (2016) Fiber-reinforced polymer composite members with adhesive bonded sleeve joints for space frame structures. J Mater Civ Eng 29(2)
25. Yang X, Bai Y, Ding F (2015) Structural performance of a large-scale space frame assembled using pultruded GFRP composites. Compos Struct 133:986–996
26. Luo FJ, Bai Y, Yang X, Lu Y (2015) Bolted sleeve joints for connecting pultruded FRP tubular components. J Compos Constr 20(1)
27. Luo FJ, Yang X, Bai Y (2015) Member capacity of pultruded GFRP tubular profile with bolted sleeve joints for assembly of latticed structures. J Compos Constr 20(3)
28. Wu C, Zhang Z, Bai Y (2016) Connections of tubular GFRP wall studs to steel beams for building construction. Compos Part B: Eng 95:64–75
29. Volkersen O (1938) Die Nietkraftverteilung in zugbeanspruchten Nietverbindungen mit konstanten Laschenquerschnitten. Luftfahrtforschung 15(1/2):41–47
30. Goland M, Reissner E (1944) The stresses in cemented joints. J Appl Mech 11(1):A17–A27
31. Hart-Smith LJ (1973) Adhesive-bonded single-lap joints. National Aeronautics and Space Administration, Hampton
32. Hart-Smith LJ (1973) Adhesive-bonded double-lap joints. National Aeronautics and Space Administration, Hampton
33. Ranisch E-H (1982) Zur Tragfähigkeit von Verklebungen zwischen Baustahl und Beton: geklebte Bewehrung. Inst. für Baustoffe, Massivbau und Brandschutz der Techn. Univ.
34. Brosens K, Van Gemert D (1998) Plate end shear design for external CFRP laminates. Fract Mech Concr Struct 3:1793–1804
35. Yuan H, Teng J, Seracino R, Wu Z, Yao J (2004) Full-range behavior of FRP-to-concrete bonded joints. Eng Struct 26(5):553–565
36. Wu Z, Yuan H, Niu H (2002) Stress transfer and fracture propagation in different kinds of adhesive joints. J Eng Mech 128(5):562–573
37. Holzenkämpfer P (2002) Ingenieurmodelle des Verbunds geklebter Bewehrung für Betonbauteile1997
38. Xia S, Teng J (2005) Behaviour of FRP-to-steel bonded joints. In: Proceedings of the international symposium on bond behaviour of FRP in structures. International Institute for FRP in Construction, pp 419–426
39. Campilho R, De Moura M, Domingues J (2005) Modelling single and double-lap repairs on composite materials. Compos Sci Technol 65(12):1948–1958
40. Campilho R, Banea MD, Pinto A, da Silva LF, De Jesus A (2011) Strength prediction of single- and double-lap joints by standard and extended finite element modelling. Int J Adhes Adhes 31(5):363–372

41. Nakaba K, Kanakubo T, Furuta T, Yoshizawa H (2001) Bond behavior between fiber-reinforced polymer laminates and concrete. Struct J 98(3):359–367
42. Wu Z, Yin J (2003) Fracturing behaviors of FRP-strengthened concrete structures. Eng Fract Mech 70(10):1339–1355
43. Fawzia S, Zhao X-L, Al-Mahaidi R (2010) Bond–slip models for double strap joints strengthened by CFRP. Compos Struct 92(9):2137–2145
44. Yu T, Fernando D, Teng J, Zhao X (2012) Experimental study on CFRP-to-steel bonded interfaces. Compos Part B: Eng 43(5):2279–2289
45. Täljsten B (1996) Strengthening of concrete prisms using the plate-bonding technique. Int J Fract 82(3):253–266
46. Yuan H, Wu ZS, Yoshizawa H (2001) Theoretical solutions on interfacial stress transfer of externally bonded steel/composite laminates. J Struct Mech Earthquake Eng, Jpn Soc Civ Eng, Tokyo 675:27–39
47. Campilho R, De Moura M, Domingues J (2008) Using a cohesive damage model to predict the tensile behaviour of CFRP single-strap repairs. Int J Solids Struct 45(5):1497–1512
48. Alfano G (2006) On the influence of the shape of the interface law on the application of cohesive-zone models. Compos Sci Technol 66(6):723–730
49. Bai Y, Zhang C (2012) Capacity of nonlinear large deformation for trusses assembled by brittle FRP composites. Compos Struct 94(11):3347–3353
50. Kadin Y, Kligerman Y, Etsion I (2006) Multiple loading–unloading of an elastic–plastic spherical contact. Int J Solids Struct 43(22):7119–7127
51. Tsai M, Oplinger D, Morton J (1998) Improved theoretical solutions for adhesive lap joints. Int J Solids Struct 35(12):1163–1185
52. DoD (USA) (2002) Structural behaviour of joints. In: Military handbook – MIL-HDBK-17-3F: composite materials handbook, volume 3 – polymer matrix composites materials usage, design, and analysis. US Dept of Defense, Washington DC
53. Al-Shawaf A (2013) Understanding and predicting interfacial stresses in advanced fibre-reinforced polymer (FRP) composites for structural applications. Adv Fibre-Reinf Polym (FRP) Compos Struct Appl :255

Chapter 13
Axial Performance of Splice Connections for Fibre Reinforced Polymer Columns

Chengyu Qiu, Chenting Ding, Xuhui He, Lei Zhang, and Yu Bai

Abstract A splice connection is introduced in this chapter for connecting tubular fibre reinforced polymer (FRP) members. This connection consists of a steel bolted flange joint (BFJ) and two steel-FRP bonded sleeve joints (BSJs). The BFJ connects two steel hollow sections, each of which is telescoped into the targeted tubular FRP member through adhesive bond, forming a BSJ. To evaluate the performance of the proposed splice connection under axial loadings, BSJs of four different bond lengths and BFJs of two bolt configurations are tested individually. Finite element (FE) models are developed which feature a bilinear bond-slip relation, contact behaviours and bolt pre-tensioning. Comparisons are made between experimental and FE results in terms of load–displacement behaviours, ultimate capacities and strain responses. Besides being capable of identifying an effective bond length for the BSJ and modelling the yielding process of the BFJ, FE analysis provides insight into the distribution of adhesive shear stress over the bond area of the BSJs, and the steel yield line pattern on the flange-plate of the BFJs. Verified by experimental results, the FE modelling technique is then utilised to understand the integrated axial behaviours of a complete splice connection.

Reprinted from Composite Structures, 189, Chengyu Qiu, Chenting Ding, Xuhui He, Lei Zhang, Yu Bai, Axial performance of steel splice connection for tubular FRP column members, 498–509, Copyright 2018, with permission from Elsevier.

C. Qiu · C. Ding
Department of Civil Engineering, Monash University, Clayton, Australia

X. He · L. Zhang
School of Civil Engineering, Central South University, Changsha, China

Y. Bai (✉)
Department of Civil Engineering, Monash University, Clayton, Australia
e-mail: yu.bai@monash.edu

13.1 Introduction

Fibre reinforced polymer (FRP) composites are increasingly used in civil engineering structures, thanks to their high specific strength, superior corrosion resistance and availability in various geometries [1–3]. In particular, glass fibre reinforced polymer (GFRP) composites are credited with sufficient strength and stiffness at moderate cost. Advances in the pultrusion manufacturing technique [4, 5] have enabled mass production of GFRP profiles at reduced cost with satisfactory quality control, motivating research into their application as bridge decks [6, 7], reinforcement [8], roof structure [9], trusses [10–12] and floor systems [13–15]. Compared to open section profiles (i.e. I or channel profiles), closed section profiles (i.e. circular or rectangular tubular profiles) exhibit better resistance against torsional and global buckling [16]. Yet these merits of tubular GFRP members coexist with the difficulty of connecting the members into truss and frame assemblies, due to the material anisotropy and the closed section shapes.

Extensive research has been conducted in pursuit of viable connection forms for tubular GFRP members. Imitating practices in steel structures, early efforts to connect GFRP tubular members utilised bolted-through web-gusset plates or flange-angle cleats for beam-column connections [17]. In the development of a connection form for axially loaded tubular GFRP profiles, the benefits of using a steel tubular sleeve connector which was inserted into and bolt-fastened to the tubular GFRP profile were underlined in [18, 19]. This steel sleeve connector facilitated versatile connection forms to adjacent members. Despite the convenience of in-situ installation, bolt fastening requires hole-drilling on the composite material, creating problems such as damaged fibre architecture, stress concentration, and exposure of the weak in-plane shear strength of the FRP composites. Adhesive bonding, as an alternative, offers structural integrity, reduced stress concentration and also improved fatigue resistance [20–22]. Combining the benefits of the adhesive bond and the steel sleeve connector, bonded sleeve connections for joining circular GFRP truss members to nodal joints were proposed and examined in [11, 12]. In an experimental investigation [23] and the numerical study [24], the FRP-steel tubular bonded sleeve connection concept was utilised in FRP beam to steel column (or FRP stud to steel beam) configurations. Such connections exhibited significant improvement in both rotational strength and stiffness over steel angle connections and bolted sleeve connections.

Despite the aforementioned works, research into developing a column-splice connection for FRP tubular members remains scarce. In steel structures, a widely used splice connection for rectangular/square hollow sections (RHS/SHS) is the bolted flange joint (BFJ), which possesses the benefits of simple constitution and convenient in-situ installation [25]. Combining the FRP-steel bonded sleeve joint and the steel hollow section bolted flange joint, a column-splice connection for GFRP tubular members is proposed as illustrated in Fig. 13.1. This steel-GFRP connection system consists of two components, i.e. a bonded sleeve joint (BSJ) coupling GFRP and steel tubes adhesively, and a bolted flange joint (BFJ) connecting two steel SHSs through fillet weld. The BSJ reduces stress concentration in the GFRP compared to

13 Axial Performance of Splice Connection for Fibre ...

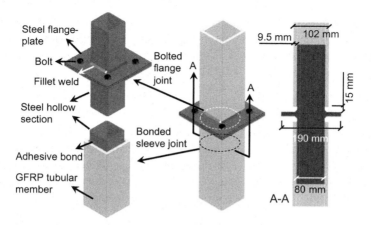

Fig. 13.1 Proposed column-splice connection for tubular GFRP members

bolt fastening, while the BFJ enables convenient installation and imparts ductility to the system (through steel yielding). As presented in Fig. 13.1, for a pultruded FRP box section of 102 × 102 × 9.5 mm, a steel SHS of outer dimension of 80 × 80 mm is selected in this study to achieve an adhesive thickness of 1.5 mm for practical assembly. To accommodate the fillet weld and to remove rotational constraint, a gap of 15 mm exists between the GFRP tubular member and the steel flange-plate. Steel flange-plate of 190 × 190 mm is selected for practical installation of the fastener (M12 bolt) and to fulfil the requirements for the edge distances of the fastener according to AS4100 [26]. Different loading scenarios may be found for the proposed splice connection depending on its potential applications. When used as a column splice, the connection is mainly subjected to a combination of axial (as a combination of tension and compression) and flexural loadings. When used as a beam splice, it would be under flexural and shear loadings. This chapter focuses on investigating the performance of the proposed connection under axial loadings for example in column applications.

Axial loading on the BSJ is resisted by shear between the GFRP and steel adherends. Such a load transfer mechanism has been extensively studied in the form of single or double lap joints. A bilinear relationship has been widely utilised to model the bond-slip behaviour due to its simple constitution and hence convenient analytical and numerical implementation. This bilinear relationship is characterised by a linearly ascending stage between the adhesive shear stress and relative slip between the adherends, followed by a linear decrease to zero shear stress at the debonding slip.

Early analytical study took advantage of this relationship to solve for the axial capacities of single lap joints in several configurations [27, 28] and also to identify the shear stress distribution [29, 30] and debonding process [31]. Experiments were carried out to measure the shear stress-slip relationship along the bond line in FRP-concrete [32, 33] and FRP-steel bonded lap joints [34–36], verifying that the bond-slip curves resembled a bilinear shape when brittle adhesives were used. For ductile adhesives, meanwhile, a trapezoidal shape was found suitable [36, 37]. The bilinear bond-slip relationship was incorporated into numerical analyses through the technique of cohesive zone modelling [38–40], in which a crack interface was predefined for the bond area and the tangential traction-slip relationship was defined as the bilinear shape. In light of previous work on single and double lap joints, an analytical solution, verified by experiments and FE analysis, was recently developed for the joint capacity of bonded sleeve connections consisting of circular steel and FRP tubular members [41]. In contrast to the earlier studies where adhesive shear stress was uniform in the transverse direction of single/double lap joints or in the circumferential direction of a circular section, the proposed connection, involving square or rectangular tubular adherends (Fig. 13.1), is distinguished by varying shear stress distribution in the transverse direction.

The failure mechanism of the bolted flange joint (BFJ) in axial loading of tension (more critical than in compression) is governed by a yield line mechanism in the steel flange-plate or/and tensile failure of the bolts under prying action. Taking into account both types of failure, design models were developed in [42] for configurations where one or two bolts were positioned at each side of the SHS. A similar design approach was later employed to solve for the layout with bolts at two sides of the RHS [43]. Focusing on the bolt failure under tensile loading, well-instrumented experiments were conducted to investigate the prying action on the bolts [44], before a modified AISC design procedure [45] was formulated. Design method is provided in Sect. 6 of Eurocode 3 Part 1–8 [46] to calculate the moment resistance of bolted flange splices or beam-column connections. Three modes of local yield lines are considered around the bolt holes at the tensile flange. Although this method is intended for I or H sections under bending loading, adaptation can be made as suggested in [47] and by mirroring the yielding lines at the tensile flange to calculate the tensile resistance of RHS/SHS bolted flange joints with bolts at two sides of the hollow section. The aforementioned experimental and analytical studies centred on the ultimate load-carrying capacities, and very limited description was presented of stiffness, failure processes and strain responses. Finite element (FE) methods were successfully utilised to understand the bending behaviour of these BFJ with [48] and without stiffeners [49]. Yet modelling results for the tensile behaviour of BFJs are still limited.

In the following this chapter experimentally and numerically investigates the mechanical performance of the proposed column-splice connection in axial loadings. Design parameters for the specimens include four different bond lengths and two types of bolt configuration (four and eight bolts). The detailed three-dimensional FE models feature utilising the bilinear bond-slip relationship for bond behaviour, contact between assembled parts and bolt pre-tensioning. Experimental results are then discussed and compared with FE modelling with respect to load–displacement behaviours, strain data, effect of bond length and yield line patterns of the connections. Finally, design recommendations are provided for optimising the performance of the proposed column-splice connection.

13.2 Experimental Program

13.2.1 Specimens

As the proposed connection consisted of bonded sleeve joints (BSJs) and bolted flange joint (BFJ) in series as described in Fig. 13.1, experiments were conducted individually on each component. Considering shear failure within the adhesive layer (cohesive failure) as the failure mode of interest for the BSJs, tensile and compressive loadings would theoretically induce identical adhesive shear stress distribution along the bond length and thus identical joint behaviour [41]; therefore, only compression tests were conducted on the BSJ specimens. For the BFJs, compressive failure would be preceded by member failure of the connected steel hollow section member, for which reason only the tensile behaviours of the BFJs were investigated.

Each of the BSJ specimens, illustrated in Fig. 13.2, was fabricated from a pultruded GFRP square tube (102 × 102 × 9.5 mm), a grade 355 steel square hollow section (SHS, 80 × 80 × 6 mm) and Sikadur-30, a two-component epoxy-based structural adhesive. The BSJs were coded '$BSJ\text{-}x\text{-}y$' where 'x' represents the bond length in mm ($x = 50, 100, 140$ or 180) and 'y' refers to the index of the repeating specimen in each bond length ($y = 1$ or 2). Bonded surfaces of the steel and GFRP were pre-treated by a procedure of 'degreasing—sandblasting—acetone cleaning' before adhesive was applied, as recommended in [50]. Fabrication of the specimens was followed by a two-week curing under room temperature before testing.

Two types of bolt configurations were adopted for the BFJs, with their geometries shown in Fig. 13.3. These two configurations were doubly symmetric and were efficient in resisting both tensile and bending actions. The BFJs were coded '$BFJ\text{-}a\text{-}b$' where 'a' represents the bolt number ($a = 4$ or 8) and 'b' the index of the repeating specimen ($b = 1$ or 2). BFJ-4 and BFJ-8 specimens shared the same geometries except for the number of bolts and their positions. Fillet welds of approximately 6 mm joined the steel SHS (80 × 80 × 6 mm, same as those in the BSJs) to the 6 mm-thick grade 250 steel flange-plate as shown in Fig. 13.3. During the welding, the flange-plate was clamped firmly against a rigid flat base to prevent likely deformation caused by heat distortion. A gusset plate was slot-welded into each steel SHS to enable application of tensile loading through gripping (Fig. 13.3a). The bolts were M12 grade 8.8 hex

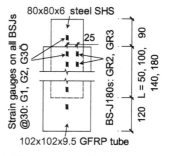

Fig. 13.2 Geometries of bonded sleeve joint specimens BSJ-50 to 180 and positions of strain gauges (all units in mm)

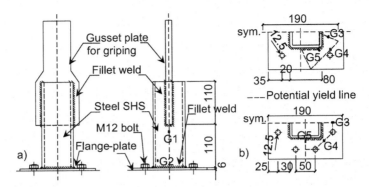

Fig. 13.3 **a** Front and side view of a BFJ-4 specimen **b** plan views of BFJ-4 and BSJ-8 and positions of strain gauges (all units in mm)

bolts with washers and nuts, pre-tensioned to around 52 kN (70% of the nominal proof load of the bolt) by a torque wrench before testing according to [45].

13.2.2 Material Properties

The pultruded GFRP square tubes were composed of a polyester matrix (volume fraction 53.3%) and E-glass fibres (volume fraction 46.7%) [51]. The strength and elastic modulus properties of the GFRP material were determined in [51, 52] according to relevant standard methods [53–55] as summarised in Table 13.1. The material properties of the steel SHS and flange-plate, tested from tensile coupons in accordance with [56], are summarised in Table 13.2. The M12 bolt was reported to have a yield strength of 1043 MPa and Young's modulus of 235 GPa [24]. The Sikadur-30 adhesive, tested in [36] in accordance with ASTM D638-10 [57], exhibited linear brittle behaviour with tensile strength of 22.3 MPa and elastic modulus of 11.3 GPa.

Table 13.1 Strength and stiffness of the GFRP material

Orientation and component	Strength (MPa)	Stiffness (GPa)	Method
Longitudinal tensile	306.5 ± 18.0	30.2 ± 1.4	ASTM D3039 [53]
Transverse tensile	–	5.5 ± 0.7	ASTM D3039 [53]
Interlaminar shear	26.7 ± 0.2	–	ASTM D2344 [54]
In-plane shear	14.9 ± 1.3	3.5 ± 0.7	10° off-axis tensile test [55]

13 Axial Performance of Splice Connection for Fibre ...

Table 13.2 Strength and stiffness of the steel materials

Steel component	Yield strength (MPa)	Ultimate strength (MPa)	Young's modulus (GPa)	Poisson's ratio
80*80*6 mm SHS [a]	420.1 ± 5.9 [b]	519.4 ± 8.4	209.5 ± 3.9	0.277 ± 0.07
6 mm-thick flange	313.6 ± 1.0	458.5 ± 0.0	200.6 ± 2.2	0.277 ± 0.01

Note [a] Tensile coupons cut from walls of the tube; [b] 0.2% offset yield strength

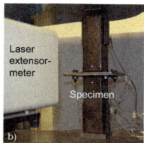

Fig. 13.4 Experimental test setup **a** compression test on BSJs **b** tension test on BFJs

13.2.3 Instrumentation and Experimental Setup

Compressive loading on the BSJ specimens was implemented under displacement control at 0.4 mm/min loading rate (Fig. 13.4a). Axial shortening of the specimens was measured by two linear variable differential transducers (LVDT). As shown in Fig. 13.2a, strain gauges were installed on the surface of the GFRPs along the centre bond line at 30 mm intervals; two additional strain gauges were installed on the BSJ-180s offset 25 mm from the centre bond line, as indicated. The BSJs were each loaded past their peak load.

Tensile loading on the BFJ specimens was carried out at a 0.5 mm/min loading rate (Fig. 13.4b). A laser extensometer was used to gauge axial elongation of the specimens. Besides strain gauges G1 and G2 on the steel SHS (Fig. 13.3a), strain gauges G3 to G5 were installed on the flange-plate where, from trial FE analysis, yield lines were likely to form (Fig. 13.3b). The BFJs were each loaded until substantial yielding deformation of the steel was observed in the load–displacement curves.

13.3 Finite Element Modelling

13.3.1 Geometric Modelling and Material Definitions

FE modelling of the specimens under axial loadings was performed using Ansys. Fig. 13.5 shows representative meshed models of the BSJ and BFJ specimens, for each of which half of the geometry was constructed due to symmetry. All the GFRP

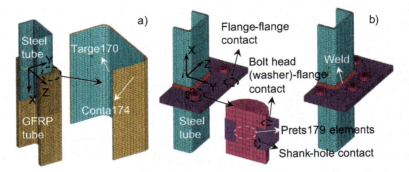

Fig. 13.5 Meshed FE models **a** BSJ specimens **b** BFJ-4 and BFJ-8 specimens

and steel components were meshed with Solid185, a 3D 8-node element in Ansys. The GFRP was modelled as an orthotropic linear elastic material whose longitudinal direction aligned with the x-axis shown in Fig. 13.5a. The walls of the tubular GFRP members were idealised as transversely isotropic composite laminates, resulting in identical interlaminar and in-plane shear moduli (Table 13.1). The longitudinal and transverse elastic moduli of the GFRP were defined according to the values in Table 13.1. The steel SHS and flange-plate were modelled as isotropic multilinear work-hardening materials, representing the stress–strain curves measured from the tensile coupon tests. The fillet welds were considered to be the same material as the steel flange-plate.

13.3.2 Modelling of Bond Behaviour in BSJs

Bond behaviour in the FE analysis was established through the cohesive zone modelling (CZM)) approach. This was enforced by a contact pair on the bonded surfaces, i.e. Conta174 and Targe170, a pair of 3D 8-node surface-to-surface contact elements, applied on the bonded surface of the GFRP and steel respectively (Fig. 13.5a). A 1.5 mm-gap existed between the contact surfaces, representing the thickness of the adhesive layer. The normal stiffness of the contact interaction (both opening and closing) was input as the product of the elastic modulus and the thickness of the adhesive. The tangential traction between the contact surfaces was modelled as a bilinear function of the shear slip between the contact pair, as shown in Fig. 13.6. This bilinear bond-slip relationship, consisting of a linear ascending elastic stage, a linear descending softening stage and also a debonding stage with zero shear stress, has been deemed appropriate for modelling cohesive failures in bonded lap joints when brittle adhesives are used [34–36]. It should be noted that, in the CZM approach, the distribution of shear stress was considered uniform through the thickness of the adhesive layer.

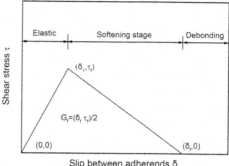

Fig. 13.6 Bilinear bond-slip relationship for GFRP-steel bond

To define this bilinear relationship in Ansys, three material parameters were required—the peak shear stress (τ_f), the stiffness of the elastic stage ($K_a = \tau_f/\delta_1$) and the critical fracture energy ($G_f = \tau_f \delta_f/2$). For the same adhesive (Sikadur-30) with identical layer thickness ($t_a = 1.5$ mm), the peak shear stress ($\tau_f = 18.4$ MPa) and critical fracture energy ($G_f = 1.25$ N/mm) had been determined in a literature through experimental investigation of steel-FRP single lap joints [36]. Assuming uniform shear stress thus shear strain through the thickness of the adhesive layer, the stiffness K_a was calculated by:

$$K_a = G_a t_a \tag{13.1}$$

where G_a is the adhesive shear modulus calculated from the elastic modulus with a Poisson's ratio of 0.3, and t_a is the thickness of the adhesive layer.

Before applying the axial loading, the GFRP end was constrained in all directions and symmetric constraint was applied on the longitudinal cut plane (xz plane in Fig. 13.5a). A load step which displaced the steel end in tension or compression was solved with automatic time-stepping for better convergence. It was further confirmed that tensile and compressive loadings generated identical load–displacement behaviour and shear stress distribution between the bonded surfaces.

13.3.3 Modelling of Contact and Pretension in BFJs

In the modelling of the BFJs, contact between the assembled steel components was considered by the contact pair Conta174 and Targe170, with a steel-to-steel friction coefficient of 0.44 [58]. Three contact pairs were identified as shown in Fig. 13.5b, i.e. those between the two flange-plates, between the bolt washer and flange-plate, and between the bolt shank and hole. Bolt pretension was applied via Prets179 elements defined at the midsection of each bolt shank (Fig. 13.5b). In terms of boundary conditions, one end of the specimen was constrained in all directions and symmetry constraint was applied on the longitudinal cut plane (xz plane in Fig. 13.5b). Application of the bolt pretension was solved in a first load step before a second load step axially displaced the free end of the specimen in the tensile direction.

13.4 Results and Discussion: BSJ Specimens

13.4.1 Failure Modes and Load–Displacement Responses

Experiments revealed brittle cohesive failure within the adhesive layer of all the BSJs. Fig. 13.7 shows typical load–displacement curves of the BSJs, characterised by a linear increase to peak load before brittle failure. Post-failure residual strength, provided by friction on the fracture surfaces, was recorded between 10 to 45 kN among specimens. Likewise, as also shown in Fig. 13.7 linear brittle load–displacement behaviours were obtained by FE modelling, except that the experimental residual strength could not be captured. Of all the BSJ specimens, discrepancy in the stiffness is found no more than 18% between experiment and FE modelling. When the peak load was imminent, cracking in the adhesive layer was observed as shown in Fig. 13.8a. In BSJ-100-1 and BSJ-140-1, post-failure loading resulted in cracks at the web-flange junction of the GFRPs, as indicated in Fig. 13.8b, possibly due to the confining pressure generated through sliding of the adherends over the uneven crack surface. Fig. 13.8c, representative of all BSJs, shows the separated adherends and the crack surface where bond failure occurred; the attachment of adhesive to both the steel and GFRP indicates cohesive failure located within the adhesive layer but closer to the GFRP.

Fig. 13.7 Typical experimental and FE load–displacement curves of BSJs

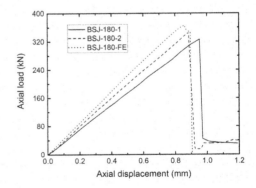

Fig. 13.8 Failure modes of BSJs **a** adhesive failure at peak load **b** cracking of GFRP after peak load (BSJ-100-1 and BSJ-140-1) **c** bond failure surface

13.4.2 Joint Capacity Versus Bond Length

Table 13.3 summarises the experimental joint capacities ($P_{u,E}$) of all BSJs and the corresponding FE estimates ($P_{u,FE}$). Except for the BSJ-50 specimens (where $P_{u,E}/P_{u,FE} = 0.761$), the FE modelling compared well to the experimental results with differences less than 13%. For all BSJs, the overestimation by FE modelling may be due to the fact, that the bilinear bond-slip relationship adopted in the FE analysis was derived from a plate-to-block single lap joint configuration [36], while the square tubular geometry of the BSJs imposed higher levels of through-thickness stress within the adhesive layer. This complex stress state in the adhesive, although dominated by shear stress, may advance expected bond failure. To account for the existence of the through-thickness stress in addition to the shear stress in the adhesive layer, application of a mixed-mode (mode I and mode II) bilinear bond behaviour in the numerical modelling [59] may represent the experiment scenario more accurately. However, this requires determination of the cohesive parameters of both mode I and mode II [60]; while further experimental calibration is needed to acquire the mode I parameters for this study. It should also be noted that implementation of the mixed-mode model in the contact behaviour often encounters with convergence difficulty for the 3D modelling as in this study, although such mixed-mode bond behaviour was successfully applied in 2D FE modelling [59]. The overestimation of joint capacities can also be attributed to the thicker adhesive layer at the corners (due to the round corners of the steel SHS) where local bond strength may be compromised. Joint capacities $P_{u,E}$ and $P_{u,FE}$ versus bond length (L) are plotted in Fig. 13.9.

Table 13.3 Comparison of experimental and FE joint capacity of BSJs

Bond length L (mm)	Experimental joint capacity $P_{u,E}$ (kN)			FE joint capacity $P_{u,FE}$ (kN)	Accuracy ($P_{u,E}/P_{u,FE}$)
	1	2	Average		
50	155	166	160.5 (± 3.4%)	211	0.761
100	303	266	284.5 (± 6.5%)	323	0.881
140	327	278	302.5 (± 8.1%)	346	0.875
180	326	346	336.0 (± 3.0%)	350	0.960

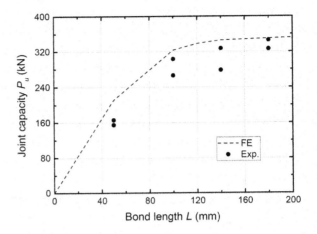

Fig. 13.9 Experimental and FE results of axial joint capacity versus bond length

The increase in $P_{u,E}$ with L slowed dramatically after L of around 100 mm. This was successfully captured by the FE modelling, indicating an effective bond length of 100 mm.

13.4.3 Strain Responses

The specimen BSJ-180-2 sustained the greatest axial load of all the BSJs. Its load-strain responses recorded by strain gauges outside the bond region (G1 and G8) are plotted in Fig. 13.10. The linear responses and the strain values indicated that both the steel and the GFRP adherends were within elastic range. From the same specimen, axial strains on the outer surface of the GFRP along the centre bond line, at four different load levels from FE modelling as well as experiment, are plotted

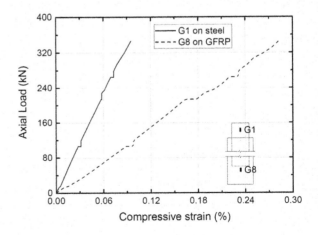

Fig. 13.10 Load-strain behaviour from strain gauges installed outside bond area (BSJ-180-2)

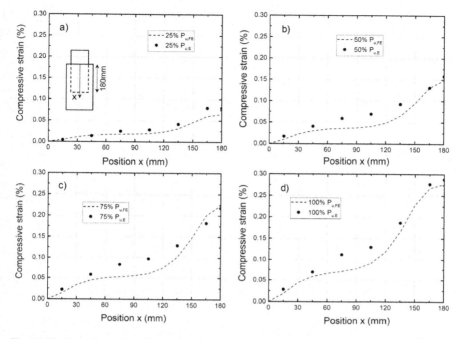

Fig. 13.11 Experimental and FE results of axial strains on the outer surface of the GFRP along centre bond line (BSJ-180-2, positions of strain gauges indicated in Fig. 13.2) at **a** 25% P_u **b** 50% P_u **c** 75% P_u **d** 100% P_u

in Fig. 13.11. It is evident that the axial strain distribution at mid-portion of the bond length features a flatter gradient, revealing lower levels of shear stress in this region than at the ends of the bond length. In general, the axial strain distribution near the GFRP end of the bond length ($x = 180$ mm) exhibits a steeper gradient than at the steel end ($x = 0$ mm), matching anticipation that greater shear slip (δ) would occur near the more flexible adherend (GFRP in this case). Also noteworthy is that in Fig. 13.11d (corresponding to 100% P_u) the strain distribution flattens near the GFRP end, indicating a drop of adhesive shear stress as a result of the GFRP end being loaded into the softening stage (see Fig. 13.6). The discrepancy between the experimental and FE strain data derives from two main sources. One is approximation of the true bond-slip relationship into a bilinear shape; the other source is that the GFRP was approximated as uniform through its wall thickness, instead of its actual mat-roving-mat layered structure.

13.4.4 Adhesive Shear Stress Distribution

Fig. 13.12 shows, from FE analysis, the distribution of adhesive shear stress along the centre bond line at four different load levels of BSJ-180 as a representation for all

Fig. 13.12 FE results of shear stress distribution along centre bond line at different load levels of BSJ-180

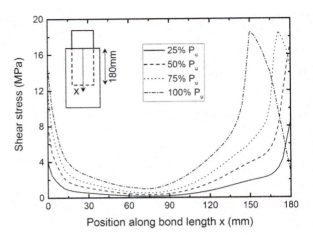

the BSJs. At 25% and 50% P_u, the full bond length is within the elastic stage of the bilinear bond-slip curve (Fig. 13.6); the majority of the shear resistance is provided near the two ends of the bond line and the maximum shear stress (τ_f) appears at the GFRP end ($x = 180$ mm, as the more flexible adherend). The increase of load from 50 to 75% P_u introduces a drop in shear stress near the GFRP end, due to its moving past the peak shear stress point (δ_1, τ_f) in the bilinear curve before entering the softening stage. As P increases further to 100% P_u, the softening length at the GFRP side extends inwards while the rest of bond length remains in the elastic stage.

Fig. 13.13 presents, at peak loads, the FE results of shear stress distribution over the entire bond area for all the BSJs. In contrast to single/double lap joints or circular bonded sleeve joints, non-uniform adhesive shear stress in the BSJs is found in the transverse direction (z direction). Generally, higher shear slip (δ) is induced further from the centre bond line, due to the greater rigidity at the corners of the tubular adherends than in the flat wall regions. The gradient of the transverse variation is

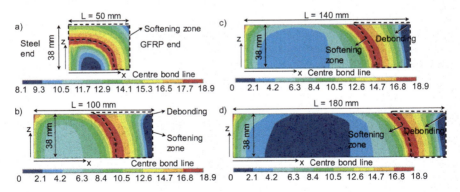

Fig. 13.13 Adhesive shear stress distribution from FE modelling at peak load: **a** BSJ-50; **b** BSJ-100; **c** BSJ-140; **d** BSJ-180

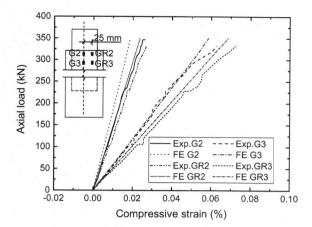

Fig. 13.14 Comparison of strain responses from 25 mm-offset and centre strain gauges (BSJ-180-1)

flatter near the two longitudinal ends ($x = 0$ mm *and 180* mm) and steeper inwards. The transverse variation of shear stress is implicitly evidenced by the experimental load-strain curves (BSJ-180-1) shown in Fig. 13.14; consistent with the FE output, the GFRP axial strains closer to the transverse corner of the adherend (GR2 and GR3) are greater than those along the centre bond line (G2 and G3).

In consideration of the shear stress distribution at peak load (100% P_u), the bond areas of the four BSJs could each be divided into two regions: a softening zone as highlighted in Fig. 13.13 and the remaining elastic zone, the boundary between the regions being defined by a line of the peak shear stress τ_f. Figure 13.13a presents a typical shear stress distribution at peak load with a short bond length (BSJ-50). Near the location where $z = 38$ mm (the adherend transverse corner), the full bond length is loaded into the softening stage of the bilinear curve. Figure 13.13b, c and d share typical stress distributions for longer bond lengths (BSJ-100, 140, 180). Stress distribution along the x direction can generally be characterised by an elastic zone near the steel side and a softening zone near the GFRP side, with the debonding point of the bilinear curve (δ_f, 0) attained at the GFRP end close to the transverse corners of the steel SHS (i.e. $z = 38$ mm). In Fig. 13.13b, c, and d, as the bond length increases, the softening zone does not exhibit notable change whereas the low-stress region in the elastic zone expands markedly. This observation signifies an inefficient increase in bond strength, thus implying the existence of an effective bond length.

13.5 Results and Discussion: BFJ Specimens

13.5.1 *Failure Modes and Load-Displacement Behaviours*

All the BFJ specimens failed through steel yielding with notable deformation of the flange-plates; no bolt deformation was visible but the nuts were found to loosen

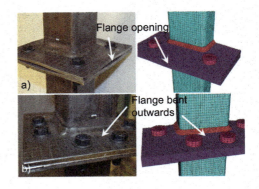

Fig. 13.15 Experimental and FE deformed shapes of BFJ specimens **a** BFJ-4 at 6 mm extension **b** BFJ-8 at 5 mm extension

after loading. As shown in Fig. 13.15, elongation of BFJ-4 (i.e. BFJ with four bolts) resulted in opening gaps between the two flange-plates in the region away from the bolts; while in BFJ-8, the flange-plates were bent outwards near the location welded with the SHS. Fig. 13.16 presents the tensile load-displacement behaviours of the BFJs. After steel yielding, the load-displacement curves kept increasing gradually at near constant slopes until the loading process was ceased at 9 mm elongation for the BFJ-4s or at 8 mm for the BFJ-8s.

The initial stiffness of the BFJs (S_i) is defined as the slope of the linear stage of the load-displacement curves, and yield capacity (P_y) is determined by the intersection of the elastic and post-yield tangent lines. Both experimental and FE values of S_i and P_y are summarised in Table 13.4. From BFJ-4 to BFJ-8, an increase of four bolts improved the experimental stiffness ($S_{i,E}$) by 233% and the yield capacity ($P_{y,E}$) by 77%. Comparison between the FE and experimental results, presented in Fig. 13.15 and Fig. 13.16, shows that both the deformed shapes and the load-displacement behaviours were accurately captured by the FE modelling. As presented in Table 13.4, the FE results of stiffness ($S_{i,FE}$) are within 16% from experimental values,

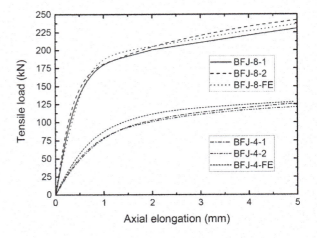

Fig. 13.16 Experimental and FE tensile load-displacement behaviours of **a** BFJ-4 **b** BFJ-8

Table 13.4 Comparison of yield capacity (P_y) and initial stiffness (S_i) of BFJs from experiment and FE modelling

Specimen	$P_{y,E}$ (kN) a, b	$P_{y,FE}$ (kN) b	$P_{y,E} / P_{y,FE}$	$S_{i,E}$ (kN/mm) a, c	$S_{i,FE}$ (kN/mm) c	$S_{i,E} / S_{i,FE}$
BFJ-4	104	113	0.920	104	117	0.889
BFJ-8	184	192	0.958	346	305	1.134

Note [a] Average value of the two repeating specimens
[b] intersection point of the load-displacement curve's elastic and post-yield tangent lines
[c] slope of the initial linear part of the load-displacement curve

and the FE yield capacities ($P_{y,FE}$) within 8%. The slightly higher yield capacities obtained from the FE modelling may be a consequence of welding residual stress in the flange-plates of the experimental specimens.

13.5.2 Stress Distribution and Load-Strain Responses

Figure 13.17 depicts from FE modelling the distributions of stress state ratio (defined as the ratio of von Mises stress to yield stress) for BFJ-4 at 114 kN and BFJ-8 at 198 kN, where a ratio larger than 1.0 indicates yielding of steel at the position. These stress distributions reveal that, as the loads increased beyond the yield capacities of the BFJs, the high stress areas formed into a pattern, allowing continuous yielding and thereby causing large elongation deformation of the specimens. Figure 13.17 also presents the stress state of the bolts, indicating that the bolts were subjected to a combination of bending and tensile action, due to the prying effect from the deformed flange-plates. The bolts of BFJ-4 were found to be under greater stresses than those of BFJ-8, yet yielding was limited to less than 16% of the cross-section area of the bolt shank.

Both experimental and FE load-strain curves are plotted in Fig. 13.18. As the strain gauges G3 to G5 were installed at highly stressed locations and orientations, their readings can indicate the yield initiation of the specimens. For BFJ-4 (Fig. 13.18a), the experimental G3 curve deviates from initial linearity at around 50 kN tensile load, which resembles the beginning of nonlinearity in the load-displacement curve

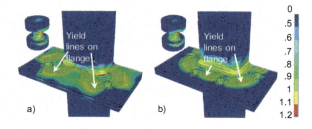

Fig. 13.17 von-Mises stress state distributions for **a** BFJ-4 at 114 kN **b** BFJ-8 at 198 kN

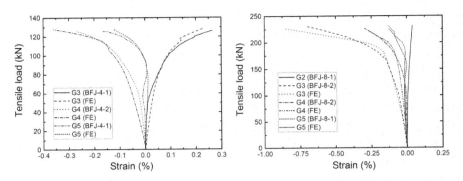

Fig. 13.18 Experimental and FE load-strain responses of **a** BFJ-4 **b** BFJ-8 (refer to Fig. 13.2 for positions of strain gauges)

(Fig. 13.16a). Likewise for the BFJ-8, both the experimental G3 curve (Fig. 13.18b) and load-displacement curve (Fig. 13.16b) begin to exhibit nonlinearity at around 86 kN tensile load. The linear behaviour of G2 shown in Fig. 13.18b demonstrates that the SHS was within the elastic range for all the BFJs. Further, the experimental and FE load-strain curves show satisfactory agreement, especially when steel yielding has occurred.

13.6 Integrated Performance of Proposed Splice Connection

Given the bonded sleeve joint (BSJ) and bolted flange joint (BFJ) investigated in prior sections, the experiment-validated FE modelling approaches can aid understanding of the performance of the proposed splice connection as an integration of BSJ and BFJ under axial loadings. Fig. 13.19 presents the meshed FE model of a design example SC-180-8, namely a splice connection using 180 mm bond length and the BFJ-8 bolt configuration. The FE model of SC-180-8 incorporates all the features and details of BSJ-180 and BFJ-8 described in Sect. 13.3, i.e. geometries (with the 15 mm-gap distance between the GFRP tube and steel flange-plate as described in Fig. 13.1), element types, material properties and contact behaviours. The load-carrying performances of SC-180-8 in both tension and compression are also indicated in Fig. 13.19.

The tensile load-displacement behaviour of the design connection SC-180-8 is plotted as the black curve in Fig. 13.19. It is confirmed that the yielding and the subsequent hardening behaviour of SC-180-8 coincide with those of BFJ-8 (FE shown as the black dashed curve and experimental as the black dotted curve), due to the same failure mode of flange-plate yielding. The initial stiffness of SC-180-8 differs from that of BFJ-8 due to the inclusion of the GFRP tube. With a gauged length of 640 mm used in Fig. 13.19, SC-180-8 exhibits slightly lower initial stiffness than

13 Axial Performance of Splice Connection for Fibre ...

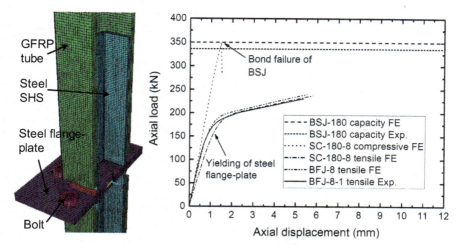

Fig. 13.19 Meshed FE model of a tubular FRP splice connection SC-180-8 and indication of its axial load-carrying performance

BFJ-8. The ultimate load capacities of SC-180-8, both in tension and compression, are bounded by the axial joint capacity of BSJ-180, which is presented in Fig. 13.19 as the horizontal dashed line from FE modelling and the horizontal dotted line from the experimental result. Generally, for other bond lengths and bolt configurations used with the splice connection, the tensile performance may show either brittle or ductile responses. The former happens when axial capacity of the BSJ is attained before yielding of the BFJ; and the resulting load-displacement curve of the splice connection develops linearly before the brittle failure of the BSJ component. The latter takes place when yielding of the BFJ precedes failure of the BSJ as in the case of SC-180-8. In this case, the load-displacement curve flattens upon yielding of the BFJ component, and continues to develop at reduced stiffness until axial capacity of the BSJ is attained. Behaviour of the connection SC-180-8 under compression loading is presented as the solid grey line in Fig. 13.19. The compressive load-displacement curve ascends linearly until the axial capacity of BSJ-180 is reached, where the bond failure of the BSJ component occurs. Furthermore, design of the compressive performance of FRP tubular members with such a connection may also involve buckling of the FRP member or yielding of the steel SHS, depending on their geometries and material properties. The tensile behaviour of the designed splice connection, which is governed by that of BFJ-8, as discussed above, utilises yielding of the steel component, and therefore ductile performance before brittle bond failure is achieved.

13.7 Conclusions

A steel splice connection was developed to join tubular FRP members and its axial performance was investigated through experimental study and FE modelling. The two components of the splice connection, namely the bonded sleeve joint (BSJ) and bolted flange joint (BFJ), were first experimentally examined. FE modelling was performed to study and understand the failure modes, load-displacement behaviours and strain responses. Verified by experimental results, the FE modelling technique was applied to understand the axial performance of the proposed splice connection integrated by the BSJ and BFJ. From the present study, the following conclusions can be drawn:

1. Under axial loading, the bonded sleeve joints (BSJs) failed in a brittle manner within the adhesive layer. Their load-displacement behaviour was linear up to peak load, followed by a sudden drop to residual strength between 10 to 45 kN. Among the four bond length groups (50, 100, 140 and 180 mm), FE analysis employing the bilinear bond-slip relationship produced estimates of joint capacity mostly within 15%. An effective bond length of around 100 mm was identified both experimentally and numerically for this joint configuration.

2. For the BSJ specimens, strain distribution on the GFRP surface indicated softening of the bond length at the GFRP end as the load increased and transverse variation of adhesive shear stress. The consistency between experimental and FE results for strain distribution further validated the FE modelling approach. Therefore the FE analysis provided insights into the adhesive shear stress distribution over the bond area. In the longitudinal direction, the GFRP (as the more flexible adherend) end of bond length sustained higher shear stress initially and, as the external load increased, entered the softening stage while the steel end remained elastic. In the transverse direction, a higher level of slip between the adherends occurred closer to the adherend transverse corner, resulting in earlier entry to the softening stage and ultimately longer softening length. At peak loads, two types of shear stress distribution were identified. For the BSJ specimen with 50 mm bond length (BSJ-50), the entire bond length near the adherend transverse corner was loaded into the softening stage. For other specimens with longer bond lengths (BSJ-100, 140 and 180), softening occurred near the GFRP side and the transverse corner of the GFRP end was loaded to the debonding point of the bilinear bond-slip relationship, while the rest of the bond area remained in the elastic stage.

3. Failure of the bolted flange joints (BFJs) was ductile through yielding of the flange-plates. The load-displacement responses could all be characterised as a 'linear ascending—yielding—linear hardening' process. The eight-bolt configuration (BFJ-8), compared to the four-bolt one (BFJ-4), improved the initial stiffness by 225% and the yield capacity by 82%. The FE modelling successfully captured the deformed shapes and produced load-displacement and load-strain responses that were in good comparison with experimental ones. The FE estimates of initial stiffness were within 14% of the experimental ones and those of

yield capacity were within 8% of the experimental estimates. From review of the FE von Mises stress distributions on the flange-plates, the improvement in stiffness and yield capacity from BFJ-4 to BFJ-8 was revealed to be associated with a more optimal yield line mechanism which involved an enlarged yielding area.

4. The axial behaviours of a complete splice connection, integrating the BSJ and BFJ, could be studied through the experiment-validated FE modelling approach. A FE model of a splice connection (SC-180-8), integrating the BSJ with 180 mm bond length and the BFJ with eight bolts, was developed, with its load-carrying performance verified by experimental data. The tensile behaviour of SC-180-8, involving yielding of the BFJ below the ultimate load of the BSJ, demonstrated that ductile failure could be achieved by utilising the yielding of the steel component.

References

1. Keller T (2001) Recent all-composite and hybrid fibre-reinforced polymer bridges and buildings. Prog Struct Mat Eng 3(2):132–140
2. Bakis C, Bank LC, Brown V, Cosenza E, Davalos J, Lesko J et al (2002) Fiber-reinforced polymer composites for construction-state-of-the-art review. J Compos Constr 6(2):73–87
3. Hollaway L (2010) A review of the present and future utilisation of FRP composites in the civil infrastructure with reference to their important in-service properties. Constr Build Mater 24(12):2419–2445
4. Bank LC (2006) Composites for construction: structural design with FRP materials. Wiley
5. Meyer R (2012) Handbook of pultrusion technology. Springer Science & Business Media
6. Bank LC, Gentry TR, Nuss KH, Hurd SH, Lamanna AJ, Duich SJ et al (2000) Construction of a pultruded composite structure: case study. J Compos Constr 4(3):112–119
7. Keller T (2003) Use of fibre reinforced polymers in bridge construction. Zurich: International Association for Bridge and Structural Engineering (IABSE).
8. Benmokrane B, El-Salakawy E, El-Gamal S, Goulet S (2007) Construction and testing of an innovative concrete bridge deck totally reinforced with glass FRP bars: Val-Alain bridge on highway 20 east. J Bridg Eng 12(5):632–645
9. Keller T, Haas C, Vallée T (2008) Structural concept, design, and experimental verification of a glass fiber-reinforced polymer sandwich roof structure. J Compos Constr 12(4):454–468
10. Hagio H, Utsumi Y, Kimura K, Takahashi K, Itohiya G, Tazawa H (2003) Development of space truss structure using glass fiber reinforced plastics. In: Proceedings: advanced materials for construction of bridges, buildings, and other structures III, Davos, Switzerland
11. Bai Y, Zhang C (2012) Capacity of nonlinear large deformation for trusses assembled by brittle FRP composites. Compos Struct 94(11):3347–3353
12. Yang X, Bai Y, Ding F (2015) Structural performance of a large-scale space frame assembled using pultruded GFRP composites. Compos Struct 133:986–996
13. Awad ZK, Aravinthan T, Zhuge Y (2012) Experimental and numerical analysis of an innovative GFRP sandwich floor panel under point load. Eng Struct 41:126–135
14. Satasivam S, Bai Y, Zhao X-L (2014) Adhesively bonded modular GFRP web–flange sandwich for building floor construction. Compos Struct 111:381–392
15. Satasivam S, Bai Y (2014) Mechanical performance of bolted modular GFRP composite sandwich structures using standard and blind bolts. Compos Struct 117:59–70

16. Wardenier J, Packer J, Zhao X, Van der Vegte G (2010) Hollow sections in structural applications. Cidect Zoetermeer, Netherlands
17. Smith S, Parsons I, Hjelmstad K (1999) Experimental comparisons of connections for GFRP pultruded frames. J Compos Constr 3(1):20–26
18. Luo FJ, Bai Y, Yang X, Lu Y (2015) Bolted sleeve joints for connecting pultruded FRP tubular components. J Compos Constr 04015024
19. Luo FJ, Yang X, Bai Y (2015) Member capacity of pultruded GFRP tubular profile with bolted sleeve joints for assembly of latticed structures. J Compos Constr 20(3):04015080
20. Kwakernaak A, Hofstede J (2008) Adhesive bonding: providing improved fatigue resistance and damage tolerance at lower costs. SAMPE J 44(5):6–15
21. Keller T, Gürtler H (2005) Quasi-static and fatigue performance of a cellular FRP bridge deck adhesively bonded to steel girders. Compos Struct 70(4):484–496
22. Zhang Y, Vassilopoulos AP, Keller T (2008) Stiffness degradation and fatigue life prediction of adhesively-bonded joints for fiber-reinforced polymer composites. Int J Fatigue 30(10):1813–1820
23. Wu C, Zhang Z, Bai Y (2016) Connections of tubular GFRP wall studs to steel beams for building construction. Compos B Eng 95:64–75
24. Zhang Z, Wu C, Nie X, Bai Y, Zhu L (2016) Bonded sleeve connections for joining tubular GFRP beam to steel member: numerical investigation with experimental validation. Compos Struct 157:51–61
25. Kurobane Y (2004) Design guide for structural hollow section column connections. Verlag TUV Rheinland
26. Standards Australia (1998) AS4100 steel structures. Sydney (SA)
27. Ranisch E-H (1982) Zur Tragfähigkeit von Verklebungen zwischen Baustahl und Beton: geklebte Bewehrung: Inst. für Baustoffe, Massivbau und Brandschutz der Techn. Univ
28. Brosens K, Van Gemert D (1998) Plate end shear design for external CFRP laminates. AEDIFICATIO Publ Fract Mech Concr Struct 3:1793–1804
29. Yuan H, Wu ZS, Yoshizawa H (2001) Theoretical solutions on interfacial stress transfer of externally bonded steel/composite laminates. J Struct Mech Earthq Eng Jpn Soc Civ Eng Tokyo 675:27–39
30. Wu Z, Yuan H, Niu H (2002) Stress transfer and fracture propagation in different kinds of adhesive joints. J Eng Mech 128(5):562–573
31. Yuan H, Teng J, Seracino R, Wu Z, Yao J (2004) Full-range behavior of FRP-to-concrete bonded joints. Eng Struct 26(5):553–565
32. Nakaba K, Kanakubo T, Furuta T, Yoshizawa H (2001) Bond behavior between fiber-reinforced polymer laminates and concrete. Struct J 98(3):359–367
33. Wu Z, Yin J (2003) Fracturing behaviors of FRP-strengthened concrete structures. Eng Fract Mech 70(10):1339–1355
34. Xia S, Teng J (2005) Behaviour of FRP-to-steel bonded joints. Proceedings of the international symposium on bond behaviour of FRP in structures. In: International institute for FRP in construction, pp 419–426
35. Fawzia S, Zhao X-L, Al-Mahaidi R (2010) Bond–slip models for double strap joints strengthened by CFRP. Compos Struct 92(9):2137–2145
36. Yu T, Fernando D, Teng J, Zhao X (2012) Experimental study on CFRP-to-steel bonded interfaces. Compos B Eng 43(5):2279–2289
37. Campilho R, De Moura M, Domingues J (2008) Using a cohesive damage model to predict the tensile behaviour of CFRP single-strap repairs. Int J Solids Struct 45(5):1497–1512
38. Alfano G, Crisfield M (2001) Finite element interface models for the delamination analysis of laminated composites: mechanical and computational issues. Int J Numer Meth Eng 50(7):1701–1736
39. Alfano G (2006) On the influence of the shape of the interface law on the application of cohesive-zone models. Compos Sci Technol 66(6):723–730
40. Campilho R, Banea MD, Pinto A, da Silva LF, De Jesus A (2011) Strength prediction of single- and double-lap joints by standard and extended finite element modelling. Int J Adhes Adhes 31(5):363–372

41. Qiu C, Feng P, Yang Y, Zhu L, Bai Y (2017) Joint capacity of bonded sleeve connections for tubular fibre reinforced polymer members. Compos Struct 163:267–279
42. Kato B, Mukai A (1985) Bolted tension flanges joining square hollow section members. J Constr Steel Res 5(3):163–177
43. Packer JA, Bruno L, Birkemoe PC (1989) Limit analysis of bolted RHS flange plate joints. J Struct Eng 115(9):2226–2242
44. Willibald S, Packer J, Puthli R (2002) Experimental study of bolted HSS flange-plate connections in axial tension. J Struct Eng 128(3):328–336
45. AISC (1997) Hollow structural sections connections manual. Chicago
46. EN 1993-1-8 (2005) Eurocode 3: design of steel structures, part 1–8: design of joints. Bryssels: CEN
47. Heinisuo M, Ronni H, Perttola H, Aalto A, Tiainen T (2012) End and base plate joints with corner bolts for rectangular tubular member. J Constr Steel Res 75:85–92
48. Wang Y, Zong L, Shi Y (2013) Bending behavior and design model of bolted flange-plate connection. J Constr Steel Res 84:1–16
49. Wheeler A, Clarke M, Hancock G (2000) FE modeling of four-bolt, tubular moment end-plate connections. J Struct Eng 126(7):816–822
50. Teng J, Fernando D, Yu T, Zhao X (2011) Treatment of steel surfaces for effective adhesive bonding. Advances in FRP composites in civil engineering. Springer, pp 865–868
51. Satasivam S (2015) Modular FRP sandwich structures for building floor construction. Monash University, Melbourne, Australia
52. Xie L, Bai Y, Qi Y, Caprani C, Wang H (2017) Compressive performance of PFRP SHS columns: effects of width-thickness ratio. In: Proceedings of the institution of civil engineers-structures and buildings in-press
53. ASTM International (2014) Standard test method for tensile properties of polymer matrix composite materials. ASTM D3039. West Conshohocken, PA
54. ASTM International (2016) Standard test method for short-beam strength of polymer matrix composite materials and their laminates. ASTM D2344. West Conshohocken, PA
55. Lee S, Munro M, Scott R (1990) Evaluation of three in-plane shear test methods for advanced composite materials. Composites 21(6):495–502
56. ASTM International (2016) Standard test methods and definitions for mechanical testing of steel products. ASTM A370-16. West Conshohocken, PA
57. ASTM International (2010) Standard test method for tensile properties of plastics. ASTM D638-10. West Conshohocken, PA
58. Trahair N, Bradford M (1998) The behaviour and design of steel structures to AS 4100, 3rd edn. Taylor & Francis
59. Da Silva LF, Campilho RD (2012) Advances in numerical modelling of adhesive joints. In: Advances in numerical modeling of adhesive joints. Springer, pp 1–93
60. Lee MJ, Cho TM, Kim WS, Lee BC, Lee JJ (2010) Determination of cohesive parameters for a mixed-mode cohesive zone model. Int J Adhes Adhes 30(5):322–328

Chapter 14
Cyclic Performance of Splice Connections for Fibre Reinforced Polymer Members

Chengyu Qiu, Yu Bai, Zhenqi Cai, and Zhujing Zhang

Abstract This study investigates the cyclic performance of splice connections developed for hollow section fibre reinforced polymer (FRP) members. Splice connection specimens, each consisting of a steel bolted flange joint between two hollow section steel-FRP bonded sleeve joints, are prepared in three configurations with difference in bolt arrangement or bond length. Correspondingly, detailed finite element (FE) models are constructed with consideration of yielding of the steel components, damage in the adhesive bond, pre-tensioning of the bolts and contact between the bolt-fastened parts. Tested under a cyclic flexural loading, the specimens experience different levels of yielding in the steel flange-plates before ultimate failure in the FRP member or in the steel flange-plate. Ductility and energy dissipation capacity are demonstrated in a specimen where plastic deformation of the steel flange-plates is fully developed. The strain responses are also analysed to identify damage in the adhesive bond and yielding in the flange-plates. The FE modelling agrees with the experimental results in terms of moment-rotation and load-strain responses, and can also predict the initiation of the ultimate failure in the FRP using the Tsai-Wu failure criterion.

14.1 Introduction

Fibre reinforced polymer (FRP) composites, compared to traditional construction materials, have unique features such as high strength-to-weight ratio and corrosion resistance. Majorly used in rehabilitation and strengthening of concrete [1] and

---

Reprinted from Composite Structures, 243, Chengyu Qiu, Yu Bai, Zhenqi Cai, Zhujing Zhang, Cyclic performance of splice connections for hollow section fibre reinforced polymer members, 112222, Copyright (2020), with permission from Elsevier.

C. Qiu · Z. Cai · Z. Zhang
Department of Civil Engineering, Monash University, Clayton, Australia

Y. Bai (✉)
Department of Civil Engineering, Monash University, Clayton, Australia
e-mail: Yu.Bai@monash.edu

steel structures [2], they are also finding applications in new constructions thanks to development of the pultrusion manufacturing technique [3] and the moderate cost of FRPs with glass fibres (GFRPs). Examples of field applications and laboratory studies include bridges decks [4, 5], space trusses [6, 7], floor slabs [8, 9] and wall panels [10]. In attempts to exploit the advantages of pultruded FRPs in frame structures, structural connections have been developed to accommodate the brittle and anisotropic characteristics of the material.

In frame structures, the beam-column and splice connections are two fundamental forms to facilitate the assembly. To date, beam-column connections of FRP structures have received considerable study. Bank et al. conducted pioneering works in which FRP bolts, seated angles and web cleats were used to connect FRP I-section beam and column similar as in steel structures [11, 12]. Unique failure modes of pultruded FRP composites were revealed such as web-flange separation and delamination of the seated angles. To mitigate these types of failure, through-bolts that engaged both flanges of the I-section were examined [11], before universal connector [13] and wrapped connector [14] were developed. Further works were conducted in [15–17] to explore the effect of bolt arrangements, angle cleats positions and the materials of connectors (i.e. steel vs. FRP). Later on, beam-column connections were developed for hollow section members where steel bolts, adhesive, flange angles and web plates were used [18]. In search of further improvement in strength, a cuff connector that integrated a hollow section FRP beam and column into a monolithic unit using adhesive bond was studied [19, 20]. The ultimate moment capacity was increased by up to 57% compared to specimen with bolted cuff. More recently, bonded sleeve connection with extended steel endplate was developed into stud connection [21] and beam-column connection [22] for hollow section FRP members. In this configuration, the extended endplate enabled easy assembly with bolts and ductile failure through steel yielding. In another recent work, innovative steel tube connectors with welded bolt nuts were designed inside rectangular hollow section FRP members for easy connection of beam and column, and to impart ductility by yielding of these steel connectors [23].

Apart from the aforementioned works which focused on the monotonic behaviour of beam-column connections, the cyclic performance was also investigated in a few studies. Early cyclic studies were conducted on connections consisting of FRP angle and T-shaped connectors [24] or universal connectors [25] that were bonded and/or bolted to I-section beams and columns. The bonded cuff connection for hollow section members was also examined under cyclic loading [20, 26]. These cyclic studies revealed limited ductility and energy dissipation due to brittle fracture or delamination of the connector elements at the beam flanges. In the more recent cyclic studies, i.e. the bonded sleeve connection with extended endplate [27] and the connection comprised of steel tubular connectors bolted inside the hollow section beam and column [28, 29], improved energy dissipation capacities were demonstrated through yielding of steel components and also bearing failure of bolted joints in FRP instead of shear-out failure [30].

In comparison, splice connections for FRP members have received less research. Splice connections are often used in beams or columns to enable their continuity over long span or multiple storeys. In one of the earliest studies, beam splices formed by bolted and bonded FRP lap plates were applied to wide flange I-sections and square hollow sections [31]. This study focused on the fatigue performance and reported the realization of 60-80% (I-section) and 46–60% (hollow section) of the intact beam fatigue life. Hybrid FRP I-beams with double lap splice joints at midspan were examined in [32], where bonded and bolted joints proved enhanced stiffness compared to bolted-only joints. In another work, steel splice plates were used instead of FRP plates, and the effect of bolt row number was investigated [33]. Later, the serviceability behaviour of beam splices formed by bonded-only [34] and bolted-only [35, 36] lap plates for pultruded wide flange I-section was investigated.

Among the aforementioned works of splice connections, few was conducted for hollow section FRP members. Therefore, a splice connection was introduced in the previous chapter, consisting of a steel bolted flange joint between two hollow section steel-FRP bonded sleeve joints [37] as shown in Fig. 14.1. The bonded sleeve joint, formed by a steel square hollow section (SHS) member coupled into an FRP SHS member with a high-strength adhesive, was intended to be prefabricated off-site to attain manufacturing accuracy. The bonded sleeve joint, by integrating the FRP and steel member into a monolithic unit, provided a high degree of interaction between the two components and a reduced stress concentration in the FRP member (compared to bolt-fastening).

Furthermore, a rotational stiffness up to 114% of a continuous member was realized, and an ultimate moment capacity of 47% of the FRP section could be attained [38]. The steel flange-plate, designed for easy on-site assembly, also enabled ductile failure by steel yielding [38], suggesting promising energy dissipation performance of the splice connection under cyclic loading. To protect the exposed steel parts (the flange-plates and the bolts) against corrosion, galvanized or stainless-steel parts can be used instead, or surface coating can be easily applied. In FRP frame structures, the energy dissipation capacity of the splice connections can be of important concern under seismic loading. Therefore, this chapter investigates the cyclic performance of the splice connection, by both full-scale experimental testing and finite element

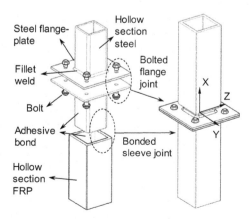

Fig. 14.1 Configuration of the proposed splice connection for hollow section FRP member

(FE) modelling. Splice connection specimens with difference in bolt arrangement or bond length are studied under cyclic flexural loading to reveal their failure modes, moment-rotation behaviours and cyclic performances. Local strain responses that indicate the interaction between the bonded members and yield status of the steel components are also reported and discussed.

14.2 Experimental Program

14.2.1 Specimens and Fabrication

The overall configuration and dimensions of the splice connection specimens are illustrated in Fig. 14.2a. The connection specimens were prepared in three configurations, namely C-170-8, C-120-8 and C-170-4. The first letter "C" refers to the cyclic loading; the first number (170 or 120) denotes the steel-FRP coupling bond length in mm (Fig. 14.2b-d); and the second number (4 or 8) represents the number of bolts on the steel flange-plate (Fig. 14.2e). One specimen per configuration was prepared considering the large scale and further verification from FE modelling. Fabrication of the specimens started with welding a 6 mm-thick steel flange-plate to one end of a steel SHS (80 × 80 × 6 mm) member by fillet weld approximately 6 mm in leg length. The other end of the steel SHS member was co-axially coupled into and bonded to a pultruded GFRP member with a hollow section of 102 × 102 × 9.5 mm. Before the adhesive was applied, the to-be-bonded surfaces of the steel and GFRP members were treated following a procedure of degreasing, grit blasting for steel/sandpaper abrading for GFRP, air nozzle blow-off cleaning, and solvent cleaning. Each of the bonded assembly was cured for two weeks under room temperature, before a pair of them was bolted together at the flange-plates with grade 8.8 M12 bolts and nuts with washers. The fasteners were pre-tensioned to around 65 kN by a torque wrench to meet the specification of "tension tight" according to AS4100 [39].

14.2.2 Material Properties

The pultruded GFRP SHS members were comprised of E-glass fibres embedded in a polyester resin matrix. The strengths and moduli of the GFRP material are summarised in Table 14.1 where the relevant test methods [40–44] are also indicated. The stress–strain behaviours of the steel SHS and flange-plate were characterised from tensile coupon tests following [45]; the key strength and modulus properties are summarised in Table 14.2. Reported in an earlier study [46], the grade 8.8 M12 bolts had a yield strength of 1043 MPa and Young's modulus of 235 GPa. The adhesive used to bond the steel and GFRP members was Sikadur-330, a two-component

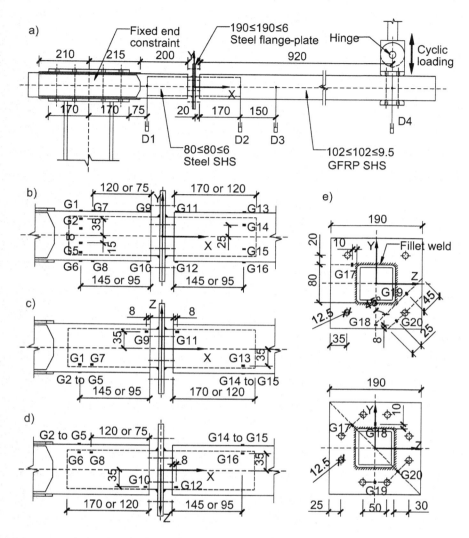

Fig. 14.2 Dimensions of the splice connection specimens (all units in mm; G1 to G20: strain gauges; D1 to D4: displacement gauges) **a** overall side view **b** side view of the connection **c** top view of the connection **d** bottom view of the connection **e** front view of the flange-plates with four and eight bolts

epoxy-based adhesive. Tested under tensile loading as per [47], the structural adhesive showed a linear stress–strain behaviour up to an average ultimate strength of 33.7 MPa; the elastic modulus was recorded as 4.09 GPa and the Poisson's ratio as 0.28.

Table 14.1 Strengths and moduli of GFRP

Component of properties	Modulus (GPa)	Strength (MPa)	Test method
Longitudinal tensile	25.18 ± 12.80[a]	330.6 ± 19.4[b]	ASTM D3039 [40]
Longitudinal compressive	23.77 ± 5.46	307.7 ± 4.3[b]	ASTM D695 [41]
Transverse flexural	6.24 ± 1.18[a]	88.5 ± 6.5	ASTM D7264 [42]
Transverse tensile	5.52	48.3[b]	From manufacturer's datasheet[c]
Transverse compressive	8.78 ± 2.37	127.9 ± 7.6[b]	ASTM D695 [41]
Interlaminar shear	–	31.2 ± 1.9	ASTM D2344 [43]
In-plane shear	3.02 ± 0.31[a]	27.6 ± 1.7[b]	10° off-axis tensile test [44]

Note[a] Adopted as the elastic moduli in the FE modelling

Table 14.2 Strengths and moduli of steel

Steel material	Yield strength (MPa)	Ultimate strength (MPa)	Young's modulus (GPa)	Poisson's ratio
6 mm-thick flange	372.8 ± 4.0	467.2 ± 3.1	205.1 ± 2.0	0.243 ± 0.08
80 × 80 × 6 mm SHS	420.1 ± 5.9[a]	519.4 ± 8.4	209.5 ± 3.9	0.277 ± 0.07

Note[a] 0.2% offset yield strength
[b] Adopted as the strength parameters in the Tsai-Wu failure criterion
[c] Transverse sample too short for tensile testing

14.2.3 Test Setup and Instrumentation

A cantilever flexural setup was adopted as shown in Fig. 14.2a where the cyclic loading was applied vertically at the free end of the specimen. The test setup, designed to subject the splice connection to a combination of shear and moment cyclic loading, simulates the load scenario of a beam or column splice when a building frame is under seismic or other lateral sway loadings. To approximate a fixed boundary condition, the constrained end of the specimen was clamped over a 340 mm length to a short steel column which was anchored to the strong floor (Figs. 14.2a and 14.3). The clamped area of the GFRP member was strengthened against failure by externally bonded steel plates. The cyclic loading was applied by a hydraulic actuator with a jack stroke limit of ± 125 mm and a load cell of ± 250 kN capacity. Shown in Fig. 14.3, a spread plate with a hinge joint was used at the loaded end of the specimen to ensure freedom of rotation and verticality of the load. In each specimen, sixteen strain gauges (G1 to G16) were attached on the GFRP member (Fig. 14.2b–d) to evaluate the interaction between the bonded steel and GFRP members. Four strain gauges (G17 to G20) were attached on the steel flange-plate (Fig. 14.2d) in positions and orientations where yielding would initiate according to the FE analysis. Four

14 Cyclic Performance of Splice Connections for Fibre … 315

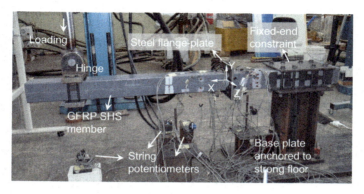

Fig. 14.3 Experiment setup for cyclic testing

displacement gauges (D1 to D4, string potentiometers) were deployed along the specimen to measure vertical deflections.

14.2.4 Cyclic Loading Program

In the absence of a cyclic loading protocol for FRP structures, the ATC-24 protocol [48] for components of steel structures was based on to devise the loading sequence in this study. The loading sequence, illustrated in Fig. 14.4, used the yield displacement (δ_y) as the reference to increase the amplitude of cycles. The magnitude of δ_y, defined as the loaded end displacement when yielding of the steel flange-plate began, was estimated by the FE modelling. For a direct comparison of cyclic performance among specimens, the δ_y value of 25 mm from specimen C-170-4 was adopted for all cyclic testings. The loading sequence, applied in a displacement-control mode

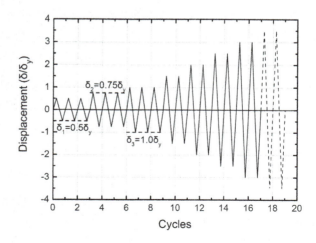

Fig. 14.4 Cyclic loading sequence applied to the specimens ($\delta_y = 25$ mm)

(12 mm/min), consisted of nine elastic cycles (amplitude $\leq \delta_y$) followed by a series of cycle pairs with amplitude increment of $0.5\delta_y$. In a cyclic testing, the loading sequence was implemented until a complete failure of the specimen, i.e. sudden drop in load resistance, or until the stroke limit of the hydraulic jack (± 125 mm) was reached.

14.3 Finite Element Modelling

14.3.1 Geometries, Element Types, Material Models and Boundary Conditions

Finite element (FE) modelling of the specimens was conducted using Ansys. The geometry of the models was constructed with symmetric simplification about the xy-plane (Fig. 14.5a). The steel and GFRP components were meshed by a 3D 8-node hexagonal element (Solid185). The GFRPs were modelled as an orthotropic elastic member with the longitudinal direction aligned parallel to the x-axis. The elastic material properties of the GFRP (longitudinal, transverse and shear moduli) were input as those obtained from the material characterisation tests (Table 14.1). In the post-processing, the Tsai-Wu failure criterion [49] was applied to evaluate the stress status of the GFRP member, and the strength parameters adopted to define the Tsai-Wu criterion are highlighted in Table 14.1. For the steel components, material models with the von-Mises yield criterion and kinematic multi-linear hardening behaviour

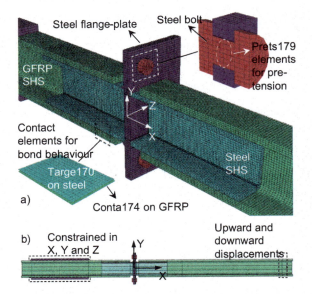

Fig. 14.5 FE modelling of the splice connection specimens **a** a typical meshed model **b** boundary conditions

were adopted; the stress–strain hysteresis behaviour was defined by fitting its envelop to the stress–strain curve obtained from the tensile coupon tests (Sect. 14.2.2).

Contact between the bolted parts was considered by defining a pair of 4-node surface-to-surface contact elements (Conta173 and Targe170) onto the potential contact surfaces. Three pairs of contact were included, i.e. between the flange-plates, between the bolt shank and the flange-plate, and between the bolt head (washer) and the flange-plate. For the contact interaction, free separation was allowed in the normal direction. While in compression the contact behaviour was governed by the augmented Lagrange algorithm where the contact pressure would be adjusted during equilibrium iterations to keep the penetration below an allowable tolerance. The tolerance was defined as 10% of the underlying element depth considering a balance between accuracy and computation time. In trial FE analyses with monotonic loading, the 10% tolerance did not result in a distinguishable difference in specimen load–deflection response compared to a 5% tolerance. Contact behaviour in the tangential direction was governed by a friction coefficient of 0.44 between the steel surfaces [50]. Pre-tensioning of the bolts was modelled by defining Prets179 elements at the mid-section of the bolt shanks.

In terms of boundary conditions, a symmetric constraint was applied on the xy cut plane of the specimen models. The top and bottom nodes at one end of the models were constrained from translation to simulate the fixed end condition (Fig. 14.5b), while the nodes at the other end were displaced upwards and downwards to simulate the cyclic loading. To solve for the mechanical response, pre-tensioning of the bolts was applied in the first load step, before the cyclic displacements were applied in the subsequent load steps.

14.3.2 Modelling of Steel-GFRP Bond

In adhesive bonded joints, local damage could be initiated in the high stress regions of the bonded area, while this does not necessarily lead to an immediate loss of load-carrying capacity. To account for the damage initiation and propagation in the steel-GFRP bonded sleeve joint of the connection specimens, the bond behaviour was modelled through the interaction between a pair of contact elements. As indicated in Fig. 14.5a, the adhesive layer was not represented by solid elements; instead, Conta174 and Targe170, a pair of 8-node surface-to-surface contact elements, were superposed onto the bonded surface of the steel and the GFRP respectively. Because of the thickness of the adhesive layer (and thus the distance between the steel and GFRP surfaces), the pair of contact elements were 1.5 mm apart from each other initially but with contact interaction.

Considering the presence of both shear and normal stresses on the adhesive layer when the bonded sleeve joint was under flexural loading, damage in the adhesive bond was defined to be initiated when:

$$\left(\frac{\sigma_p}{\sigma_p^{cr}}\right)^2 + \left(\frac{\sigma_t}{\sigma_t^{cr}}\right)^2 = 1 \tag{14.1}$$

where σ_p or σ_t is the peel or tangential stress between the contact pair; σ_p^{cr} or σ_t^{cr} is the corresponding critical stress that would lead to damage initiation in the state of pure peel or tangential stress. Compressive damage was not included in the modelling as the adhesive exhibited a 115% higher strength in compression than in tension [51], and also considering that compressive stress could still be transferred when debonded adherends were in contact. As illustrated in Fig. 14.6, once damage was initiated, the stress-separation (σ-δ) relation of the contact pair entered the debonded stage in a brittle manner, ideally with a $\delta_{p,t}^f/\delta_{p,t}^{cr}$ ratio of 1.0. However, in the numerical implementation, a debonding slope was defined ($\delta_{p,t}^f/\delta_{p,t}^{cr} > 1.0$) to overcome convergence difficulty. From trial FE analyses, a $\delta_{p,t}^f/\delta_{p,t}^{cr}$ ratio of 1.15 was found to allow substantial propagation of the adhesive bond damage, and was therefore adopted. To complete the definition of the bond behaviour, the state of ($\delta_{p,t}^f$, 0) in Fig. 14.6 was reached when:

$$\frac{G_p}{G_p^{cr}} + \frac{G_t}{G_t^{cr}} = 1 \tag{14.2}$$

where G_p or G_t is the work done by the peel (σ_p) or tangential (σ_t) stress with the corresponding separation (δ_p or δ_t); G_p^{cr} or G_t^{cr}, as the area under the corresponding σ-δ curve, was calculated by Eqs. (14.3) or (13.4).

$$G_p^{cr} = \sigma_p^{cr} \delta_p^f / 2 \tag{14.3}$$

$$G_t^{cr} = \sigma_t^{cr} \delta_t^f / 2 \tag{14.4}$$

It should be noted that under cyclic loading, damage in the adhesive bond was deemed to be cumulative that any unloading and reloading beyond the elastic stage would occur with a reduced slope as shown in Fig. 14.6. Moreover, at damaged regions, compressive normal stress could still be transferred with the initial stiffness.

The critical peel stress (σ_p^{cr}) was input as the tensile strength of the adhesive, i.e. $f_{t,a} = 33.7$ MPa. The critical tangential stress (σ_t^{cr}) was input as the shear strength of the adhesive, i.e. $\tau_a = 0.8 f_{t,a} = 27.0$ MPa based on the empirical correlation concluded in [52] where the shear-slip behaviour of steel-FRP bonded joints was investigated using linear brittle epoxy adhesives with 1–2 mm in thickness. Between the contact pair, the contact stiffness in the normal ($\sigma_p^{cr}/\delta_p^{cr}$ and σ_c/δ_c) and tangential ($\sigma_t^{cr}/\delta_t^{cr}$) directions were calculated based on a linear elastic and isotropic deformation of the adhesive layer:

$$\sigma_p^{cr}/\delta_p^{cr} = \sigma_c/\delta_c = E_a/t_a \tag{14.5}$$

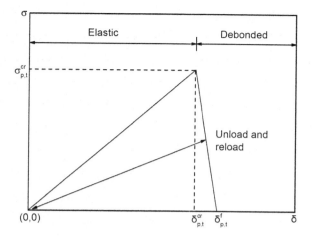

Fig. 14.6 Stress-separation (σ-δ) relation for steel-GFRP adhesive bond

$$\sigma_t^{cr}/\delta_t^{cr} = G_a/t_a \tag{14.6}$$

$$G_a = \frac{E_a}{2(1+v_a)} \tag{14.7}$$

where σ_c and δ_c are the stress and relative displacement between the contact pair in the normal compressive direction; t_a (1.5 mm) is the thickness of the adhesive; G_a is the shear modulus of the adhesive; E_a (4.09 GPa) and v_a (0.28) are the elastic modulus and Poison's ratio.

14.4 Results and Discussion

14.4.1 Moment-Rotation Responses and Failure Modes

The moment-rotation (M-θ) hysteresis responses of all the specimens are presented in Fig. 14.7 where the bending moment M was calculated considering a lever arm of 946 mm measured horizontally from the loading position to the centre of the steel flange-plates. The rotation angle θ was derived from the vertical deflections measured by the displacement gauges (Fig. 14.2a):

$$\theta = \frac{D_3 - D_2}{L_{2,3}} - \frac{D_1}{L_1} \tag{14.8}$$

where D_i = deflection measured by displacement gauge i, $L_{2,3}$ = 150 mm, and L_1 = 75 mm.

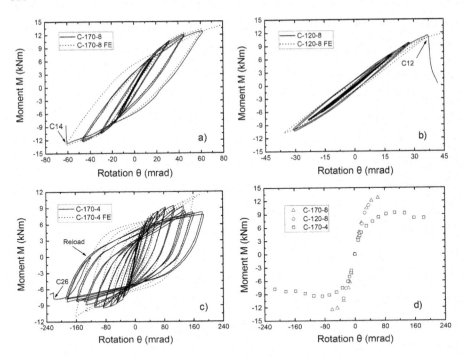

Fig. 14.7 Moment-rotation (M-θ) responses **a** specimen C-170-8 **b** C-120-8 **c** C-170-4 **d** envelop curves of all specimens

The M-θ curve of specimen C-170-8 (Fig. 14.7a) started to show identifiable nonlinearity and residual rotation (upon unloading to 0 kNm) in the 10th cycle (C10). The nonlinearity, due to yielding of the steel-flange plate (to be discussed in Sect. 14.4.3), continued to develop as the cycle amplitude increased. In the upward excursion of C14, the specimen attained its peak bending moment (M_u) of + 12.7 kNm; in the downward excursion of the same cycle ultimate failure occurred at − 12.5 kNm as cracking happened near the lower web-flange junctions of the GFRP SHS member (Fig. 14.8a), which was instantly followed by debonding between the steel and the GFRP SHS member at the upper face. Similar to in C-170-8 and because of the same bolt configuration, the M-θ curve of C-120-8 started its nonlinearity in C10 due to yielding of the steel flange-plate. But with a shorter bond length of 120 mm, C-120-8 was unable to withstand C12 and failed in the upward excursion at M_u of + 11.0 kNm. The failure mode of C-120-8, shown in Fig. 14.8b, was similar to that of C-170-8, except that the GFRP web-flange cracking and the debonding occurred at the reversed sides compared to that in C-170-8. Presented in Fig. 14.7c, specimen C-170-4 exhibited the most ductile M-θ hysteresis response because it experienced the most substantial yielding in the steel flange-plate, as visualised by the opening between the flange-plates highlighted in Fig. 14.8c. In Fig. 14.7c, C-170-4 started to develop residual rotation in C10, and in the downward excursion of C16 reached its peak bending moment (M_u) of − 9.5 kNm. After that the peak M in each cycle started

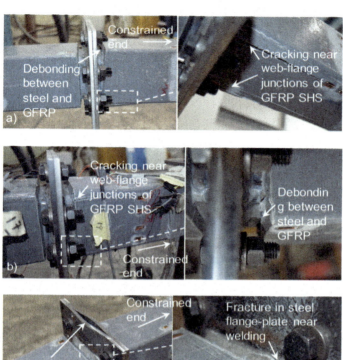
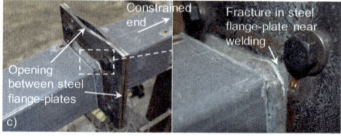

Fig. 14.8 Failure modes **a** specimen C-170-8 **b** C-120-8 **c** C-170-4

to gradually deteriorate, possibly due to the initiation of fracture in the steel flange-plates. After C25 during which the stroke limit of the hydraulic jack (±125 mm) was attained, an extension bar was attached to the jack and C-170-4 was reloaded downwards until occurrence of ultimate failure at rotation θ of -221 mrad. The ultimate failure was highlighted in Fig. 14.8c as the fracture of the steel flange-plates near the corner welding of the steel SHS.

The envelops of the M-θ responses of the three specimens are plotted together in Fig. 14.7d with the key performance results summarised in Table 14.3. A comparison of C-170-8 and C-120-8 shows that the increase of bond length from 120 to 170 mm improved the initial rotational stiffness S_i (defined as the average secant stiffness of the first 3 cycles) by 13% (from 404 to 456 kNm/rad) and the peak moment capacity M_u by 15% (from 11.0 to 12.8 kNm). Comparing C-170-8 and C-170-4, the change of bolt arrangement (from the 8-bolt to 4-bolt) resulted in 22% decrease in S_i (from 456 to 355 kNm/rad) and 26% decrease in M_u (from 12.7 to 9.5 KNm), but prevented the

Table 14.3 Mechanical performance of all specimens under cyclic loading

Specimen	Rotational stiffness (kNm/rad) [a, b]		Peak moment (kNm) [b, c]		Failure cycle	Ultimate rotation (mrad)[c]	Accumulated dissipated energy (kJ)
	S_i	$S_{i,FE}$	M_u	$M_{u,FE}$			
C-170-8	456	425	12.7 (+)	14.5 (+)	14	61 (−)	2.36
C-120-8	404	385	11.0 (+)	12.9 (+)	12	37 (+)	0.57
C-170-4	355	333	9.5 (−)	11.4 (+)	26	221 (−)	21.9

Note [a] Average secant stiffness of the first 3 cycles
[b] Subscript 'FE' means result from FE modelling
[c] ' + ' means in the upward excursion, ' − ' means downward

brittle failure of GFRP cracking and better utilized the yielding of steel, improving the ultimate rotation from 61 to 221 mrad.

The M-θ responses from the FE modelling are also plotted in Fig. 14.7a to c alongside the experimental results. Overall, the FE modelling is able to well capture the hysteresis responses in terms of the nonlinear behaviours and residual rotations. More detailed comparisons of stiffness and strength evolution are presented in the following section. For C-170-8 and C-120-8, the peak moment from FE modelling ($M_{u,FE}$) was determined by checking the stress state of the GFRP member against the Tsai-Wu failure criterion, as further discussed in Sect. 14.4.4. For C-170-4, from C20 onwards the FE M-θ curve started to show evident deviation from the experimental curve with overestimation of the bending moment. This happened because fracture might have been initiated in the steel flange-plates in the experimental scenario but such steel fracture was not considered in the FE modelling. In FE modelling, the ultimate failure and $M_{u,FE}$ of C-170-4 (Table 14.3) was deemed to be attained when the von-Mises stress in the steel flange-plate reached the steel ultimate strength of 467.2 MPa.

14.4.2 Cyclic Performance

In this study, the cyclic performance of the specimens is evaluated in regard to their stiffness and strength evolutions and energy dissipation capacities. The evolutions of stiffness and strength versus the loading cycles are plotted in Fig. 14.9a and b respectively. Herein the stiffness and strength of each cycle are the secant stiffness (S_s) and bending moment (M) when the loading displacement reversed in direction. Of each cycle the average values (S_s and M) of the upward and downward excursions were adopted, considering the reasonable symmetry of the M-θ responses (Fig. 14.7). In Fig. 14.9a, generally for all the specimens, the stiffness fluctuated slightly in C1 to C6 before declining continuously afterwards as yielding developed in the steel flange-plate. The most notable drop in stiffness occurred between C9 and C10, where the cycle amplitude increased from $1.0\delta_y$ to $1.5\delta_y$ (Fig. 14.4). Comparing the stiffness

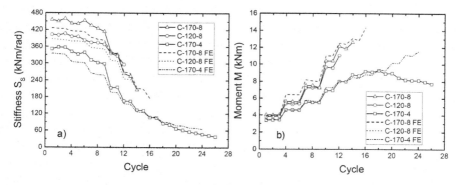

Fig. 14.9 Evolution of **a** stiffness (S_s) **b** strength (M) of the specimens

evolution of specimens C-170-8 (presented as triangles) and C-120-8 (circles) which had the same bolt arrangement, C-120-8 showed slightly lower stiffness in C1 to C9 because of the shorter bond length. However, as the stiffness degraded further (yielding of the steel flange-plates became more substantial) the stiffness of the two specimens converged towards each other. Compared to C-170-8 and C-120-8, the overall lower S_s of C-170-4 (squares) shows that the bolt arrangement has a greater effect on the stiffness than the bond length. Among the specimens, C-170-4 sustained the most extended stiffness degradation because of the most substantial steel yielding before ultimate failure. In Fig. 14.9b which presents the strength evolution, the bending moment (M) sustained by C-170-8 (triangles) and C-120-8 (circles) shows steady increase with the cycle amplitude, until the brittle ultimate failure of GFRP. In contrast, the M of C-170-4 stabilized and deteriorated slowly after peaking at C16, because the brittle GFRP failure was preceded by substantial yielding of the steel flange-plates.

The evolutions of stiffness and strength from FE modelling are also plotted in Fig. 14.9a and b as the dashed lines. In Fig. 14.9a, the FE modelling was able to capture the stiffness degradation, although the stiffness in the first nine cycles was underestimated (within 15%). The underestimation may be attributed to the neglect of the minor initial bowing deformation of the steel flange-plates induced through welding to the steel SHS member. This deformation resulted in minor gaps between the steel flange-plates and these were subsequently closed when the bolts were pre-tensioned. The initial deformation of the flange-plates was shown in [53] to slightly increase the initial flexural stiffness of the bolted flange joint. The strength evolution produced by FE modelling (Fig. 14.9b) also agreed well with the experimental results (difference within 13%) before occurrence of ultimate failure or strength deterioration. The overestimation of the ultimate strength of C-170-8 and C-120-8 is to be discussed in Sect. 14.4.4. In C-170-4, while the experimental result of strength stabilized and deteriorated after C16, due to excessive yielding and possibly fracture in the steel flange-plates, the FE results kept increasing steadily. This was, again, due to the absence of a fracture behaviour for the steel material in the FE modelling.

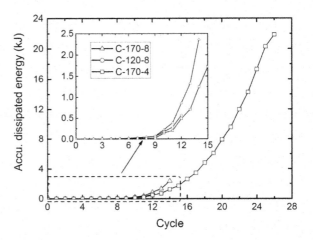

Fig. 14.10 Accumulated energy dissipation of specimens

The energy dissipation performance of the specimens is shown in Fig. 14.10 as the accumulated dissipated energy versus the loading cycles. In each cycle the dissipated energy was calculated as the area enclosed by the M-θ hysteresis loop. For all the specimens, energy dissipation was hardly visible until C10 where a notable drop in stiffness happened compared to C9 (Fig. 14.9a). Up to C14 where its ultimate failure occurred, C-170-8 dissipated the most energy due to its highest stiffness (highest bending moment at similar imposed rotation). Beyond C14, C-170-4 obviously outperformed the others with a total dissipated energy of 21.9 kJ as it withstood the most loading cycles by full utilization of the plastic deformation of the steel flange-plates. In comparison to C-170-8, C-120-8 exhibited inferior cyclic performance with 13% lower rotational stiffness (first three cycles S_i), 15% lower peak moment (M_u) and 76% lower energy dissipation. C-170-4 presented a remarkable higher (826%) energy dissipation than C-170-8, while was 22% and 26% lower in S_i and M_u respectively. Nevertheless, C-170-4 still had a S_i that amounted to 109% of a continuous member (considering a connection length of 392 mm as per Fig. 14.2) and a M_u equal to 29% of the theoretical elastic moment capacity of the connected section. For the configuration of C-170-4, the excellent energy dissipation capacity, along with the satisfactory rotational stiffness and strength, suggests its potential for application in seismic regions with brittle pultruded FRP members.

14.4.3 Local Strain Responses

Mechanical responses that are not distinguishable in the M-θ results may be reflected in the strain responses. To indicate the interaction between the bonded steel and GFRP SHS members, the strain profiles of the bonded sections are plotted at the maximum positive M of different cycles in Fig. 14.11, based on the measurement of strain gauges G1 to G6 installed on the surface of the GFRP member (Fig. 14.2b-d). In

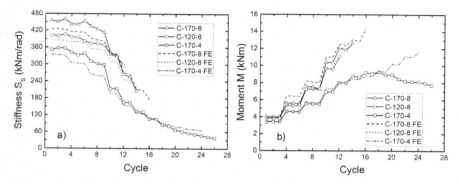

Fig. 14.11 Strain profiles of steel-GFRP bonded sections at different cycles **a** C-170-8 **b** C-120-8

specimen C-170-8 (Fig. 14.11a), the strain profile was close to linear in the early cycles (e.g. C6, C9 and C11) but showed apparent nonlinearity at C13 and C14, implying a degradation in the interaction between the steel and the GFRP sections and thus possible damage in the adhesive bond in later stage of the cyclic loading. In C-120-8 with a shorter bond length (Fig. 14.11b), the deviation of strain profile from linearity could be identified at C11 and C12.

For strain gauges G1, G2, G5 and G6, the load-strain responses are plotted in Fig. 14.12. In both C-170-8 and C-120-8, in the first nine cycles (C1 to C9) the load-strain responses were almost linear. In C-170-8 (Fig. 14.12a) the first appearance of nonlinearity was observed in the G5 and G6 curves at − 9.1 kNm of C10; then at + 10.1 kNm of C12, nonlinear responses also occurred in the G1 and G2 curves. Likewise, in C-120-8 (Fig. 14.12b) similar nonlinearity first appeared in the G5 and G6 curves at − 8.5 kNm of C10 before in the G1 and G2 curves at + 8.4 kNm of C12. These nonlinearities in the load-strain responses, most likely caused by damage in the adhesive bond, were all initiated at or near the compression flange (negative strain). The strain gauges in Fig. 14.12 (G1, G2, G5 and G6) were positioned near the GFRP end of the bonded region, where the compression flange was characterised with an

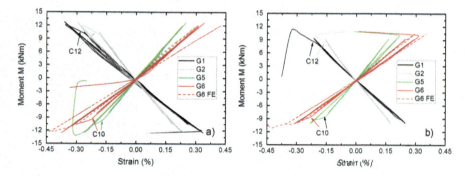

Fig. 14.12 Load-strain responses from strain gauges G1, G2, G5 and G6 **a** C-170-8 **b** C-120-8

adhesive stress state of peel-and-shear while the tension flange was characterised with a state of compressive-and-shear. The initiation of the load-strain nonlinearities at the compression flange suggested that damage in the adhesive bond was more likely to be induced by a stress state of peel-and-shear than compressive-and-shear. This finding was consistent with the damage criterion of Eq. 14.1 adopted in the FE modelling. The load-strain responses of G6 by FE modelling are also presented in Fig. 14.12 as the dashed curves. It can be seen that, with the modelling of debonding behaviour, the FE results were able to capture the nonlinear responses, although at a relatively later stage, i.e. at − 10.5 kNm of C12 for C-170-8 and at −9.2 kNm of C10 for C-120-8.

The damage in adhesive bond was further evidenced by the load-strain responses from G9 as presented in Fig. 14.13. In C10 the tensile strain in both C-170-8 and C-120-8 experienced a release when the bending moment increased beyond − 9.5 kNm and − 8.7 kNm respectively. This response, also captured in the FE modelling at −10.9 kNm of C12 for C-170-8 and at − 8.4 kNm of C10 for C-120-8, was associated with the onset of debonding at the position as exemplified in Fig. 14.14 for C-120-8. Figure 14.14 presents the total contact stress, as the vector sum of the

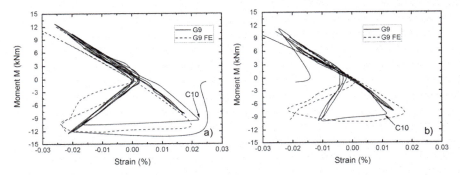

Fig. 14.13 Load-strain responses from strain gauges G9 for **a** specimen C-170-8 **b** C-120-8

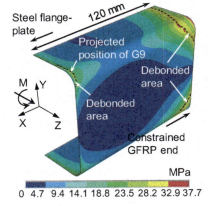

Fig. 14.14 Distribution of total contact stress over bonded area from FE modelling (C-120-8 at M of − 8.40 kNm for C10)

normal and tangential stresses, over the steel-GFRP bonded area, at the load level where the strain release (G9) was captured in the FE modelling. The highlighted debonded areas, representing the regions where the contact status was loaded into the debonded stage (Fig. 14.6), mainly appeared at the GFRP end because of its smaller stiffness than the steel member [54]. At this load level the debonded area appeared and expanded over the projected position of G9, resulting in the release of tensile strain in the detached region of the GFRP. Total contact stress beyond a trivial level was noted in the debonded areas. This may be resulted from i) the existence of normal compressive stress which could still be transferred in the debonded regions; and ii) the debonding slope defined for the stress-separation behaviour (Fig. 14.6) to overcome convergence difficulty (Sect. 14.3.2), resulting in a quick reduction but not immediate loss of the peel and tangential stresses.

The strain data from the strain gauges on the steel flange-plates (G17 to G20 in Fig. 14.2e) can provide a direct indication of steel yielding, as a verification of the failure modes and M-θ responses. The envelop curves of the load-strain responses are plotted in Fig. 14.15 with indication of the steel yield strain (0.18%). In both C-170-8 (Fig. 14.15a) and C-170-4 (Fig. 14.15b), the yield strain was first attained in C10, at + 8.3 kNm (G19) and at + 6.0 kNm (G18) respectively, echoing the observation of residual rotations in C10 from the M-θ responses. For C-170-8 where the deformation of the steel flange-plate was difficult to visually notice, the strain responses presented in Fig. 14.15a can evidence the yielding of the steel flange-plate. Figure 14.15b shows that the yielding in C-170-4 was more substantial. Besides this, in the curve of G17 which was positioned near the ultimate steel fracture, the envelop strain of C22, compared to that of C21, experienced a decrease although with an increased cycle displacement amplitude, signalling possible initiation of the steel fracture.

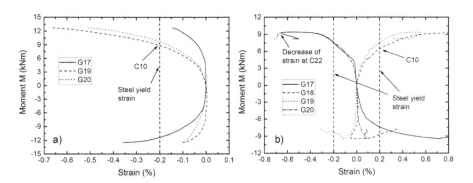

Fig. 14.15 Envelop curves of load-strain response from strain gauges on steel flange-plates of **a** specimen C-170-8 **b** C-170-4

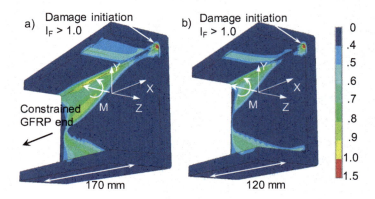

Fig. 14.16 Distribution of Tsai-Wu failure index (I_F) on GFRP from FE modelling: **a** C-170-8 at $M_{u,FE}$ of +14.5 kNm; **b** C-120-8 at $M_{u,FE}$ of +12.9 kNm

14.4.4 GFRP Failure

The ultimate failure of specimens C-170-8 and C-120-8, i.e. cracking of the GFRP SHS member near the web-flange junctions, was predicted in the FE modelling by detecting the exceedance of Tsai-Wu failure criterion. For these two specimens, the distributions of Tsai-Wu failure index (I_F) on the GFRP members (at the constrained side) are presented in Fig. 14.16 where $I_F > 1.0$ indicates exceedance of the failure criterion. Due to the brittle failure mode, in FE modelling the ultimate bending moments ($M_{u,FE}$) of C-170-8 and C-120-8 were deemed to be reached when $I_F > 1.0$. As shown in Fig. 14.16 a and b, $M_{u,FE}$ of both C-170-8 and C-120-8 was reached in an upward loading excursion as failure was initiated ($I_F > 1.0$) at the inside of the upper web-flange junctions. This failure initiation signalled an imminent cracking of the GFRP member near the web-flange junctions as observed in the experiments. As the crack development of the GFRP was not modelled, the major debonding (Fig. 14.8a and b) that instantly followed was not reflected in the modelling either. The ultimate bending moments of C-170-8 and C-120-8 from experimental testing (M_u) and FE modelling ($M_{u,FE}$) are listed in Table 14.3. The FE modelling overestimated the ultimate moment by less than 18%, resulting in the occurrence of failure one cycle later than the experimental result for C-170-8 and two cycles later for C-120-8. A possible reason for the overestimation is that pultruded FRP members usually exhibit weaker strength at or near the web-flange junctions than in the flanges or webs [55, 56].

14.5 Conclusions

This chapter investigated the cyclic flexural performance of splice connections developed for hollow section FRP members. The splice connection specimens, each comprised of a steel bolted flange joint between two hollow section steel-GFRP bonded sleeve joints, were prepared in three configurations with difference in bond length or bolt arrangement. Based on the results from the experiments and the corresponding FE modelling, the following conclusions can be drawn:

1. Under the cyclic flexural loading, the specimens experienced different levels of yielding in the steel flange-plates before the ultimate failure in the mode of GFRP cracking near the web-flange junctions, i.e. in C-170-8 (the specimen with 170 mm steel-GFRP bond length and eight bolts at the steel flange-plate) and C-120-8, or fracture in the steel flange-plate, i.e. in C-170-4. Specimen C-170-8 exhibited the highest rotational stiffness (456 kNm/rad) and ultimate moment capacity (12.7 kNm); while C-170-4 presented the most ductile moment-rotation behaviour and the highest rotation capacity (221 mrad).
2. In terms of cyclic performance, through yielding of the steel flange-plate, the specimens began to show energy dissipation as their cycle secant stiffness underwent a steady degradation. The energy dissipation capacity of C-170-8 (with 170 mm-bond length and eight bolts) and C-120-8 was limited by the brittle failure of GFRP cracking prior to substatial yielding in the steel flange-plate. Specimen C-170-4, without significant compromise in the rotational stiffness (22% lower than C-170-8) and peak bending moment (26% lower), showed remarkable improvement in total energy dissipation (826% higher) by fully exploiting the plastic deformation of the steel flange-plate before failure in the GFRP.
3. In the steel-GFRP bonded sleeve joints of the splice connections, damage in the adhesive bond could be identified from the nonlinear strain profile of the bonded section, and further evidenced from the nonlinearity and strain release in the load-strain responses. The strain data of the steel flange-plates can indicate steel yielding which may be difficult to identify from the deformations and moment-rotation responses. From the steel strain data of specimen C-170-4 (with 170 mm-bond length and four bolts), the decrease of cycle envelope strain may signal the initiation of the ultimate steel fracture.
4. The FE modelling produced moment-rotation responses that agreed well with the experimental results. Before failure or deterioration of strength, for all the specimens the difference in stiffness evolution was within 15% and that in strength evolution was within 13%. With the definition of a debonding behaviour, the FE modelling was able to capture the nonlinear load-strain responses in the steel-GFRP bonded sleeve joints. The ultimate failure of web-flange cracking in the GFRP members (in C-170-8 and C-120-8 with eight bolts but different bond length) was predicted by examining the GFRP stress state

against the Tsai-Wu failure criterion. Using this modelling approach, the location of failure initiation could be predicted, and the ultimate moments were overestimated within 18%.

The ductility and energy dissipation capacity are important considerations in the design of a frame structure, especially one with brittle FRP members. The cyclic performance of specimen C-170-4 demonstrated the possibility of substantial energy dissipation at connection level using the developed splice connections. To realize this favourable performance, it is important that the splice connection is designed with a strength governed by the plastic deformation of the steel flange-plates, thereby delaying or preventing the brittle FRP web-flange junction cracking during the course of a seismic loading. The geometries of the flange-plates (e.g. thickness, number and position of bolts etc.) and the steel-FRP bond length are design variables that could be optimized to fulfill the objective of energy dissipation as well as other design requirements such as connection stiffness and strength.

References

1. Teng J, Chen J, Smith ST, Lam L (2003) Behaviour and strength of FRP-strengthened RC structures: a state-of-the-art review. Proc Inst Civ Eng-Struct Build 156(1):51–62
2. Zhao X-L, Zhang L (2007) State-of-the-art review on FRP strengthened steel structures. Eng Struct 29(8):1808–1823
3. Meyer R (2012) Handbook of pultrusion technology: Springer Science & Business Media.
4. Kumar P, Chandrashekhara K, Nanni A (2004) Structural performance of a FRP bridge deck. Constr Build Mater 18(1):35–47
5. Hayes MD, Ohanehi D, Lesko JJ, Cousins TE, Witcher D (2000) Performance of tube and plate fiberglass composite bridge deck. J Compos Constr 4(2):48–55
6. Zhang D, Zhao Q, Huang Y, Li F, Chen H, Miao D (2014) Flexural properties of a lightweight hybrid FRP-aluminum modular space truss bridge system. Compos Struct 108:600–615
7. Yang X, Bai Y, Ding F (2015) Structural performance of a large-scale space frame assembled using pultruded GFRP composites. Compos Struct 133:986–996
8. Satasivam S, Bai Y, Zhao X-L (2014) Adhesively bonded modular GFRP web–flange sandwich for building floor construction. Compos Struct 111:381–392
9. Satasivam S, Bai Y, Yang Y, Zhu L, Zhao X-L (2018) Mechanical performance of two-way modular FRP sandwich slabs. Compos Struct 184:904–916
10. Xie L, Qi Y, Bai Y, Qiu C, Wang H, Fang H et al (2019) Sandwich assemblies of composites square hollow sections and thin-walled panels in compression. Thin-Walled Struct 145:106412
11. Bank LC, Mosallam AS, McCoy GT (1994) Design and performance of connections for pultruded frame structures. J Reinf Plast Compos 13(3):199–212
12. Bank L, Mosallam A, Gonsior H (1990) Beam-to-column connections for pultruded FRP structures. ASCE, Serviceability and durability of construction materials, pp 804-813
13. Mosallam AS, Abdelhamid MK, Conway JH (1994) Performance of pultruded FRP connections under static and dynamic loads. J Reinf Plast Compos 13(5):386–407
14. Bank LC, Yin J, Moore L, Evans DJ, Allison RW (1996) Experimental and numerical evaluation of beam-to-column connections for pultruded structures. J Reinf Plast Compos 15(10):1052–1067
15. Mottram J, Zheng Y (1999) Further tests on beam-to-column connections for pultruded frames: web-cleated. J Compos Constr 3(1):3–11

16. Mottram J, Zheng Y (1999) Further tests of beam-to-column connections for pultruded frames: flange-cleated. J Compos Constr 3(3):108–116
17. Mottram J, Turvey GJ (2003) Physical test data for the appraisal of design procedures for bolted joints in pultruded FRP structural shapes and systems. Prog Struct Mat Eng 5(4):195–222
18. Smith S, Parsons I, Hjelmstad K (1998) An experimental study of the behavior of connections for pultruded GFRP I-beams and rectangular tubes. Compos Struct 42(3):281–290
19. Smith S, Parsons I, Hjelmstad K (1999) Experimental comparisons of connections for GFRP pultruded frames. J Compos Constr 3(1):20–26
20. Singamsethi S, LaFave J, Hjelmstad K (2005) Fabrication and testing of cuff connections for GFRP box sections. J Compos Constr 9(6):536–544
21. Wu C, Zhang Z, Bai Y (2016) Connections of tubular GFRP wall studs to steel beams for building construction. Compos B Eng 95:64–75
22. Zhang ZJ, Bai Y, Xiao X (2018) Bonded sleeve connections for joining tubular glass fiber-reinforced polymer beams and columns: experimental and numerical studies. J Compos Constr 22(4):04018019
23. Martins D, Proença M, Correia JR, Gonilha J, Arruda M, Silvestre N (2017) Development of a novel beam-to-column connection system for pultruded GFRP tubular profiles. Compos Struct 171:263–276
24. Bruneau M, Walker D (1994) Cyclic testing of pultruded fiber-reinforced plastic beam-column rigid connection. J Struct Eng 120(9):2637–2652
25. Mosallam A (1997) Design considerations for pultruded composite beam-to-column connections subjected to cyclic and sustained loading conditions. Marketing technical regulatory sessions of the composites institutes international composites expo 14-B
26. Carrion JE, LaFave JM, Hjelmstad KD (2005) Experimental behavior of monolithic composite cuff connections for fiber reinforced plastic box sections. Compos Struct 67(3):333–345
27. Zhang Z, Bai Y, He X, Jin L, Zhu L (2018) Cyclic performance of bonded sleeve beam-column connections for FRP tubular sections. Compos B Eng 142:171–182
28. Martins D, Proença M, Gonilha JA, Sá MF, Correia JR, Silvestre N (2019) Experimental and numerical analysis of GFRP frame structures. Part 1: cyclic behaviour at the connection level. Compos Struct 220:304–317
29. Martins D, Sá MF, Gonilha JA, Correia JR, Silvestre N, Ferreira JG (2019) Experimental and numerical analysis of GFRP frame structures. Part 2: monotonic and cyclic sway behaviour of plane frames. Compos Struct 220:194–208
30. Bank LC (2012) Progressive failure and ductility of FRP composites for construction. J Compos Constr 17(3):406–419
31. Nagaraj V, Gangarao HV (1998) Fatigue behavior and connection efficiency of pultruded GFRP beams. J Compos Constr 2(1):57–65
32. Manalo A, Mutsuyoshi H (2012) Behavior of fiber-reinforced composite beams with mechanical joints. J Compos Mater 46(4):483–496
33. Hai ND, Mutsuyoshi H (2012) Structural behavior of double-lap joints of steel splice plates bolted/bonded to pultruded hybrid CFRP/GFRP laminates. Constr Build Mater 30:347–359
34. Turvey G (2014) Experimental and analytical investigation of two-and six-plate bonded splice joints on serviceability limit deformations of pultruded GFRP beams. Compos Struct 111:426–435
35. Turvey GJ, Cerutti X (2015) Flexural behaviour of pultruded glass fibre reinforced polymer composite beams with bolted splice joints. Compos Struct 119:543–550
36. Turvey GJ, Cerutti X (2015) Effects of splice joint geometry and bolt torque on the serviceability response of pultruded glass fibre reinforced polymer composite beams. Compos Struct 131:490–500
37. Qiu C, Ding C, He X, Zhang L, Bai Y (2018) Axial performance of steel splice connection for tubular FRP column members. Compos Struct 189:498–509
38. Qiu C, Bai Y, Zhang L, Jin L (2019) Bending performance of splice connections for assembly of tubular section FRP members: experimental and numerical study. J Compos Constr 23(5):04019040

39. Australia S (1998) AS4100 steel structures. Australia, Sydney
40. ASTM International (2014) Standard test method for tensile properties of polymer matrix composite materials. ASTM D3039. West Conshohocken, PA
41. ASTM International (2015) Standard Test Method for Compressive Properties of Rigid Plastics. ASTM D695–15. West Conshohocken, PA.
42. ASTM International (2015) Standard test method for flexural properties of polymer matrix composite materials. ASTM D7264. West Conshohocken, PA
43. ASTM International (2016) Standard test method for short-beam strength of polymer matrix composite materials and their laminates. ASTM D2344. West Conshohocken, PA
44. Lee S, Munro M, Scott R (1990) Evaluation of three in-plane shear test methods for advanced composite materials. Composites 21(6):495–502
45. ASTM International (2010) Standard test methods and definitions for mechanical testing of steel products. ASTM A370–10. West Conshohocken, PA
46. Zhang Z, Wu C, Nie X, Bai Y, Zhu L (2016) Bonded sleeve connections for joining tubular GFRP beam to steel member: Numerical investigation with experimental validation. Compos Struct 157:51–61
47. ASTM International (2010) Standard test method for tensile properties of plastics. ASTM D638–10. West Conshohocken, PA
48. Krawinkler H (1992) Guidelines for cyclic seismic testing of components of steel structures. Appl Technol Council
49. Tsai SW, Wu EM (1971) A general theory of strength for anisotropic materials. J Compos Mater 5(1):58-80
50. Trahair N, Bradford M (1998) The behaviour and design of steel structures to AS 4100. Third Edition ed: Taylor & Francis
51. de Castro J, Keller T (2008) Ductile double-lap joints from brittle GFRP laminates and ductile adhesives, part I: experimental investigation. Compos B Eng 39(2):271–281
52. Xia S, Teng J (2005) Behaviour of FRP-to-steel bonded joints. Proceedings of the international symposium on bond behaviour of FRP in structures. In: International institute for FRP in construction, pp 419–426
53. Wheeler AT, Clarke MJ, Hancock GJ (2000) FE modeling of four-bolt, tubular moment end-plate connections. J Struct Eng 126(7):816-822
54. Qiu C, Feng P, Yang Y, Zhu L, Bai Y (2017) Joint capacity of bonded sleeve connections for tubular fibre reinforced polymer members. Compos Struct 163:267–279
55. Turvey GJ, Zhang Y (2005) Tearing failure of web–flange junctions in pultruded GRP profiles. Compos A Appl Sci Manuf 36(2):309–317
56. Turvey GJ, Zhang Y (2006) Shear failure strength of web–flange junctions in pultruded GRP WF profiles. Constr Build Mater 20(1–2):81-89

Chapter 15
Fire Performance of Loaded Fibre Reinforced Polymer Multicellular Composite Structures

Lei Zhang, Yiqing Dai, Yu Bai, Wei Chen, and Jihong Ye

Abstract Multicellular web-flange composite structures were assembled using glass fibre reinforced polymer (GFRP) box sections as web sections and plates as face sheets, further with glass magnesium (GM) or gypsum plaster (GP) panels on the surface for fire protection. The structures were subjected to ISO 834 fire curve from underside where the GM or GP panels were installed and a constant load on top to introduce bending during fire exposure. Experimental results showed that the fire endurance times of the structures before failure were extended from 54 min without protective panels to 83 min or 103 min by a single layer of GP or GM panel. When double layers were used, the fire endurance time increased to 113 min for GP panels and 158 min for GM panels. Numerical modelling was further established to estimate the temperature distribution in the specimens. Effects of the GM and GP layers on the thermal and mechanical performances of loaded specimens in fire can be clarified. The fire performance of GFRP multicellular web-flange composite structures enhanced by fire resistance panels may also be well demonstrated experimentally and numerically.

15.1 Introduction

Fibre reinforced polymer (FRP) composites have been used as structural components in construction for their high strength-to-weight ratio and superior resistance

---

Reprinted from Construction and Building Materials, 296, Lei Zhang, Yiqing Dai, Yu Bai, Wei Chen, Jihong Ye, Fire performance of loaded fibre reinforced polymer multicellular composite structures with fire-resistant panels, 123733, Copyright 2021, with permission from Elsevier.

L. Zhang
School of Civil Engineering, Xuchang University, Xuchang, China

Y. Dai · Y. Bai (✉)
Department of Civil Engineering, Monash University, Clayton, Australia
e-mail: yu.bai@monash.edu

W. Chen · J. Ye
China University of Mining and Technology, Xuzhou, China

© The Author(s), under exclusive license to Springer Nature Singapore Pte Ltd. 2023
Y. Bai (ed.), *Composites for Building Assembly*, Springer Tracts in Civil Engineering,
https://doi.org/10.1007/978-981-19-4278-5_15

to corrosion [1–4]. For example, glass fibre reinforced polymer (GFRP) sections have been used in bridge decks [5–8], building structures [9–11], roof structures [12], reinforcements for concrete [13–16], supporting frame structures [17, 18], transmission tower structures [19], cooling towers [20] and piles [21]. In general, the elastic modulus of GFRP composites is about 10–20% of steel [22], therefore GFRP structural components can be designed as multicellular web-flange sandwich sections which may provide improved bending stiffness despite of their relatively low elastic modulus [23–25]. The effects of elevated and high temperatures also degrade mechanical properties of GFRP composites [26–34]. Existing researches have indicated that the most rapid degradation in elastic modulus of GFRP materials happens at their glass transition temperature (T_g) of about 100–120 °C [35, 36]. In fire scenarios, building structures may be exposed to much higher temperatures according to the ISO 834 fire curve [37]. Since the elastic modulus and strengths of GFRP composites decrease significantly at the elevated and high temperatures, several methods have been proposed to improve the endurance time of GFRP structures in fire scenarios. For example, a water-cooling system was proposed and used for GFRP tubular components [38] to extend their endurance time in fire, and the effectiveness was experimentally verified showing considerable increase in the fire endurance time of loaded GFRP components with water-cooling system. Furthermore, the post-fire performance of the GFRP components was also investigated [39].

Another solution was to install fire-resistant panels at the structural surface. The enhancement of structural fire endurance time from fire-resistant panels is also well evidenced in a series of fire exposure experiments on cold-formed steel structures with fire resistant panels, including gypsum plasterboard, bolivian magnesium panel and calcium silicate panel [40, 41]. The results indicated that the bolivian magnesium panels provided better performance with a fire endurance time of over 90 min for the wall structures subjected to mechanical loading and ISO 834 fire curve. Applications of the fire-resistant panels and materials were also introduced for GFRP structures to improve their performance in fire scenarios [42–45]. The protective effects of such fire-resistant layers assembled with GFRP components or structures have been investigated in several studies. For example, different approaches including a calcium silicate (CS) board, a vermiculite/perlite (VP) based mortar and an intumescent coating were applied respectively to a side of GFRP square hollow beams with a cross section of $100 \times 100 \times 8$ mm and a length of 1.5 m. The specimens were then subjected to a service load by four-point bending and simultaneously exposed to fire with reference to ISO 834 from the underside with the fire protection [45]. It took about 34 min before the upper flange reaching the glass transition temperature for the specimen without protection; while this became 61 min or 74 min for the specimen with the CS panel or VP mortar respectively. In another study, three sides of GFRP square hollow beams with or without CS boards were exposed to fire, where the cross section of the GFRP beams was also $100 \times 100 \times 8$ mm and the total length was 1.3 m. The CS boards increased the fire endurance duration from 38 min to about 70 min [32]. Intumescent coatings were used to protect a sandwich structure with carbon fibre reinforced polymer (CFRP) face sheets, and the coating effectively prolonged the exposure time before failure [46]. In a recent study [42], GFRP multicellular web-flange structures were installed with different fire-resistant panels and exposed to fire from underside. All specimens were subjected to no loading. When the fire insulation performance was defined by the temperature increase of no more than 140 °C at the upper surface of the specimens, an adequate fire insulation time of over 90 min was provided by glass magnesium (GM) panels or gypsum plaster (GP) panels, and 72 min by CS panels, with the same thickness of 12 mm. The specimens were then cooled to room temperature and submitted to four-point bending tests, and specimens with GM panels presented the highest post-fire stiffness and flexural capacity [43], where about 59% stiffness and 52% flexural capacity remained in comparison to those without fire exposure.

Finite element (FE) modelling has also been used to understand the thermal responses of GFRP structures with or without fire-resistant layers exposed to fire. In several early researches, modelling approaches were developed for GFRP including thermo-physical properties (e.g. thermal conductivity, specific heat capacity and thermal conductivity [35]) and stiffness [36]. An FE model was established to predict the thermal responses of GFRP square hollow sections (SHS) in fire [33] and the results were compared with those from experiments [32]. It suggested that the heat convention in the cavity of SHS presented noticeable effects on the heat transfer behavior in the modelling. A recent state-of-the-art paper systematically reviewed experimental and numerical research concerning fire performance of FRP composites [47]. It was found that most results were mainly focused on the material scale and small-scale experiments and numerical approaches; while only very limited full-scale fire experiments were conducted.

The above results on the application of fire-resistance panels for improvement of fire performance for GFRP components further suggest this as an effective way to increase fire endurance time for GFRP multicellular web-flange sandwich structures. Such structures have shown their potential for building floor and decking applications while limited studies are reported about their fire endurance performance under loading with installation of fire resistance panels. Indeed, mechanical loading needs to be taken into account and the effects of thickness and type of the fire-resistant panels need to be understood to help in determination of appropriate type and thickness of the panels in practice. Finally, it is also expected to achieve a fire endurance time of more than 120 min as considered for further applications in multistory building structures [48, 49].

In this chapter, five large-scale GFRP multicellular structures were fabricated using GFRP square tubes as web sections and plates as face sheets and the connections between them were achieved by epoxy adhesive with a thickness of 1 mm. Glass magnesium (GM) or gypsum plaster (GP) panels, in either single or double layers, were then assembled to the underside of the structures. One GFRP multicellular web-flange structure was unprotected (i.e. without fire resistant panel) as a reference. The specimens were loaded with a constant load at the midspan and the undersides of specimens were then exposed to fire with reference to ISO 834. The temperature variations inside the specimens and the deflections at the mid span were continuously recorded, and the fire performance (considering the load carrying capacity in addition to insulation and integrity performances) of the loaded specimens enhanced by fire resistance panels was evaluated. FE modelling based on temperature-dependent properties of GFRP, GM and GP materials was developed to estimate the temperature responses of specimens subjected to fire. The modelling results were then verified based on the experimental scenarios and further used to investigate the effects of the panel thickness on the thermal responses of the GFRP structures. Results from the specimens exposed to fire without mechanical loading were also comparatively discussed to investigate the effects of mechanical loading on the thermal responses of the specimens.

15.2 Experimental Investigation

15.2.1 Specimen Preparation

Each of the GFRP multicellular web-flange structures used in this study is composed of three pultruded GFRP box sections (i.e. rectangular hollow sections) with a dimension of 3600 mm (length) × 75 mm (width) × 100 mm (height) × 10 mm (wall thickness) and two pultruded GFRP panels with a dimension of 3600 × 475 × 8 mm as shown in Fig. 15.1. The GFRP box sections and GFRP panels consisted of glass fibres and polyester resin. According to the microscopy performed on the GFRP box sections and panels [42], the flat panel consisted of five mat layers, four 90° roving layers and three 0° roving layers in the through-thickness direction; and each wall of the box sections consisted of four mat layers and five roving layers (two in 0°, one in 90°, one in 45° and one in −45°). The tensile strength in the longitudinal direction of the GFRP panels and box sections were 378.1 MPa and

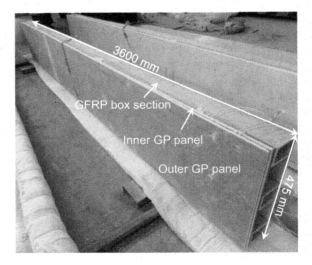

Fig. 15.1 Configuration of specimens with two layers of GP panels as an example

360.0 MPa measured from tensile experiments; and the in-plane shear strength of the box sections was 20.4 MPa evaluated by the 10° off-axis tensile experiments [43]. Based on the results obtained from the thermogravimetric analysis (TGA) on the samples from the GFRP materials [42], the mass fraction of fibre was 63.3% for the GFRP panels and 69.5% for the GFRP box sections; while the volume fraction of fibre was 45.3% for the GFRP panels and 52.3% for the box sections. Furthermore, the glass transition temperature (T_g) was determined as 127 °C (based on the peak of the loss modulus curve) through dynamic mechanical analyses (DMA) [42]. The GFRP box and panel sections were bonded together by an epoxy structural adhesive with a thickness of 1 mm. The epoxy adhesive had a glass transition temperature (T_g) of 80 °C and lap shear strength of about 24 MPa according to the manufacturer data specification [43]. The specimens were prepared and cured for two weeks at room temperature (about 25 °C) and indoor humidity. In total, five multicellular web-flange specimens were prepared with a curing time of at least two weeks, for the fire experiments including four with fire resistant panels and one without. To evaluate the load capacity of the multicellular web-flange structure at room temperature, a specimen with identical GFRP sectional configuration was prepared and examined under four-point bending with the same span length of 3300 mm and spacing of 1100 mm between the load points. The ultimate load capacity of the specimen was measured as 230 kN at room temperature.

Glass magnesium (GM) panels and gypsum plaster (GP) fire-resistance panels were used in this study and both had a thickness of 12 mm for a single layer. Two of the GFRP structures were assembled with a single layer of the GM or GP panel, and another two were assembled with two layers of GM or GP panels, i.e. with a total panel thickness of 24 mm as shown in Figs. 15.1 and 15.2. The fire-resistant panels were installed to the underside of the GFRP multicellular web-flange structures by steel self-drilling screws (ST 4.8, see Fig. 15.2). The screws were drilled all through the fire-resistant panels, the lower GFRP face sheet, the adhesive layer and a lower side of the GFRP box sections, to provide the connection between the fire-resistant panels and the GFRP structures as shown in Fig. 15.2. The screws were only located along the central axes of the box sections with a spacing of 300 mm between adjacent screws. The specimen with a single layer of GM or GP panel is referred as specimen SGM or SGP, where S denotes a single layer. Similarly, the specimen with two layers

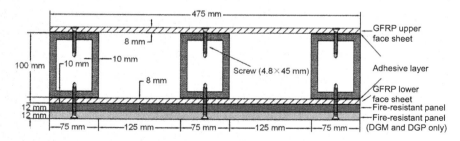

Fig. 15.2 Section view for components and dimensions of specimens

of GM or GP panels is referred as specimen DGM or DGP. In addition, one GFRP specimen was unprotected, i.e. without fire resistant panel, as a reference specimen and is named as specimen NN.

15.2.2 Experimental Setup

Fire experiments were conducted using a horizontal furnace as shown in Fig. 15.3. The furnace had a width of 3000 mm and a depth of 1500 mm, and the specimens were placed above the furnace with two roller supports at 150 mm from each end of the specimens resulting in a span length of 3300 mm (i.e. a span to depth ratio of about 28). The gap between the specimen and furnace at each end caused by the roller supports was sealed by aluminium silicate wool. Meanwhile the aluminium silicate wool was also used to seal the gaps between the specimens and the furnace lids as shown in Fig. 15.4. The critical sections of the specimen have been marked by dash lines in Fig. 15.3 and numbered as S1 to S5, where S1 and S5 are the sections at 150 mm from the supporting points, S3 the midspan, and S2 and S4 are the 1/3 and 2/3 spans respectively. The mechanical load was applied by a block of weight supported on two rollers at S2 and S4 with a spacing of 1100 mm as shown in Fig. 15.3. It should be noted that for safety consideration the weight block was linked to a reaction frame using extended steel cables to prevent it from falling down if structures failed. The load of 3.5 kN/m^2 equivalent was applied as the mechanical load in fire tests according to the design standards of building structures [50], where the live load ranges between 1.5 and 3.0 kN/m^2 for general structures such as residential and office buildings. The total weight was applied as 5.49 kN (2.4% of the ultimate load capacity at room temperature) and the resulting bending moment at the midspan was then 3.02 kN.m. The initial axial stress in the top and bottom faces at the moment

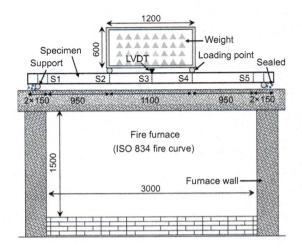

Fig. 15.3 Furnace configuration and loading setup (unit in mm)

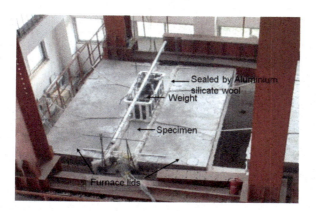

Fig. 15.4 Experimental system and installation of specimen

region was 6.32 MPa (1.7% of the tensile strength of the GFRP panels); and the initial shear stress in the mid-height of webs at the shear regions of 0.58 Mpa (2.9% of the in-plane shear strength of the box sections). The initial mid-span deflection was 3.7–3.9 mm for all specimens (see Sect. 5.4.2). The deflection at the midspan (S3) was measured by a linear variable differential transformer (LVDT) and recorded every two seconds. A high temperature camera was equipped within the fire furnace to record the conditions of the underside surface of the specimens.

A series of K type thermocouples was installed in the midspan section (i.e. section S3 in Fig. 15.3) of the specimens to obtain the temperature profiles there. For example, the layout of thermocouples for specimens with two layers of fire resistant panels (i.e. specimens DGM and DGP) is shown in Fig. 15.5, where T1 and T22 are thermocouples installed on top of the upper GFRP face sheet and others were also installed inside the specimens within drilled holes of a depth of 2 mm (T2 and T23) and 4 mm (T3 and T24) measured from the upper surface of the GFRP face sheet (see Fig. 15.5). Similarly, thermocouples were also installed in the adhesive layers (T4 and T16), different depths in the web of GFRP box sections (T5-15) and the lower

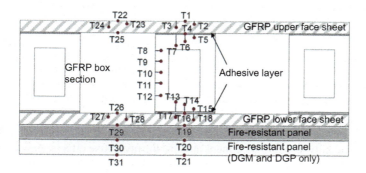

Fig. 15.5 Layout of thermocouples for specimens

GFRP face sheet (T17-18 and T27-28), as well as the surfaces of the fire-resistant panels (T19-21 and T29-31) as shown in Fig. 15.5.

Mechanical load through the weight was first applied. After about 20 min and the midspan deflection of specimens became stable, the fire load was initiated within the gas furnace and the resulting temperature increase followed the ISO 834 fire curve [37] as described by Eq. (15.1):

$$T(t) = T_a + 345 lg(8t + 1) \tag{15.1}$$

where $T(t)$ is the time-dependent furnace temperature; T_a is the room temperature of about 22 °C; t is the fire exposure time.

The average temperature of the furnace was monitored via eight internal thermocouples at the mid-height of the four furnace walls. Temperature results from the thermocouples were recorded by a data acquisition system every two seconds during the fire experiments. With reference to existing standards [37], the following four criteria were defined as the failure of specimens and the fire exposure experiments would stop accordingly: (i) the temperature at T1 or T22 rises for more than 180 °C, or both of them rise for over 140 °C from room temperature; (ii) flames developed on the surface of the upper face sheet for more than ten seconds; and (iii) loss of load carrying capacity of the structures, and (iv) any unsafe situation noticed.

15.3 Numerical Modeling

15.3.1 Material Properties

A finite element (FE) heat transfer model was established for the specimens with fire-resistant panels exposed to fire by Abaqus to understand their thermal responses in fire. Temperature-dependent material properties (including the thermal conductivity, specific heat capacity and density) of the involved materials were used in the modelling since such properties of GFRP profiles and fire-resistant panels change with temperatures during fire exposure.

The thermal conductivities (see Fig. 15.6a), specific heat capacities (see Fig. 15.6b) and densities (see Fig. 15.6c) of the GFRP face sheet and web box sections were obtained according to the results reported in [35, 42] based on similar GFRP materials with close mass and volume fractions of glass fibres. The multilinear models for the temperature-dependent thermal conductivities of GP and GM were experimentally determined in the literature [51], where the GP and GM specimens were similar with those used in the present research. Besides, since GP panels collapsed gradually from the specimens and lost the protective effects at about 950 °C as witnessed in the experiments, the thermal conductivity of GP panels at the temperature over 900 °C was amended to increase linearly to 10 W/m K from 900 to 950 °C (see Fig. 15.6a). Also, the specific heat capacities and the densities of GP and GM

15 Fire Performance of Loaded Fibre Reinforced Polymer ...

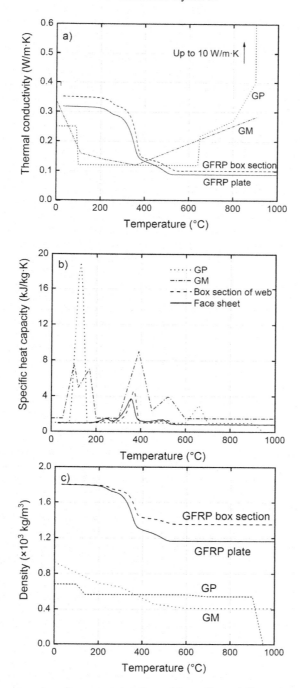

Fig. 15.6 Temperature-dependent properties of materials used in modeling: **a** thermal conductivity, **b** specific heat capacity and **c** density

panels shown in Fig. 15.6b, c were in reference to those reported in [51], as determined by differential scanning calorimetry (DSC). Considering the gradual collapse of the GP panels at the temperature over 900 °C in fire, the specific heat capacity and the density of GP panel were amended to decrease linearly to 1.0 J/kg K and 0.1 kg/m^3 from 900 to 950 °C (see Fig. 15.6b, c). The peaks of specific heat capacities at about 100 °C for GP as shown in Fig. 15.6b are caused by water evaporation; and those at other certain temperatures are due to decomposition of different constituents as an endothermic process. For example, about 350 °C for GFRP plates and box sections because of the decomposition of polymer matrix.

The temperature-dependent thermophysical properties of the epoxy adhesives on the overall thermal responses of the specimens were not considered in the heat transfer modelling due to their thin thicknesses (1 mm). Constant thermal conductivity of 0.2 W/m K [52], specific heat capacity of 1200 J/kg K [53] and density of 1250 kg/m^3 [53] were used. Though the thermal conductivity of the steel screws was about 50 W/m K [54], the effect of the screws on the heat transfer of the specimens was not taken into account in the modelling because the screws were only located along the central axes of the box sections with a spacing of 300 mm between adjacent screws (see Sect. 15.2.1). The section area of the screw heads was relatively small (in total 3.06×10^{-3} m^2), implying very limited effects on the heat transfer behavior of the specimens with fire exposure surface area of 1.57 m^2.

15.3.1.1 Structural Modelling and Thermal Boundary Conditions

Based on the thermophysical properties of the GFRP components, fire resistant panels and adhesive layers, FE models for the five specimens investigated in the experiments were established. Only the midspan cross-section (i.e. S3 in Fig. 15.3) was simulated as 2D modelling for computational efficiency. Figure 15.7 shows the FE model for specimen SGP as an example. The sectional dimensions of the models are the same as those in the experiments, where the width is 475 mm and the thickness for one layer of GP or GM panel is 12 mm (see Fig. 15.2).

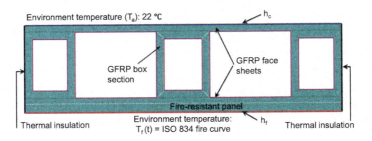

Fig. 15.7 FE model and thermal boundary conditions for specimen SGP as an example where h_c or h_f is the convection coefficient at top or bottom surface

As shown in Fig. 15.7, the furnace environment temperature for the underside of the specimens follows the ISO 834 fire curve as a time-dependent temperature as described in Eq. (15.1). The total fire exposure times in the modelling are according to the experiments. Heat transfer from the furnace environment to the unprotected and protected specimens considers heat convection and radiation, where the convection coefficient is 25 W/m² K and the emissivity is 0.95 [30, 44]. The heat received at the specimen underside is then transferred into the GM or GP fire-resistant panels (if any) and the inner GFRP face sheet and web sections through conduction, and further transferred through the cavities by thermal radiation [33]. Convection within the cavity was insignificant according to existing modelling studies [33, 52]. It should be noted that thermal radiation considered for the aforementioned surfaces is associated with the emissivity of 0.75 within 20 °C and 0.95 above 353 °C, and other values by interpolation in-between these two temperatures [55]. The surfaces at the left and right sides of the specimens are thermally insulated since they are well sealed by aluminium silicate wool in the experiments. At the upper surface of the specimens, heat is also transferred to the indoor environment by convection and the convection coefficient (h_c) is defined by Eq. (15.2) [55]:

$$h_c = 0.14k \left[P_r \frac{g\beta}{\nu^2} (T_c - T_a) \right]^{1/3} \qquad (15.2)$$

where k is the conductivity of air (0.03 W/m K); Pr is the Prandtl number and taken as 0.71; g is the gravitational acceleration, i.e. 9.81 m/s²; β is the coefficient of thermal expansion of air as 3.43×10^{-3} K⁻¹; ν is the dynamic viscosity of air as 1.57×10^{-5} m²/s; T_a is the room temperature, i.e. 22 °C. Therefore, h_c depends only on one variable T_c as the temperature at the upper surface of the specimen.

It should be noted that mechanical loading was not considered in the 2D FE modelling, because mechanical responses almost have no influence on the thermal responses of GFRP materials in the specimens [30, 33]. The effect of the mechanical loading on the failure of the GP panels can be considered through the change of the temperature-dependent material properties from 900 to 950 °C (see Fig. 15.6). The 2D model was then meshed by 8-node quadratic heat transfer quadrilateral elements (DC2D8) as shown in Fig. 15.7, and the dimension of each element is 1 mm. The solving method of transient heat transfer step using automatic incrementation was applied in the FE modelling with initial increment of one second, minimum increment of 0.05 second and maximum increment of ten seconds.

15.4 Results and Discussion

15.4.1 Temperature Responses

The temperature according to Eq. (15.1) as per ISO 834 [37] is plotted in Fig. 15.8 together with the temperature results recorded by thermocouples in the fire furnace from the experiment of specimen DGM, associated with the longest duration (158 min) of fire endurance time in the experiments. It can be seen that the differences between the ISO 834 fire curve and furnace temperature are minor and within 2% during the entire experimental process. As shown in Fig. 15.8, the temperature in the furnace increased rapidly in the first 6 min to more than 600 °C after the fire was applied, and reached 1000 °C within 83 min. The maximum temperature in the furnace recorded in the experiment was 1100 °C at approximately the end of the fire exposure of 158 min for specimen DGM. The temperature progressions of fire furnace from other experimental scenarios were similar to specimen DGM as presented in Fig. 15.8.

It also can be seen from Fig. 15.8 that the temperature at the outer surface of the specimen exposed to fire was close to the furnace temperature (and the ISO 834 fire curve) and slightly higher with the maximum temperature reaching about 1200 °C for specimen DGM at about 158 min of fire exposure. This is likely because the furnace temperature was measured through the thermocouples at middle height of the furnace walls, while higher temperatures occurred in the upper zone of the furnace (which was closer to the fire exposed surface of the specimens) considering the upward movement of hot air [42]. In terms of the other specimens, the fire-resistant panels extended the fire endurance time from about 54 min for specimen NN to 83 min for specimen SGP with single layer of GP panel, to 103 min for specimen SGM with single layer of GM panel, to 113 min for specimen DGP with double layers of GP panels.

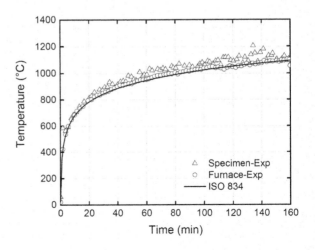

Fig. 15.8 Furnace temperature in comparison to ISO 834 fire curve for experiment on specimen DGM and corresponding temperature at bottom of fire-resistant panel

The temperature variations at different locations of all the five specimens obtained from the experiments and FE modelling are presented in Fig. 15.9. Among them, Fig. 15.9a shows the temperature variation with time at mid depth of the lower GFRP face sheets as obtained from the average temperature of thermocouples T18 and T28 (see Fig. 15.5). As shown in Fig. 15.9a, at the end of the fire endurance time (54 min, where the temperature rose at T3 and T24 reached 140 °C) of specimen-NN without fire-resistant panels, the temperature at mid depth of its lower GFRP face sheets was 657 °C. When a single layer of GP or GM panel with a thickness of 12 mm was applied, the fire endurance time was considerably extended to 83 min for specimen SGP with GP panel or 103 min for specimen SGM with GM panel, where the temperature rise at T3 and T24 reached 140 °C. The corresponding temperature at mid depth of its lower GFRP face sheets was 779 °C for specimen SGP and 662 °C for specimen SGM. In terms of the temperature progression, the average temperature rise was 12.0 °C/min for specimen NN, 9.4 °C/min for specimen SGP and 6.4 °C/min for specimen SGM.

As shown in Fig. 15.9a, the temperature progressions at the lower GFRP face sheets for specimens SGP and SGM are highly consistent in the first 45 min, suggesting similar protective effects provided by the single layer of GP and GM panels in this stage. After that, the temperature at the lower GFRP face sheet of specimen SGP then increased sharply and became about 200 °C higher than that of specimen SGM at the 83 min of fire exposure (i.e. the end of fire exposure for specimen SGP). The results recorded from the high temperature camera are shown in Fig. 15.10 for the underside of specimen SGP. In comparison to the surface before fire exposure (Fig. 15.10a), it can be noticed that cracks developed in the GP panels of specimen SGP at about 26 min of the fire exposure as shown in Fig. 15.10b (corresponding to a furnace temperature of about 870 °C). Parts of the GP panels collapsed gradually from the GFRP structure at the fire exposure of about 40 min (Fig. 15.10c) and the corresponding temperature at the underside of the GP panel reached about 950 °C. The cracking and partial collapse of the GP panel from SGP specimen may be responsible for its temperature sharp increase after the first 45 min of fire exposure, in comparison to specimen SGM.

When the double layers of GP or GM panel with a total thickness of 24 mm were installed, the fire endurance time was further extended to 113 min for specimen DGP or 158 min for specimen DGM. Their temperature progressions during fire exposure are also shown in Fig. 15.9a where the average temperature rise was 6.7 °C/min for specimen DGP and 4.9 °C/min for specimen DGM. Specimens DGP and DGM presented similar temperatures in the mid-depth of the lower GFRP face sheet in the first 50 min of fire exposure. After that, the temperature in specimen DGP increased rapidly from about 900 to 950 °C, again due to partial collapse of the GP panels. This led to higher temperatures of about 300 °C than those of specimen DGM after the fire exposure of 90 min. It should be noted that the GM panels in both specimens SGM and DGM connected well to the GFRP structures in the experiments as supported by the results from the high temperature camera. In terms of the fire endurance time and temperature development within GFRP structures, the GM panels also present better fire insulation performance than the GP panels. When the fire-resistant

Fig. 15.9 Experimental and modelling results of temperature progressions with time for all specimens at mid-depth of: **a** GFRP lower face sheet, **b** web box sections and **c** upper face sheet

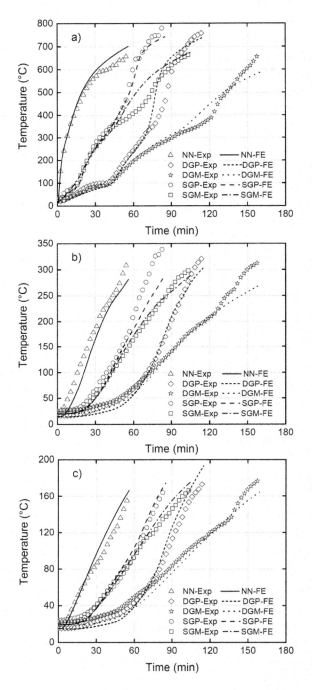

15 Fire Performance of Loaded Fibre Reinforced Polymer …

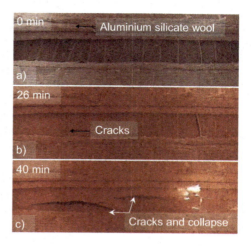

Fig. 15.10 GP panels of specimen SGP exposed to fire: **a** intact GP panel at 0th min, **b** cracks developed at 26th min, and **c** obvious cracking and collapse at 40th min of fire exposure

panels were increased from single layer to double layers, the fire endurance time was extended for 30 min for the specimens with GP panels and 55 min for GM panels. Obviously, two layers of fire-resistance panels (with a total thickness of 24 mm) offer better temperature mitigation than single layer panels (with a thickness of 12 mm) investigated in the experiments.

The modelling results concerning the temperature responses at the mid-depth of the lower GFRP face sheet are also presented in Figs. 15.9a, and their agreements with the experimental results can be witnessed where the average difference for the temperatures at the end of fire exposures is about 5% in general. The maximum difference is about 10% as noticed from specimen DGP, where the temperature at the end of fire exposure is 655 °C from the experiment and 586 °C from the modelling.

The temperatures at the mid-depth of the web box sections of the specimens exposed to fire is presented in Fig. 15.9b, where they are taken from the readout of thermocouple T10 (see Fig. 15.5). It can be seen that such temperatures at the end of fire exposure for all specimens range from 298 °C (specimen SGM) to 339 °C (specimen SGP). The corresponding temperatures at the mid-depth of the web box sections from modelling results are also present in Fig. 15.9b, showing slight underestimations, of about 8% difference in average for the temperatures at the end of fire exposures. The differences between experiments and modelling are mainly noted from the specimens SGP and DGM in the last 30 min of fire exposure.

Figure 15.9c shows the temperatures at the mid-depth of the upper GFRP face sheets of the specimens exposed to fire. It can be seen that, at the end of fire exposure although with different endurance time durations, the temperatures at the mid-depth of the upper GFRP face sheets are between 155 °C (specimen NN) and 175 °C (specimen SGP). The temperatures recorded at the upper GFRP face sheets are much lower than those at the lower GFRP face sheets, suggesting the effective thermal insulation effects within their individual fire endurance times offered from lower GFRP face sheets and single or double fire-resistant panels. It should be noted that, the glass transition temperature T_g of the GFRP materials used in the experiments is 127 °C, therefore, the mechanical properties of the upper GFRP face sheets would have been significantly impacted by the elevated temperatures at the end of their fire exposures. The modelling results are generally consistent with the experimental results, with an average difference for the temperatures at the end of fire exposure of about 7%.

15.4.2 Mid-Span Deflection

The developments of mid-span deflection for all specimens subjected to fire exposure and mechanical load are presented in Fig. 15.11. In the experiments, the mechanical load was applied before fire exposure (i.e. started at time 0 min). This resulted in the initial linear increase of deflection to be consistent within 3.7–3.9 mm for all specimens and stabilized within the following 20 min as shown in Fig. 15.11. It should be noted that the mid-span deflection was caused by the bending moment and the shear effects as well as thermal effects. It also suggests that the mechanical contributions from the fire-resistant panels to the specimens are minor before the fire exposure. After the fire exposure initiated at time 0 min in Fig. 15.11, with the rapid increase of the temperature in the furnace, the mid-span deflection in specimen NN develops rapidly from 3.8 mm to 10.6 mm at the first 4 min. It presents an approximately bilinear behavior up to the end of fire exposure. As shown in Fig. 15.9, the temperature differences between the lower and upper GFRP face sheets at 4 min of the fire exposure were about 183 °C for the unprotected specimen NN and about 22 °C for the protected specimens, suggesting greater thermal bowing effect as well as the larger stiffness loss of different components for the unprotected specimen. Both thermal bowing effect and stiffness loss contributed to the increase of deflection.

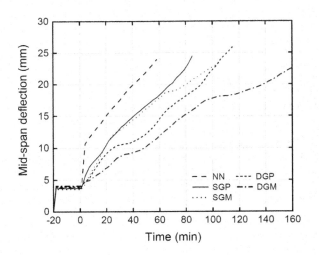

Fig. 15.11 Mid-span deflections before and during fire exposure

At the end of the fire exposure with different endurance times, the ultimate deflection was 23.9 mm for specimen NN, 24.5 mm for specimen SGP with a single layer of GP panels and 25.9 mm for specimen DGP with two layers of GP panels. The ultimate deflections for specimens with GM panels was relatively smaller as shown in Fig. 15.11, corresponding to 23.5 mm for specimen SGM with single layer of GM panels and 22.5 mm for specimen DGM with double layers of GM panels. Despite of similar ultimate deflections of the specimens, the developing progressions of the deflection curves with fire exposure time was quite different among the specimens in relation to their different fire exposure durations. Certainly, the specimen NN without fire resistant panels showed the fastest rate of deflection increase and the specimens DGM and DGP with double layers of fire-resistant panels showed much slower increase in deflection during fire exposure. For example, the deflection development was 0.37 mm/min on average corresponded to the entire period during the fire test for specimen NN, and it was obviously mitigated to 0.25 mm/min for specimen SGP and further to 0.21 mm/min for specimen DGP. The installation of GM panels reduced the development rate of deflection to 0.17 mm/min for specimen SGM and further to 0.12 mm/min for specimen DGM. Consistently, the performance of fire protection offered from the GM panels is superior to that from the GP panels. This can further be seen from the deflection curves of specimens SGM and DGM in Fig. 15.11, as their deflections slowed down after reaching about 18 mm.

15.4.3 Effects of Fire Resistance Panel on Thermal Response

The thickness of single layer of GM and GP panels is 12 mm as a common product in market. Considering other thicknesses may also be commercially available or achieved through multiple layers, the effects of thickness of GM and GP panels on the thermal performance of such GFRP structures exposed to fire can be investigated by the established modelling approach, in order to provide references for optimal thicknesses of the fire-resistant panels in such applications.

When the upper GFRP face sheets reach the glass transition temperature (T_g, 127 °C in this study) of the GFRP materials, the stiffness and carrying capacity of such composite structures degrade severely [45]. Therefore, the fire exposure time (defined as t_p in min) when the temperate at the mid-depth of the upper GFRP face sheet reach its T_g can be used to reflect the thermal insulation and fire protective performance of the fire-resistant panels with different thicknesses, in reference to the approach conducted in [45]. The results are presented in Fig. 15.12 where the thickness of fire-resistant panels ranges from 0 to 30 mm and t_p for the experimental specimens with 0, 12 and 24 mm thicknesses are also included. As shown in Fig. 15.12, t_p for specimens with different thicknesses of fire-resistant panels ranges from 44 to 102 min for the GP group and from 44 to 162 min for the GM group. The increase in fire endurance time is approximately in linear relationship with the panel thickness, although the increase of t_p with the thickness of GP panels becomes less obvious for thick ones. It can be noted again that the in terms of fire protection, GM panels outperform GP

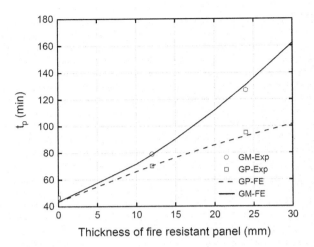

Fig. 15.12 Variation of t_p with thickness of fire-resistant panels

panels and this advantage is more obvious for thick panels. For example, when the thickness of the panels is smaller than 10 mm, the difference in t_p for such GFRP structures with GP or GM panels is within 5 min, while for the panel thickness of 30 mm this difference in t_p becomes about 60 min.

15.4.4 Mechanical Loading Effects

Mechanical loading during fire exposure in the experimental investigation introduced structural deformation to the specimens and therefore tensile stress to the fire-resistant panels from the underside. These further caused cracking and partial collapse of the GP panels as shown in Fig. 15.10. It becomes meaningful to investigate the effects of the mechanical loading in thermal performance of such structures with different types of fire-resistant panels in fire.

A comparative study can be made with the thermal responses reported in [42], where similar GFRP structures and fire-resistant panels were used as the present study while they were exposed to fire according to ISO 834 without mechanical loading. The specimens in the reference [42] were composed by similar GFRP components and fire-resistant panels using similar assembling approaches, resulting in the same thickness of the specimens as the present study. Although the specimen width in [42] was 580 mm in comparison to the 475 mm width in current study, the heat transfer mainly occurred in the depth direction because of fire from underside only and the temperature distribution was rather uniform along the width direction at the same depth. Therefore, the results from the reference [42] can be used for a comparison of thermal performance for the loaded and unloaded GFRP multicellular web-flange structures protected with different fire-resistant panels.

Figure 15.13 shows the temperature responses at mid-depth of the web box sections and upper GFRP face sheets in loaded (i.e. specimens SGP and SGM from current study) and unloaded specimens [42] with single layer of GM or GP panels. The fire exposure time was 89 min for the specimen with single layer of GM. It can be seen from Fig. 15.13a that temperature responses in the loaded and unloaded specimens with single layer of GM are comparable within the fire exposure times. For example, the temperature from current specimen SGM is only 19 °C (7.8%) higher for the mid-depth of web box sections or 11 °C (8.6%) higher for the upper face sheet at the fire exposure of 89 min. The results suggested that the mechanical loading did not introduce significant effects on temperature responses of such GFRP structures protected with GM panels in fire as evidenced also in Fig. 15.13a.

However, as seen from Fig. 15.13b, more obvious differences between the loaded and unloaded specimens with single layer of GP panels are witnessed from the

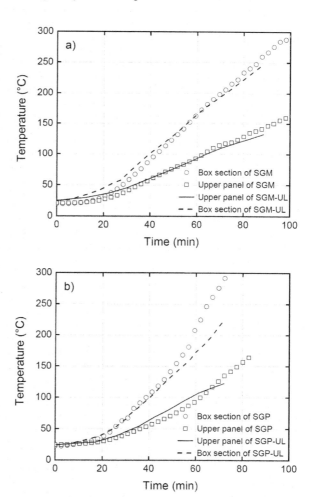

Fig. 15.13 Comparisons of temperature responses at mid-depth of web sections and upper GFRP face sheets in loaded and unloaded specimens with single layer of **a** GM and **b** GP panels

temperature responses at the mid-depths of web box sections and upper face sheet. The total experimental duration for unloaded specimen in [42] is 72 min. At the end of its fire exposure time (72 min), the temperature at the mid depth of the web box Sections (223 °C) is 66 °C lower than that (289 °C) of the loaded specimen SGP. Also, the temperature at the upper face sheet is 12 °C lower than that of loaded specimen SGP.

The temperature progressions were also much different for the loaded and unloaded specimens with single layer of GP panels as witnessed in Fig. 15.13b. From 50 min of the fire exposure to the end, the temperature increase in the GFRP web box sections was 88 °C, i.e. from 135 °C to 223 °C for the unloaded specimen. However, the temperature increased from this duration time became 143 °C for the loaded specimen SGP, i.e. from 146 °C to 289 °C. Within the same duration time of fire exposure, the temperature rise in the upper face sheet was 38 °C for the unloaded specimen but 62 °C for the loaded specimen SGP. The more significant increase in temperature in specimen SGP after about 50 min fire exposure is likely associated with the cracking and partial collapse (see Fig. 15.10) of GP panels due to the mechanical load and the resulting structural deformation. Considering the significant effects of mechanical loading on the thermal performance of the specimens assembled with GP panels, it is therefore suggested that evaluation of the fire performance of structures with GP panels should take mechanical loading effects into consideration.

15.5 Conclusions

This chapter presented an experimental and modelling study of prefabricated fire-resistant panels for improving the fire performance of modular GFRP multicellular structures subjected to mechanical load in bending. The results allow the following conclusions to be drawn:

1. The fire endurance time of the loaded GFRP multicellular web-flange structure of 118 mm depth without fire resistant panel (specimen NN) was 54 min. With the installation of 12 mm thickness fire resistance panels at the fire exposure side, the fire endurance time was effectively extended to 83 min for loaded specimens when single layer of GP panels was used, or 103 min when GM panels were used. This fire endurance time was further extended to 113 min for loaded specimens when double layers of GP panels were used or 158 min when double layers of GM panels were used, corresponding to a total thickness of 24 mm of fire-resistant panels. In terms of fire endurance time, the effectiveness of both types of fire-resistant panels was witnessed while GM panels appears outperforming GP panels.
2. The fire experiments indicated that cracking of GP panels occurred after about 26 min of fire exposure for the loaded specimen SGP or DGP with one or two layers of GP panels. When the fire furnace temperature increased to about

900 °C, partial GP panels collapsed gradually from the specimens SGP and DGP; while GM panels still attached well to the specimens SGM and DGM, until the end of fire exposure when the temperature rise at the upper GFRP face sheet reached 140 °C.
3. The results from FE modelling are generally consistent with the experimental results in terms of the temperature progressions at different locations through the depth of specimens. The validated modeling approach was further used to investigate the thermal performance of such GFRP multicellular web-flange structures assembled with GP or GM panels of different thicknesses. The fire exposure duration (t_p) when the upper GFRP face sheet reached its glass transition temperature (T_g, 127 °C) was used to compare their fire performances. It is found that the difference in t_p between GP and GM panels is insignificant when the panel thickness is less than 10 mm. While more superior fire performance was offered by GM panels when the panel thickness is more than 10 mm and this advantage becomes more significant when a thick panel is applied.
4. In comparison of the loaded and unloaded specimens of similar GFRP multicellular web-flange structure assembled with one layer of GM panels, the mechanical loading in the fire exposure scenario did not induce obvious differences (less than 10%) in the temperature responses within the GFRP web box sections or the upper GFRP face sheets. However, for the loaded and unloaded specimens with one layer of GP panels, the temperature responses in fire with mechanical loading showed higher temperature developments. This is likely due to the cracking and partial collapse of GP panels occurred for the loaded specimen during fire exposure.

References

1. Bakis C, Bank L, Brown V, Cosenza E, Davalos J, Lesko J, Machida A, Rizkalla S, Triantafillou T (2002) Fiber-reinforced polymer composites for construction—state-of-the-art review. J Compos Constr 6(2):73–87
2. Hollaway L (2010) A review of the present and future utilisation of FRP composites in the civil infrastructure with reference to their important in-service properties. Constr Build Mater 24(12):2419–2445
3. Bazli M, Jafari A, Ashrafi H et al (2020) Effects of UV radiation, moisture and elevated temperature on mechanical properties of GFRP pultruded profiles[J]. Constr Build Mater 231:117137
4. Ahmed A, Guo S, Zhang Z et al (2020) A review on durability of fiber reinforced polymer (FRP) bars reinforced seawater sea sand concrete. Constr Build Mater 256:119484
5. Keller T, Schollmayer M (2004) Plate bending behavior of a pultruded GFRP bridge deck system. Compos Struct 64(3):285–295
6. Lee J, Kim Y, Jung J, Kosmatka J (2007) Experimental characterization of a pultruded GFRP bridge deck for light-weight vehicles. Compos Struct 80(1):141–151
7. Kim HY, Lee SY (2012) A steel-reinforced hybrid GFRP deck panel for temporary bridges[J]. Constr Build Mater 34:192–200
8. Sutherland LS, Sá MF, Correia JR et al (2016) Quasi-static indentation response of pedestrian bridge multicellular pultruded GFRP deck panels[J]. Constr Build Mater 118:307–318

9. Keller T (2001) Recent all-composite and hybrid fibre-reinforced polymer bridges and buildings. Prog Struct Mat Eng 3(2):132–140
10. Evernden M, Mottram J (2012) A case for houses to be constructed of fibre reinforced polymer components. Proc ICE-Constr Mater 165(1):3–13
11. Zhu D, Shi H, Fang H, Liu W, Qi Y, Bai Y (2018) Fiber reinforced composites sandwich panels with web reinforced wood core for building floor applications. Compos B 150:196–211
12. Keller T, Haas C, Vallee T (2008) Structural concept, design, and experimental verification of a glass fiber-reinforced polymer sandwich roof structure. J Compos Constr 12(4):454–468
13. Maranan G, Manalo A, Benmokrane B, Karunasena W, Mendis P (2016) Behavior of concentrically loaded geopolymer-concrete circular columns reinforced longitudinally and transversely with GFRP bars. Eng Struct 422–436
14. Xue W, Peng F, Zheng Q (2016) Design equations for flexural capacity of concrete beams reinforced with glass fiber–reinforced polymer bars. J Compos Constr 20(3):04015069
15. Ahmed E, Benmokrane B, Sansfacon M (2017) Case study: design, construction, and performance of the La Chancelière Parking Garage's concrete flat slabs reinforced with GFRP bars. J Compos Constr 21(1)
16. Gu X, Dai Y, Jiang J (2020) Flexural behavior investigation of steel-GFRP hybrid-reinforced concrete beams based on experimental and numerical methods. Eng Struct 206:110–117
17. Bank L, Mosallam A, McCoy G (1994) Design and performance of connections for pultruded frame structures. J Reinf Plast Compos 13(3):199–212
18. Boscato G, Russo S (2013) Free vibrations of a pultruded GFRP frame with different rotational stiffnesses of bolted joints. Mech Compos Mater 48(6):655–668
19. Yeh H, Yang S (1997) Building of a composite transmission tower. J Reinf Plast Compos 16(5):414–424
20. Zhang W, Meng C, Li Y, Wu N, Liang T (2013) Optimization design of the GFRP hyperbolic cooling tower structure. Int Conf Mech Autom Eng 2013:76–79
21. Guades E, Aravinthan T, Islam M, Manalo A (2012) A review on the driving performance of FRP composite piles. Compos Struct 94(6):1932–1942
22. Bank L (2006) Composites for construction structural design with FRP materials. John Wiley & Sons
23. Kulpa M, Siwowski T, Stiffness and strength evaluation of a novel FRP sandwich panel for bridge redecking. Compos Part B Eng 167:207–220
24. Dawood M, Taylor E, Ballew W, Rizkalla S (2010) Static and fatigue bending behavior of pultruded GFRP sandwich panels with through-thickness fiber insertions. Compos B Eng 41(5):363–374
25. Mathieson H, Fam A (2016) In-plane bending and failure mechanism of sandwich beams with GFRP skins and soft polyurethane foam core. J Compos Constr 20(1):04015020
26. Khaneghahi MH, Najafabadi EP, Bazli M et al (2020) The effect of elevated temperatures on the compressive section capacity of pultruded GFRP profiles. Constr Build Mater 249:118725
27. Jarrah M, Najafabadi EP, Khaneghahi MH et al (2018) The effect of elevated temperatures on the tensile performance of GFRP and CFRP sheets. Constr Build Mater 190:38–52
28. Manalo A, Maranan G, Sharma S, Karunasena W, Bai Y (2017) Temperature-sensitive mechanical properties of GFRP composites in longitudinal and transverse directions: a comparative study. Compos Struct 173:255–267
29. Wu C, Bai Y, Mottram J (2016) Effect of elevated temperatures on the mechanical performance of pultruded FRP joints with a Single Ordinary or Blind Bolt. J Compos Constr 20(2):04015045
30. Correia J, Bai Y, Keller T (2015) A review of the fire behaviour of pultruded GFRP structural profiles for civil engineering applications. Compos Struct 127:267–287
31. Hajiloo H, Green MF, Gales J (2018) Mechanical properties of GFRP reinforcing bars at high temperatures. Constr Build Mater 162:142–154
32. Morgado T, Correia J, Silvestre N, Branco F (2018) Experimental study on the fire resistance of GFRP pultruded tubular beams. Compos B Eng 139:106–116
33. Morgado T, Silvestre N, Correia JR, Branco FA, Keller T (2018) Numerical modelling of the thermal response of pultruded GFRP tubular profiles subjected to fire. Compos B Eng 137:202–216

34. Dai Y, Bai Y, Keller T (2019) Stress mitigation for adhesively bonded photovoltaics with fibre reinforced polymer composites in load carrying applications. Compos Part B Eng 177
35. Bai Y, Vallee T, Keller T (2007) Modeling of thermo-physical properties for FRP composites under elevated and high temperature. Compos Sci Technol 67(15):3098–3109
36. Bai Y, Keller T, Vallée T (2008) Modeling of stiffness of FRP composites under elevated and high temperatures. Compos Sci Technol 68:3099–3106
37. ISO 834-1 (1999) Fire-resistance tests—elements of building construction—Part 1: General requirements
38. Bai Y, Hugi E, Ludwig C, Keller T (2011) Fire performance of water-cooled GFRP columns Part I: fire endurance investigation. ASCE J Compos Constr 15(3):404–412
39. Bai Y, Keller T (2011) Fire performance of water-cooled GFRP columns Part II: post-fire investigation. ASCE J Compos Constr 15(3):413–421
40. Chen W, Ye J, Bai Y, Zhao XL (2013) Improved fire resistant performance of load bearing cold-formed steel interior and exterior wall systems. Thin-Walled Struct 73:145–157
41. Chen W, Ye J, Bai Y, Zhao XL (2012) Full-scale fire experiments on load-bearing cold-formed steel walls lined with different panels. J Constr Steel Res 79:242–254
42. Zhang L, Bai Y, Chen W, Ding FX, Fang H (2017) Thermal performance of modular GFRP multicellular structures assembled with fire resistant panels. Compos Struct 172:22–33
43. Zhang L, Bai Y, Qi Y, Fang H, Wu B (2018) Post-fire mechanical performance of modular GFRP multicellular slabs with prefabricated fire resistant panels. Compos B Eng 143:55–67
44. Bai Y, Keller T, Correia JR, Branco FA, Ferreira JG (2010) Fire protection systems for building floors made of pultruded GFRP profiles. Part 2: Modeling of thermomechanical responses. Compos Part B Eng 41(8):617–629
45. Correia JR, Branco FA, Ferreira JG, Bai Y, Keller T (2010) Fire protection systems for building floors made of pultruded GFRP profiles Part 1: experimental investigations. Compos B Eng 41B(8):617–629
46. Schartel B, Humphrey JK, Gibson AG, Horold A, Trappe V, Gettwert V (2019) Assessing the structural integrity of carbon-fibre sandwich panels in fire: bench-scale approach. Compos Part B Eng 164:82–89
47. Nguyen KT, Navaratnam S, Mendis P, Zhang K, Barnett J, Wang H (2020) Fire safety of composites in prefabricated buildings: from fibre reinforced polymer to textile reinforced concrete. Compos Part B Eng 187:107815
48. Australian Building Codes Board (2015) Building code of Australia: Class 2 to Class 9 buildings
49. Eurocode 2 (2004) Design of concrete structures. Part 1–2: general rules—Structural fire design
50. AS/NZS 1170.1-2002(R2016) (2002) Structural design actions—permanent, imposed and other actions
51. Chen W, Ye J, Bai Y, Zhao XL (2014) Thermal and mechanical modeling of load-bearing cold-formed steel wall systems in fire. J Struct Eng 140(8):A4013002.1–A4013002.13
52. Fu Y, He Z, Mo D, Lu S (2014) Thermal conductivity enhancement of epoxy adhesive using graphene sheets as additives. Int J Therm Sci
53. Mchugh J, Fideu P, Herrmann A, Stark W (2010) Determination and review of specific heat capacity measurements during isothermal cure of an epoxy using TM-DSC and standard DSC techniques. Polym Testing 29(6):759–765
54. Nassif AY, Yoshitake I, Allam A (2014) Full-scale fire testing and numerical modelling of the transient thermo-mechanical behaviour of steel-stud gypsum board partition walls. Constr Build Mater 59:51–61
55. Holman J (2010) Heat transfer, 10th edition. McGraw-Hill Higher education

Chapter 16
Large Scale Structural Applications

Yu Bai, Sindu Satasivam, Xiao Yang, Ahmed Almutairi,
Hosea Ivan Christofer, and Chenting Ding

Abstract With the success of applications in aerospace, marine, electrical, automotive transportation industries, fiber reinforced polymer (FRP) composites have also found their places in civil engineering. The use of glass fibers (therefore GFRP) further reduces material costs and incorporates beneficial environmental aspects such as low energy consumption and low carbon dioxide emissions. Pultrusion is an automated and economical manufacturing method for continuous production of constant cross-section structural profiles with consistent material properties. However, in comparison to steel, pultruded GFRP composites are highly orthotropic materials and associated with lower elastic modulus and shear strength and lack of ductility at material level. In the previous chapters, a range of design concepts and structural solutions have been developed with experimental demonstrations for structural members and connections using pultruded GFRP composites, in consideration of their distinct material features. Experimental and modelling investigations have shown results of their mechanical performance of the developed structural members and connections in terms of stiffness, strength or even ductility. In the last chapter of this work, large scale structural applications are introduced with the use of developed structural members and connections. Further highlights of such structural applications are their ways of construction where design for manufacturing and assembly (DfMA) is practiced.

16.1 Introduction

In recent decades, fibre reinforced polymer (FRP) composites have gained considerable attentions for construction in civil infrastructure. The demand for building civil structures with FRP composites has grown due to their material properties such as high strength, light weight, and excellent corrosion resistance. In the retrofitting of existing structures, FRP fabric and plates are externally bonded and/or mechanically fastened to reinforced concrete (RC) beams, slabs and columns [1] and the use of

Y. Bai (✉) · S. Satasivam · X. Yang · A. Almutairi · H. I. Christofer · C. Ding
Department of Civil Engineering, Monash University, Clayton, Australia
e-mail: yu.bai@monash.edu

© The Author(s), under exclusive license to Springer Nature Singapore Pte Ltd. 2023
Y. Bai (ed.), *Composites for Building Assembly*, Springer Tracts in Civil Engineering,
https://doi.org/10.1007/978-981-19-4278-5_16

FRP to strengthen steel structures has also become an option [2]. Glass fibre reinforced polymer (GFRP) composites made from pultrusion process further reduce material costs and are associated with low energy consumption and low carbon dioxide emissions. Applications of GFRP composites have been reported particularly in new structural constructions [3–7], including rebars and grids for reinforced concrete [8–14], bridge deck superstructures [15–18], truss and frame structures [19–22], building structures [23–27], cooling towers [28], electricity transmission towers [29] and other infrastructure facilities [30].

GFRP composites have relatively low material stiffness due to its low modulus of elasticity in comparison to steel, and thus serviceability criteria may be more critical in design. Sandwich construction may improve the structural stiffness by increasing the second moment of area and therefore compensate for a low elastic modulus of GFRP composites. GFRP web-flange sandwich structures are composed of GFRP profiles sandwiched between two pultruded flat plates, forming modular units as developed and examined in previous Chaps. 2–5. Various geometric parameters such as the thickness of face sheets and section and spacing of web-core components can be adjusted, providing further design flexibility. The sandwich units can then be assembled in the transverse direction to form decks or slabs that are adhesively bonded or mechanically bolted onto supporting beams or girders to form a composite system. Such a solution is implemented to develop a bridge structure in Sect. 16.2 with a full-scale experimental demonstration.

A large-scale space frame structure assembled by pultruded circular hollow section (CHS) GFRP members is further presented in Sect. 16.3 based on the understanding of proposed sleeve connections in Chaps. 9–14. This structure was built with a span length of 8 m, width of 1.6 m and depth of 1.13 m, and weighing only 773 kgf. The structural details and design considerations may be further extended for applications of supporting structures for bridges and roofs. An experimental demonstration is introduced through a scenario of three-point bending to understand the structural stiffness, load-carrying capacity, and failure modes. The structure may shown its overall stiffness and load-carrying capacity for potential bridge applications. It is further found that the bending of critical compressive members may cause large nonlinear deformation of the overall structure. Such bending of the critical compressive members results in a decrease in overall structural stiffness with maintenance of the load applied and increase of deformation until the ultimate material failure, in a way similar to structural ductility.

The uses of proposed sleeve and splice connections are further implemented in the construction of building frames. A two-storey house structure assembled from pultruded GFRP profiles is introduced in Sect. 16.4. The structure consists of three space units formed by four single-bay portal frames. The two storeys configuration makes the ground space for general residential uses and the upper one for a bedroom as in typical loft design of a tiny house. The column bases are connected onto a continuous beam which can be supported by pad footings. The floor beams for the upper level are connected to the columns using the proposed sleeve connections and span between the adjacent frames. Floor slabs are to be supported onto the beams using the aforementioned GFRP web-flange sandwich structures. Space of the upper storey is formed by two steeply pitched rafters connected to the columns through splice connections at an appropriate angle. In each frame a horizontal beam member is installed between the columns at the floor level to increase in-plane rigidity of the frame under lateral loadings, in addition to the contribution from joint stiffness. Out-of-plane rigidity can be provided by the beams and continuous purlins connected to the rafters.

The concept of design for structural assembly has been further extended for construction of other large-scale civil infrastructure. Retaining wall structures in coastal regions may be benefited from the applications of pultruded GFRP composites because of their material corrosion resistance and light weight for installation in comparison to steel and steel reinforced concrete. Proper designs of connections should be developed to allow assembly of continuous retaining wall structures through individual GFRP sections. In Sect. 16.5, a modular GFRP retaining wall system is presented, consisting of web-flange plank and pile sections joined by mechanical interlocking connections. The web-flange section is double-H configuration with sectional inertial moment contributed by flange areas distant from the neutral axis. The plank section is designed with pin and eye connections in the flanges to join with other web-flange or pile sections. This therefore allows achieving continuous retaining walls with adaptation to possible changes in wall orientation.

16.2 An All-GFRP Footbridge

The mechanical performance of a 9 m length GFRP composite system is introduced in this section for pedestrian bridge applications. The structure was designed for assembly with incorporation of a bidirectional pultrusion orientation, where the pultrusion direction of the square tubes are placed perpendicular to the pultrusion direction of the flat plates. This orientation improves structural performance by enhancing the junction between the square tubes and face plates to prevent premature cracking along the transverse direction. Three different span lengths are examined including an 8.8 m and 6 m single span, and two 4.4 m continuous spans. The spans are subjected to a uniformly distributed load. Bending performance, effective width and the degree of composite action provided by both the adhesive connections and the transversely placed box profiles within the deck are investigated.

16.2.1 Materials and Structure

GFRP flat plates and box square hollow sections (SHS) with two thicknesses (6 or 9 mm) were used for the structural assembly. The tensile properties were tested in accordance with ASTM D3039. The material properties are given in Table 16.1 as an average of ten specimens for the 6 mm-thick sections (flat plates, connecting plates and I-profiles) and 9 mm-thick sections. In addition, the in-plane shear modulus G_{LT}

Table 16.1 Mechanical properties for GFRP materials in use

Property	Unit	6 mm-thick GFRP	9 mm-thick GFRP
Longitudinal elastic modulus E_L	GPa	22.99	24.62
Transverse elastic modulus E_T	GPa	10.32	10.03
Poisson's ratio v_{LT}	–	0.3	0.31
Poisson's ratio v_{TL}	–	0.15	0.14
Shear modulus G_{LT}	GPa	3.26	–

for the 6 mm-thick sections was found using a 10° off-axis tensile test on coupons of size 250 × 25 mm. A two-part epoxy was used to adhesively bond the GFRP components together.

The GFRP composite structure is shown in Fig. 16.1. The sandwich deck was constructed using pultruded GFRP box sections adhesively bonded between two 6 mm-thick GFRP flat plates. The flat plates each had a width of 500 mm and a length of either 1.5 m or 3 m, placed in a staggered configuration as shown in Fig. 16.2a. Adjacent flat plates were connected together using 6 mm-thick pultruded GFRP connecting plates, the pultrusion direction of which was placed parallel to that of the flat plates (along the x-axis, see Fig. 16.2a). The box profiles had dimensions of 76 × 76 × 9.5 mm, with their pultrusion direction placed along the y-axis of the

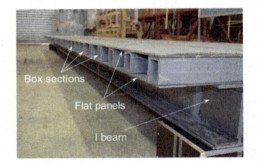

Fig. 16.1 Assembled composite structure using GFRP flat panels, box sections and I beams

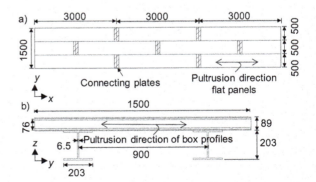

Fig. 16.2 Geometries of GFRP modular composite structure with arrangement of components in **a** plan view and **b** typical cross-section (units in mm)

bridge (see Fig. 16.2a), perpendicular to the pultrusion direction of the flat plates. The sandwich deck was adhesively bonded onto 203 × 203 × 6 mm pultruded I beams spaced 900 mm centre-to-centre. The typical cross-section of the structure is shown in Fig. 16.2b. The thickness of all bond lines was controlled by using spacer wire of 0.7 mm thickness, placed at regular intervals along the adherend surfaces.

16.2.2 Experimental Setup and Scenarios

Three span configurations were examined: 8.8 m simply supported span (Test 1, see Fig. 16.3a), 6 m simply supported span (Test 2, see Fig. 16.3b) and a 4.4 m continuous two-span (Test 3, see Fig. 16.3c). All span configurations were tested under a 4.07 kPa uniformly distributed load, which corresponds to the SLS load based on design criteria for an FRP pedestrian footbridge [31]. Loading was applied by placing weights on the surface of the deck in ten load steps and was maintained for five minutes after each load step to collect strain and displacement data. Load levels were measured using load cells, placed underneath each support. For Tests 1 and 2, two 25 kN load cells were placed at each support (LC1 to LC4, see Fig. 16.3a,b). Test 3 had two additional 50 kN load cells at the central support (LC 5 and LC6, see Fig. 16.3c).

Strain gauges were placed at four different cross-sections (CS1–1 to 4–4 as shown in Fig. 16.3). They were placed along the depth of the deck and I beams and in the longitudinal direction of the bridge at cross-section CS 1–1, which was 2.25 m from the end of the bridge. They were also placed on the deck in the transverse bridge direction at this cross section. At cross sections CS 2–2 (situated 3 m from the end of the bridge) and CS 3–3 (situated 3.75 m from the end of the bridge), longitudinal strain gauges were placed along the upper flat plate. Finally, longitudinal strain gauges were placed along the depth of the structure and along the upper flat plate, and transverse strain gauges were also placed along the upper and lower flat plates of the bridge deck at cross-section CS 4–4 (situated 4.5 m from the end of the bridge).

Deflections were measured using displacement transducers placed along the span of the bridge at either 1.5 m intervals (Test 1) or 1 m intervals (Test 2) on one side, with an additional transducer at the midspan position on the opposing side of the bridge as shown in Fig. 16.3a and b. Deflections for Test 3 were measured along one span length as shown in Fig. 16.3c, with maximum deflection measured by D3 situated 1.85 m from the support.

16.2.3 Load–Deflection Responses

Figure 16.4a shows the deformed shape for the GFRP composite structure in bending with a span length of 8.8 m and obvious deflection can be observed when the structure is subjected to uniformly distributed loading. In Fig. 16.4b, the structure presents an

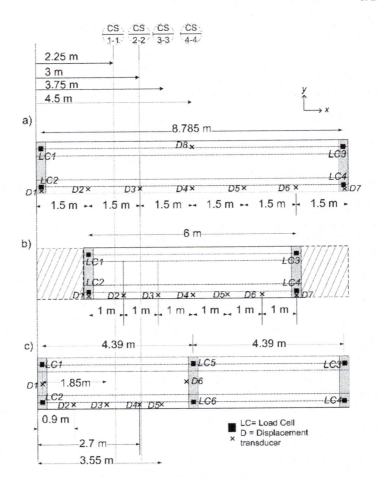

Fig. 16.3 Plan view showing cross-sections and positions of displacement transducers for **a** 8.8 m simply supported (Test 1), **b** 6 m simply supported (Test 2) and **c** 4.4 m continuous two spans (Test 3)

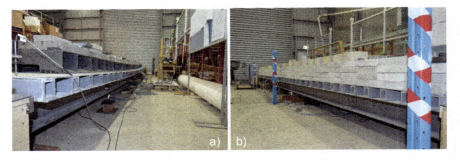

Fig. 16.4 GFRP modular composite structure subjected to uniformly distributed loading for **a** with a single span length of 8.8 m and **b** with two continuous spans of 4.4 m each

Fig. 16.5 Midspan load–deflection curves for of 8.8 m, 6 m and 4.4 m spanning composite structure under a uniformly distributed load of 4.07 kPa

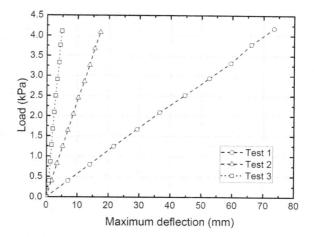

internal support at the middle length therefore forming a continuous two span configuration, where deflection was not visually noticeable with the uniformly distributed load applied.

The load–deflection responses for all experimental scenarios of the GFRP composite structure are shown in Fig. 16.5. Linear responses were observed for all specimens up to a load of 4.07 kPa. In consideration of a deflection limit of $L/500$ where L is the span length, the span length of 8.8 m did not satisfy with the deflection requirement and the maximum deflection for the span length of 6 m was still larger than the limit value; while the proposed continuous two-span design well conforms the deflection requirement of $L/500$ (experimental deflection 4.8 mm vs. the limit value of 8.8 mm). Actually, the propose continuous span configuration further takes advantage of the superior material properties in tension for the GFRP deck at the internal support, where traditional bridge deck made from concrete may be prone to crack when subjected to tension at the internal support due to negative bending moment there.

16.2.4 *Longitudinal Strain Distribution Along Specimen Depth*

The longitudinal strain distribution along the depth of the composite system is shown in Fig. 16.6 at 4.07 kPa. The strain distributions were obtained at midspan (CS 4–4, see Fig. 16.3a and b) for Tests 1 and 2, and within the positive bending region (CS 1–1 in Fig. 16.3c) for Test 3. As bending moments were low in the negative bending region, strain results were not informative from the negative bending region (CS 4–4) for Test 3. The axial strain distributions are linear from the lower flange of the I-beams to the lower flat plate, indicating that full composite action was provided by the adhesive bond between the I beams and deck. In addition, the strain distributions

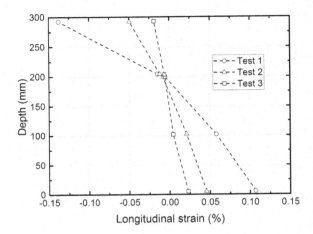

Fig. 16.6 Longitudinal strain distributions along the depth at 4.07 kPa

are also linear from the lower flat plate to the upper flat plate, indicating that the box sections as the transverse webs of the deck provided full composite action.

This behaviour was not observed in the study presented in Chap. 4 where a steel FRP composite beam also utilised built-up decks with bidirectional orientations. In that study, partial composite action was observed between the upper and lower flat plates as a result of the weak in-plane shear stiffness of the transversely oriented box sections as webs. The thickness of the box-profile webs in Chap. 4 was 6 mm, with box-profiles spaced 105 mm centre-to-centre. In comparison, the web-thickness of the box sections in this chapter was 9.5 mm and spaced at 252 mm centre-to-centre. These strain distributions indicate that a thicker transverse web can provide sufficient shear stiffness. Hence, built-up pultruded sandwich decks with bidirectional pultrusion orientations still can show full composite action between the upper and lower decks.

16.2.5 Longitudinal Strain Distribution Along Specimen Width

Figure 16.7 shows the longitudinal strain distributions of the upper flat plate along the specimen width for Test 1 (8.8 m), measured from the centre of the beam width. At CS 2–2, the compressive strains along the upper plate is the lowest at the positions where connecting plates are situated compared to the compressive strains at the centre of the beam ($y = 0$ in Fig. 16.7a). At CS 3–3, which does not have connecting plates, the compressive strains are relatively uniform along the width of the beam. At CS 4–4, the compressive stains are the lowest at the centre of the beam width where connecting plates located. These changes in strain distribution are therefore due to the presence of connecting plates. The connecting plates increase the thickness and stiffness of the flat plate locally, resulting in lower compressive strains. A change in

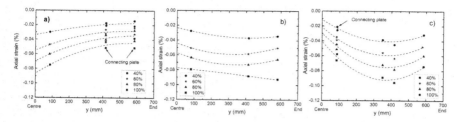

Fig. 16.7 Longitudinal strain distributions along the width of the upper flat plate for Test 1 at 40%, 60%, 80% and 100% of 4.07 kPa loading measured at **a** CS 2–2, **b** CS 3–3 and **c** CS 4–4 sections

thickness of the flat plates (caused by adhesively bonding connecting plates) therefore results in a change of effective width. This was also observed for Test 2 (6 m).

16.2.6 Conclusions

A GFRP composite structure was developed using the approach of design for manufacturing and assembly (DfMA) and presented in this chapter for pedestrian bridge applications. The built-up GFRP sandwich deck had a bidirectional pultrusion orientation, and was adhesively bonded to and supported by pultruded GFRP I-beams. Experiments were performed on the structure with a single span of 8.8 or 6 m and two continuous spans of 4.4 m. The load–deflection curves from different span configurations showed linear responses up to a load of 4.07 kPa. The proposed continuous two-span design well conforms the deflection requirement of L/500. The continuous span design for such GFRP composite structures may further use the material strength in tension for the GFRP deck at the internal support, in comparison to the weak tensile strength of traditional reinforced concrete deck.

Full composite action can be achieved at the adhesive connection between the I beams and the lower flat plate. Full composite action can also be achieved within the GFRP deck between the upper and lower flat plates because the box square hollow sections as transverse webs provide sufficient shear stiffness, allowing for the full formation of the structural bending stiffness.

The presence of connecting plates in the bridge deck results in a localized increase in the thickness of the upper flat plate, therefore introducing a reduction in compressive strains in the longitudinal direction and causing further change in effective width. However, such connecting plates can achieve full load transfer between adjacent flat plates joined together with them.

16.3 GFRP Space Frame

The sleeve connection introduced in previous chapters are used in this section for assembly of a space frame structure using pultruded GFRP members with circular hollow sections (CHS). Such members are then interconnected via Octatube plate joints by bolted connections [32] at the space nodes. The mechanical performance of the large-scale space frame is examined through three-point bending experiments. In the experiments, the load–displacement responses at the middle span and load–strain responses at the concerned members are recorded. Different failure modes are achieved through specifically designed experimental scenarios in this study. The experimental studies demonstrate the structural stiffness and load carrying capacity, as well as a mechanism for large nonlinear structural deformation induced by the bending of compressive members in the space frame structure. The dynamic and fatigue performances of the GFRP space frame structure was further investigated and introduced in [33].

16.3.1 Components and Structure

The steel connector for the sleeve connection was formed by welding a gusset plate to a slotted steel tube as shown in Fig. 16.8a. The 120 mm length unslotted section of steel tube with 73 mm outer diameter served as the bonded sleeve. The 120 mm bond length was found to be sufficient through initial bond length tests. Both ends of the CHS GFRP members were adhesively bonded with these steel connectors to form one structural member as shown in Fig. 16.8a. Then, the structural members were connected by means of bolts to a plate nodal joint known as an Octatube joint [32] (see Fig. 16.8b). The Octatube nodal joint consists of an octagonal base plate to which were welded two semi-octagonal plates placed at right angles to each other. Each Octatube nodal joint had eight joint legs. Four legs of the octagonal base plate were connected with chord members and four legs of the semi-octagonal plates were connected with diagonal web members.

The experimental space frame structure was formed by 5 × 1 grids and each grid is 1.6 × 1.6 m. As a result, the experimental structure had the span of 8 m, width of 1.6 m and depth of 1.13 m. The structure was assembled manually by two

Fig. 16.8 **a** bonded sleeve connection between steel connector and GFRP member and **b** an Octatube nodal joint for bolting with bonded sleeve connections of structural members

Fig. 16.9 a Completed space frame structure during lifting and **b** in position for testing

Table 16.2 Major experimental parameters and results of all specimens

Materials	Outer diameter (mm)	Thickness (mm)	E-modulus (GPa)	Strength (MPa)
GFRP	92	8	39.3	300
Steel	76	4	196.8	451.3/772.4[a]
Adhesive	–	–	1.9	28.6/25[b]

[a] Yielding/ultimate tensile strength for steel tube
[b] Tensile/shear strength for adhesive

technicians within half a day. The plies between bolts were brought into a snugly tightened condition using an ordinary spud wrench. The total weight of the entire space frame structure was 773 kgf. After assembly as shown in Fig. 16.9, a portal crane was employed to lift the space frame structure and place it for the testing frame.

The detailed structural geometries and material properties are summarized in Table 16.2. The strength value for GFRP was provided from the manufacturer as 300 MPa. A two-component epoxy adhesive was used, with the nominal tensile strength 28.6 MPa, tensile modulus 1.9 GPa, and shear strength 25 MPa [34].

16.3.2 Experimental Scenarios

Three loading scenarios were incurred and designated as Tests 1, 2, and 3. The loading scheme for each scenario was the same where a point load with 8 kN incremental steps was applied on the upper node at the middle of the span (see Fig. 16.10). Two corner nodes on one side were simply supported and two corner nodes on the other side were supported by rollers which can move freely along the span direction. For Test 1, the mechanical bolts utilized to connect steel connectors and nodal joints were M16 high-strength bolts with a strength grade of 8.8 (nominal shear capacity equal to 480 MPa). Such bolts were replaced by Grade 10.9 M16 high-strength bolts with nominal shear capacity of 624 MPa in Test 2. In Test 3, additional compressive forces were introduced to the two most critical compressive members (see Fig. 16.10), in

Fig. 16.10 Experimental setup for a space frame structure assembled using GFRP members and sleeve connections

order to initiate bending deformation of the two critical compressive members. It is expected that such bending of the critical compressive members may induce a change in the overall structural stiffness, and therefore the structural assembly using linear elastic GFRP members may show a nonlinear deformation capacity.

The instrumentation of the experimental structure includes strain gauges attached to the upper and lower surfaces of the most critical compressive and tensile members, and those attached around the bolt holes on the gusset plates of the critical compressive and tensile members respectively to monitor the strain variation on the steel components. Dial gauges was also mounted on the loading point and the four lower nodes near the middle span to record the vertical deflections.

16.3.3 Results and Discussion

During Tests 1, 2, and 3, no premature failures in the GFRP members and bonded sleeve connections were observed before the ultimate stage. In Test 1 or 2, the space frame structure failed due to shear failure of the Grade 8.8 bolts or Grade 10.9 bolts (see Fig. 16.11a) at the critical compressive member near the loading point. No obvious bending occurred in the critical compressive members in these two experimental scenarios. During the loading process of Test 3, bending was observed at the critical compressive members as highlighted in Fig. 16.11b, because the additional compressive forces together with their eccentricity applied to these two members could cause their bending. Meanwhile, the deflection at the loading point was observed to increase more significantly with a small increment of the applied

Fig. 16.11 Failure modes observed for **a** Tests 1 and 2 and **b** and **c** Test 3

load. Finally, one of the most critical compressive members failed due to material cleavage at the end of the GFRP tube (see Fig. 16.11c) and the structure lost its load-carrying capacity.

The load and midspan displacement curves from all three experiments are shown in Fig. 16.12. No obvious slope decrease is observed for Test 1 and a linear response is evident until the final failure at load level 144 kN. The slope of Test 2 curve shows a slight decrease at about 144 kN and finally fails at 164 kN. However as shown in Fig. 16.12, the slope of the load–displacement curve in Test 3 begins to decrease at about 96 kN due to the bending of the critical compressive members, corresponding to the accelerated increase of displacement with the sustaining of load. Such large nonlinear behaviour from Test 3 provided sufficient prior warning of structural failure in a very similar manner to the structural ductile failure process such as steel yielding. However, this time it was introduced and initiated through the bending and the resulting loss of axial stiffness of the members in compression.

Figure 16.13 shows the strain results of the critical compressive member from all experimental scenarios. Linear responses are seen from the entire loading processes for Tests 1 and 2; while only from the beginning to approximately 120 kN for Test 3. After that in Test 3, the increase of compressive strain on the upper side of the member accelerates and on the lower side the compressive strain reduces. This suggests that the critical member was under obvious bending, leading to such bending strains and further material failure as shown in Fig. 16.11b, c. In addition, all the other strains measured from the GFRP indicated that the corresponding structural members were mainly subjected to axial loading. All the strain results measured from steel suggested that no steel yielding occurred in the corresponding steel components.

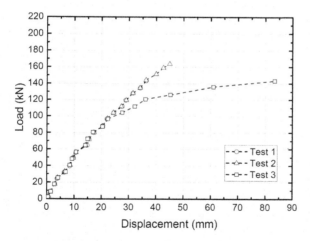

Fig. 16.12 Load–displacement curves from all experimental scenarios

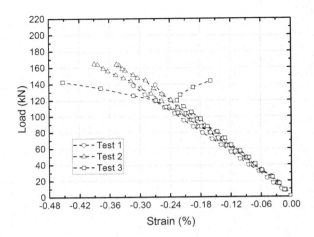

Fig. 16.13 Load–strain curves for critical compressive members from all experimental scenarios

16.3.4 Conclusions

A space frame structure that incorporates pultruded CHS GFRP components as structural members was developed using proposed sleeve connections through adhesive bonding and then bolting with steel Octatube nodal joints. The experimental study demonstrated that the bonded sleeve connections behaved well under monotonic loading and no damages were observed before the structure lost its load carrying capacity. The assembled large-scale structure had a linear response at lower load levels and showed reasonable stiffness and load carrying capacity.

Large nonlinear load–displacement responses were found for the space frame structure assembled using elastic and brittle GFRP members. Such nonlinear performance was caused by the bending of the critical compressive members. The loss of axial compressive stiffness of members in bending in a space frame configuration induced reduction in the overall structural bending stiffness, analogous to the ductile response achieved through material plasticity such as in steel yielding.

16.4 GFRP House Frames

House construction in rural or remote regions can be challenging because of limited or even no access to major machinery or equipment. As a result, heavyweight construction materials such as concrete and hot rolled steel may be either excluded from the designs or incorporated at great cost to the projects. Remote regions may also be in climates where diurnal temperature range and moisture combine to form harsh environments for building materials. Thin-walled steel and timber used in traditional housing in such environments may suffer accelerated rates of material degradation and detrimentally change structural properties. Labour and time are other critical constraints for building activities in remote regions due to their geographic nature and long distance to supplies. These constraints may further remove traditional craft-based labour-intensive construction methodology and wet-in-wet processes.

Lightweight and corrosion resistant GFRP composites appear as a potential candidate to develop structures for housing in remote regions and the approaches through design for manufacturing and assembly are also welcomed for this need. With the understandings on the sleeve connections received from Chaps. 9–12 and the splice connections from Chaps. 13, 14, such connections are used to assemble building frames in this section to develop a house structure with a loft design. Structural analysis of the frame structure is carried out through finite element (FE) modelling where the GFRP members are represented by beam elements and the connections modelled by spring elements with behaviours characterised from available experiments. Load cases are further considered including both ultimate limit state (ULS) and serviceability limit state (SLS). After all the GFRP members and steel connections are manufactured, the structural assembly may require only manpower to complete without the need of heavy machinery. It took less than one day for the structural assembly further revealing the great potential for fast construction of housing in rural and remote areas.

16.4.1 Conceptual Design

The two-storey portal frame house assembled from pultruded GFRP profiles is proposed as conceptualized in Fig. 16.14. The structure is mainly formed by four single-bay portal frames placed in parallel. Each of them is assembled by pultruded

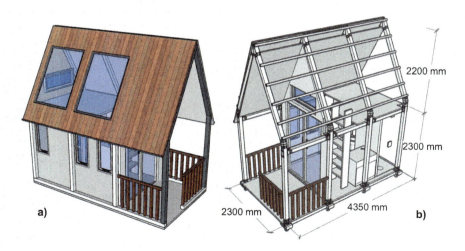

Fig. 16.14 Concept of a loft design house using pultruded GFRP composites **a** architectural and **b** interior configurations

GFRP members as columns and rafters with assistance of the proposed splice connections and proper bolting details. As shown in Fig. 16.14, the floor area is 2.3 m by 4.35 m to illustrate the design concept and also to consider the convenience for transportation and potential modification and adaptation to a tiny house or a mobile one. The four parallel portal frames form three divisions where the outer division is dedicated to a pergola space and the other two form the internal space (see Fig. 16.14). Through a loft design, the structure presents two floors. The ground floor is intended for general residential functions, including kitchen benchtop and bathroom as shown in Fig. 16.14. The upper floor is intended as a bedroom where the space is formed by two steeply pitched rafters in consideration of relevant architectural requirements.

The floor beams of the upper level are connected to the frame columns using sleeve connections as beam-column connections (see Fig. 16.15) and span between the frames. At each side of a frame, the column end with the sleeve connector and welded endplate (see Fig. 16.15) can be joined onto an edge beam using go-through bolts and then supported by footings. The developed splice connections can be used to join two rafters or the rafter and column as highlighted in Fig. 16.15. Out-of-plane rigidity of the frames can be provided by the floor beams, top bracings and purlins. GFRP sandwich wall panels as developed previously [35] can be formed with proper geometries and installed with the frames and roof panels can be attached onto the rafters. The column, rafter and beam members are GFRP tubular sections of 102 × 102 × 6.4 mm. Therefore, results in previous chapters for the structural connections developed in [36–38] with similar sections can be referred for evaluation of the mechanical performance of such a building structure.

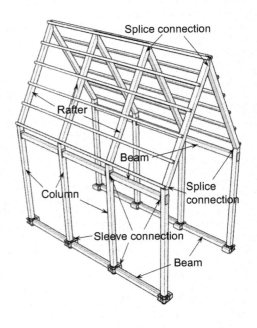

Fig. 16.15 Structural frames and connections

16.4.2 Structural Analysis

FE modelling was developed for the GFRP frame structure as shown in Fig. 16.16. The pultruded GFRP members for beams, columns, rafters and purlins were modelled using 2-node beam elements (beam188) considering the shear deformability and second-order effect. Each of such members was discretised into elements of approximately 25 mm in length. The moduli and strengths of these GFRP members measured directly from material testing as summarised in Table 16.3 in reference to previous studies [38].

As mentioned the connections between the structural members are modelled by spring elements (combin39). For the splice connections between columns and rafters or rafters and rafters (see Fig. 16.15), rotational spring elements were assigned for the in-plane (see Z-axis in Fig. 16.16) and out-of-plane rotations (see X-axis in Fig. 16.16). Moment-rotation behaviours of these spring elements were defined as multi-linear approximations (see Fig. 16.17a) based on the previous experimental results. For the beam-column connections, spring elements were defined for the rotations about X-axis (see Fig. 16.16) with their moment-rotation behaviour characterized based on the experimental results as shown in Fig. 16.17a. The rotational behaviour about Y-axis was defined with a constant stiffness of 3.33×10^5 kNm/rad

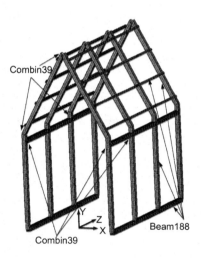

Fig. 16.16 FE modelling for the frame structure

Table 16.3 Mechanical properties of GFRP materials [38]

	Beam, column and rafter		Roof purlin	
	Modulus (GPa)	Strength (MPa)	Modulus (GPa)	Strength (MPa)
Longitudinal tensile	25.2	330.6	27.6	506
Longitudinal compressive	23.8	307.7	27.6	255
Interlaminar shear	–	31.2	–	34

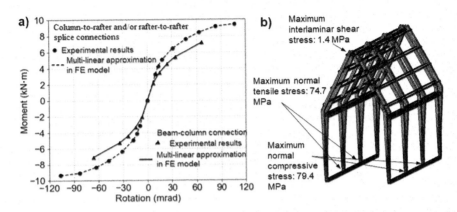

Fig. 16.17 **a** Moment–rotation relationships for connections in FE modelling; **b** Maximum stresses of the frame structure under ULS load case and deformation shape of the frame structure under SLS load case

based on a detailed FE modelling. In the above splice and beam-column connections, relative translational and torsional movements between the connected members were restrained by coupling the corresponding nodal degree of freedoms (DoFs). The connections between the purlins and rafters was assumed rigid and they were implemented by coupling the corresponding DoFs. The column base and the edge beam of ground floor were assumed rigidly connected to the footing. In the FE modelling this was realized by coupling the DoFs of the column base nodes to the corresponding beam nodes.

Two load cases were considered in the FE modelling individually. The first load case is for ultimate limit state (ULS) with $1.2G + 1.5Q + W_u$, where G is the dead load, Q is the live load and W_u is the wind load for ULS. The second load case is for serviceability limit state (SLS) with $G + 0.7Q + W_s$, where W_s is the wind load for SLS. In consideration of G, a density of 1,850 kg/m^3 was used for the GFRP members. The weight of the floor slabs was 0.617 kPa based on the adhesive bonded GFRP sandwich slabs (with about 115 mm in depth), and an additional 0.5 kPa was applied onto the floors to account for other superimposed weights. The self-weight of the roof cladding was taken as 0.2 kPa assuming also a GFRP cladding. According to AS1170.1 [39], the live load Q consists of a 2 kPa distributed load on the floors considering domestic and residential activities, and 0.25 kPa on the roofs. In the FE analysis, the loads of G and Q (in kPa) were applied and distributed as line loads (in kN/m) onto the beams and purlins considering that the floor slabs and roof claddings were simply supported. The wind load was considered in two directions (i.e. along the X- or Z-axis, see Fig. 16.16) and with or without internal pressure according to AS1170.2 [40]. The regional wind speed V_R was taken as 43 m/s for ULS and 37 m/s for SLS [40].

Under the first load case for ULS, the maximum normal tensile stress (74.7 MPa) in the structure was identified at the base of interior windward columns along their longitudinal direction (see Fig. 16.17b), in comparison to the tensile strength of 330.6 MPa (see Table 16.3). The maximum normal compressive stress is 79.4 MPa, and is located at the base of interior leeward columns, in reference to the compressive strength of 307.7 MPa (see Table 16.3). The maximum interlaminar shear stress is found to be 1.4 MPa located at the top of exterior rafters and the interlaminar shear strength is 31.2 MPa as provided in Table 16.3. The interior windward column-rafter connections are subjected to the largest moment (1.27 kNm) among all the splice connections; and the interior beam-column connections at the windward side are subjected to the largest moment (0.62 kNm) among all the beam-column connections. These are well within the elastic range of the corresponding moment-rotation curves (see Fig. 16.17a). The maximum stresses of the frame structure under ULS load case and the deformation shape of the structure under the load case for SLS is shown in Fig. 16.17b. Under SLS load case, the largest lateral displacement is identified at the middle of the windward rafters with a value of 30.7 mm, and the drift ratio of the overall structure is 0.57%. Although there are no standard values specified for GFRP building structures, this ratio looks within the reasonable range (i.e. $\leq 0.67\%$) for steel structures as suggested in [41]. However, the drift ratio of the ground floor calculated as about 1.0% becomes beyond this range. Therefore, the contribution from further wall studs and floor beams to the structural lateral stiffness can be considered in the FE modelling, or additional bracings and other lateral supports may be required.

16.4.3 Assembly Process

Considering that the developed GFRP frame structure is designed for fast assembly and erection with the absence of heavy machinery, the assembly process can be planned as several stages where components can be carried, erected, and secured safely through manual labour entirely. The national occupational health and safety commission (NOHSC) advises that load should be kept below or equal to 20 kg for lifting, lowering, and carrying loads in standing position by a single person to minimise possibility of back injury [42]. Under this principle the procedures were established for the assembly process. Moreover, all structural connections are bolted in consideration to accommodate future relocation purposes.

Prior to the assembly of building frames, the footings should have been set up appropriately with the base plate connections in place. Therefore accordingly, the GFRP edge beams of ground floor and GFRP side support member can be installed on top of the base plates (see Fig. 16.18a) and secured with bolts from the bottom (see Fig. 16.18a). Each edge beam member weights about 20 kg and hence two people may safely place the edge beams in position. As shown in Fig. 16.18a, no nuts are present on top of the edge beams and side supports, allowing the sleeve connections at the column base to slot in from the top seamlessly in a later stage. M12 bolts were utilized for all the structural connections in the structural frames in reference to the previous studies [36, 38, 43].

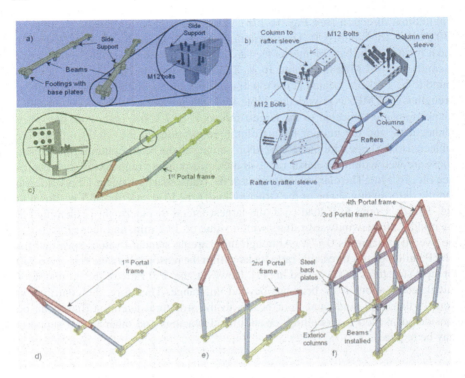

Fig. 16.18 Assembly process of proposed frame structure **a** installation of edge beams of ground floor; **b** assembly of first portal frame and **c** its position ready for erection; **d, e, f** erections of portal frames and installation of edge beams of upper floor

Individual portal frames can be assembled in their structural positions with assistance of proper working platforms. Alternatively, they may be formed on the ground, for example by firstly connecting two GFRP rafter members using the rafter to rafter sleeve connection (see Fig. 16.18b). Once the steel sleeves are inside the GFRP members, bolts and nuts can be snug-tighten with a podge spanner or wrench according to AS/NZS 5131. With a similar process, the rafters are then connected to the GFRP column members using the rafter-column steel sleeve connections. Lastly, for the portal frame assembly, the column end sleeve connections can be installed at the end of each column (see Fig. 16.18b). In total, a single portal frame weighs around 85 kg inclusive of GFRP and steel components.

To avoid the need for heavy machinery a tilt-up construction approach may be adopted to erect the individual portal frames. It is important to put the column end sleeve connections adjacent to the bolts of the edge beams as seen in Fig. 16.18c in order to locate the portal frames in position and to prevent them from slipping during the erection processes. Once the portal was upright and the bolts on the edge beam can be secured to the column end sleeve connections. Temporary props or ties may be introduced at proper locations if needed to support the portal frame and facilitate

16 Large Scale Structural Applications 377

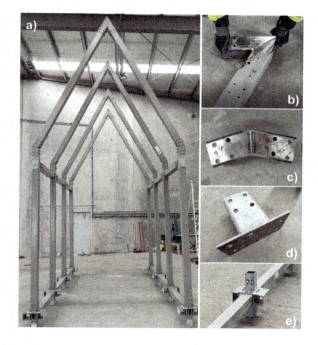

Fig. 16.19 a Assembled GFRP house frame with b rafter to rafter sleeve, c rafter to column sleeve, d beam to column sleeve and e column end sleeve connections

the erection process. The second portal frame (next to the first one) can be erected similarly as seen in Fig. 16.18e and the third and fourth then come in order and can be erected accordingly (see Fig. 16.18f).

GFRP beams for the upper floor (as indicated in Fig. 16.18f) can be prepared in place by firstly slotting in the corresponding sleeve connections into each end of the beams. After all the portal frames are erected, the beams can be lifted and installed manually as each one of them weight about only 11.5 kg (see Fig. 16.18f). Additionally, steel back plates (see Fig. 16.18f) can be placed at the outer side of the two exterior columns and bolted with the sleeve connections of the beam ends using go-through bolted. No back plates are required for the two interior columns and the sleeve connections of two adjacent beam ends can be bolted together using bolts though the columns. The main frame structure as shown in Fig. 16.19a is therefore formed in this way of GFRP members and steel connectors as shown in Fig. 16.19b, c, d, e with bolts only.

16.4.4 Further Considerations

After the main structural frame is assembled, GFRP joists can be placed on the ground level and supported by the GFRP edge beams. The spacing of the joist can be determined in reference to the results of web-flange sandwich structures presented in Chaps. 2 and 3 in order to form the floor system. The upper level joists and the corresponding sandwich floor system can be installed similarly. The roof purlins and panels may also be inspired by the concept of GFRP web-flange sandwich assemblies, where a smaller overall depth can be used because of the less live load in comparison to the floor system. Proper connection approaches and fasteners may be required though with specific waterproof considerations.

16.5 GFRP Modular Retaining Wall

Considering the relatively low elastic modulus of GFRP composites, appropriate sectional design is important to achieve adequate structural stiffness in bending for retaining wall applications. Such sections are further required to be equipped with convenient connection configurations in order to form continuous retaining walls with adaptation to possible changes in wall orientation. Figure 16.20 shows a GFRP modular retaining wall system formed with web-flange sections and round pile sections. It can be seen that an interlocking mechanism is provided with the sections where such pin and eye connections can join the flanges of the web-flange sections to other web-flange or pile sections. Furthermore, this connection configuration is provided at multiple locations around a pile section and therefore wall orientations can change at pile locations as seen in Fig. 16.20. The retaining wall system must provide sufficient bending resistance and satisfy relevant strength and deformation requirements. Especially regarding deformation, although there are no specific design specifications for GFRP structures as a retaining wall system, allowable deflections in the range of L/100 to L/60 are recommended for retaining walls in general [44, 45]. Experimental investigations were therefore conducted to evaluate the bending performance of both web-flange and round pile sections and the connections.

Fig. 16.20 A modular GFRP retaining wall system with web-flange and pile sections and mechanical interlocking connections

16.5.1 Member Performance

The web-flange sections are manufactured through pultrusion process with nominal depth, width and thickness of 260 mm, 580 mm and 6.4 mm respectively. The nominal diameter of the round pile is 300 mm and the thickness is 6.4 mm. A series of experimental studies was conducted to evaluate the bending performance of the GFRP web-flange and round pile hollow sections [46]. As shown in Fig. 16.21a, b, such specimens were loaded under four-point bending with a span length of 2.8 m. The web-flange specimen failed through shear cracking along the web-flange junction, due to the shear stress there exceeding the corresponding inter-laminar shear strength, as also observed in a few previous studies [47–50]. The specimens with pile section lost the load-carrying capacity due to local crushing failure at the loading location.

The design of retaining wall structures is often dominated by deflection limits and this requirement is even more critical for the structures made from GFRP composites because of their low elastic modulus. The hollow web-flange and pile sections further allow filling of concrete (or other backfill materials) to provide lateral constrains to the GFRP materials and improve the overall bending stiffness. Four-point bending experiments were conducted on the web-flange and pile specimens filled with concrete (see Fig. 16.22a, b respectively).

The resulting load and displacement curves for the web-flange and pile specimens without concrete filling are shown in Fig. 16.23a. The slope in the linear stage may be used to determine the bending stiffness and the pile specimen showed much higher stiffness and also a larger ultimate load. For the specimens with concrete filling, the resulting load–displacement curves (Fig. 16.23b) received from the middle span evidenced considerable improvements in bending stiffness for both sections. FE modelling was further developed in [51] showing agreement with experimental stiffness.

Fig. 16.21 Bending experiments on **a** GFRP web-flange and **b** round pile sections without concrete filling

Fig. 16.22 Bending experiments on **a** GFRP web-flange and **b** round pile sections with concrete filling

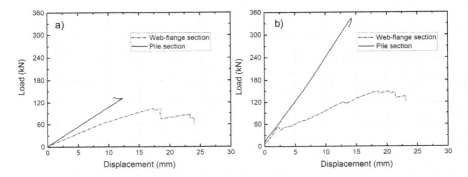

Fig. 16.23 load–displacement curves of GFRP web-flange and round pile sections **a** without and **b** with concrete filling

16.5.2 Connection Performance

To evaluate the connection performance of two adjacent sections in terms of load transfer and rotation capacity, three web-flange sections of 415 mm in length were assembled through their specific joint configurations and tested in the transverse direction (Fig. 16.24). The two ends of the specimen were fixed covering the web-flange junctions to minimize end rotation and junction rotation, so that the connection rotation could be the focus.

The load and displacement curve measured at the middle span is shown in Fig. 16.25, where an excessive deformation of more than 60 mm can be seen due to rotation of the connection. The large deflection introduced high tensile stresses at the web-flange junction and the transverse direction was also associated with weak material strengths of the pultruded GFRP composites. Initial cracking at web-flange junctions was noticed first at about 7 kN (Fig. 16.25) and the cracking developed with

16 Large Scale Structural Applications

Fig. 16.24 Experimental setup for evaluation of mechanical interlocking connection

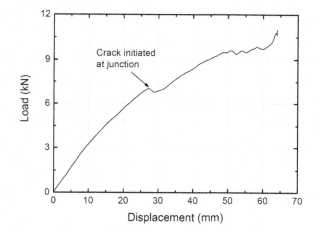

Fig. 16.25 Load–displacement responses for three web-flange sections assembled by mechanical interlocking connections

the increase in loading. However, the cracking at the junctions did not induce the ultimate failure and the load continued to increase with larger deflection achieved as seen in Fig. 16.25, until the loading was manually stopped. It was seen that the mechanical interlocking system was mechanically effective in transferring the stresses to adjacent sections up to a joint rotation of 12° or greater, and it was also convenient for assembly.

16.6 Summary

Pultruded GFRP composites have shown advantageous corrosion resistance for structural applications in aggressive environments. The material properties of GFRP composites also support the development of light weight, high strength and more durable civil structures through the methodology of DfMA. This chapter therefore focused on such practices of pultruded GFRP composites as major load-carrying members for civil construction, where DfMA was constantly exercised to deliver large-scale applications in comparison to conventional construction methods. Design considerations at structural level are further given to improve the structural stiffness in compensation of relatively low elastic modulus of GFRP composites and to provide large non-linear deformation capacity in contrast to the linear elastic and brittle feature inherent to the material. A particular design effort was made in the approaches to connect pultruded GFRP members with closed sections, enable their implementation with compatibility to the methodology of DfMA.

This chapter presented several examples of pultruded GFRP composites for construction of large-scale structures in civil applications, including i) bridge superstructure or decking system using web-flange sandwich sections, and the resulting composite structural system with the support girder, ii) space frame structures, iii) building frame structures, and iv) modular retaining wall structures. It is expected that such practices may demonstrate applicability or potential solutions for other similar needs, and provide insightful results to further assist design and industrial uptake of such GFRP applications for civil infrastructure with extra environmental benefits. Given these efforts in consideration of particular material advantages and design preferences including light weight and high strength, and structural assembly and modular solutions, further developments may be lifted toward high performance structures, design for robotic assembly and automated construction, and design for structural reusability with detachability.

References

1. Teng JG, Chen JF, Smith ST, Lam L (2003) Behaviour and strength of FRP-strengthened RC structures: a state-of-the-art review. Struct Build 156:51–62
2. Zhao XL, Zhang L (2007) State-of-the-art review on FRP strengthened steel structures. Eng Struct 29:1808–1823
3. Bakis CE, Bank LC, Brown VL, Cosenza E, Davalos JF, Lesko JJ, Machida A, Rizkalla SH, Triantafillou TC (2002) Fiber-reinforced polymer composites for construction—state-of-the-art review. J Compos Constr 6:73–87
4. Gand AK, Chan TM, Mottram JT (2013) Civil and structural engineering applications, recent trends, research and developments on pultruded fiber reinforced polymer closed sections: a review. Front Struct Civ Eng 7(3):227–244
5. Bank LC (2006) Composites for construction: structural design with FRP materials. John Wiley & Sons, Hoboken, NJ
6. Keller T (2001) Recent all-composite and hybrid fibre-reinforced polymer bridges and buildings. Prog Struct Mat Eng 3(2):132–140
7. Fang H, Bai Y, Liu WQ, Qi Y, Wang J (2019) Connections and structural applications of fibre reinforced polymer composites for civil infrastructure in aggressive environments. Compos B 164:129–143
8. Benmokrane B, El-Salakawy E, El-Gamal SE, Sylvain G (2007) Construction and testing of Canada's first concrete bridge deck totally reinforced with glass FRP bars: Val-Alain Bridge on HW 20 East. ASCE J Bridg Eng 12(5):632–645
9. Dieter DA, Dietsche JS, Bank LC, Oliva MG, Russell JS (1814) Concrete bridge decks constructed with fiber-reinforced polymer stay-in-place forms and grid reinforcing Transportation Research Record. J Transp Res Board 1:219–226

10. Rizkalla S, Hassan T, Hassan N (2003) Design recommendations for the use of FRP for reinforcement and strengthening of concrete structures. Prog Struct Mat Eng 5(1):16–28
11. Matta F, Galati N, Nanni A, Ringelstetter TE, Bank LC, Oliva MG (2005) Pultruded grid and stay-in-place form panels for the rapid construction of bridge decks. Composites, pp 1-9. Convention and Trade Show American Composites Manufacturers Association, USA
12. Tomlinson D, Fam A (2015) Performance of concrete beams reinforced with basalt FRP for flexure and shear. J Compos Constr 19(2):04014036
13. Fang H, Xin X, Liu W, Qi Y, Bai Y, Zhang B, Hui D (2016) Flexural behavior of composite concrete slabs reinforced by FRP grid facesheets. Compos B 92:46–62
14. Peng F, Xue W (2018) Design approach for flexural capacity of concrete T-beams with bonded prestressed and nonprestressed FRP reinforcements. Compos Struct 204:333–341
15. Keller T (2002) Overview of fibre-reinforced polymers in bridge construction. Struct Eng Int 12(2):66–70
16. Bank LC (2006). Application of FRP composites to bridges in the USA. In: Proceedings of the International Colloquium on Application of FRP to Bridges, pp 9–16. Japan Society of Civil Engineers (JSCE), Tokyo, Japan
17. Jeong J, Lee Y-H, Park K-T, Hwang Y-K (2007) Field and laboratory performance of a rectangular shaped glass fiber reinforced polymer deck. Compos Struct 81(4):622–628
18. Davalos JF, Chen A, Zou B (2011) Stiffness and strength evaluations of a shear connection system for FRP bridge decks to steel girders. J Compos Constr 15(3):441–450
19. Bank L, Mosallam AS, McCoy GT (1994) Design and performance of connections for pultruded frame structures. J Reinf Plast Compos 13(3):199–212
20. Mottram JT, Zheng Y (1996) State-of-the-art review on the design of beam-to-column connections for pultruded frames. Compos Struct 35(4):387–401
21. Keller T, Bai Y, Vallée T (2007) Long-term performance of a glass fiber-reinforced polymer truss bridge. J Compos Constr 11(1):99–108
22. Mosallam AS (2011) Design guide for FRP composite connections. ASCE Manuals and Reports on Engineering Practice No. 102. The American Society of Civil Engineers. ISBN 978-0-7844-0612-0
23. Keller T, Theodorou NA, Vassilopoulos AP, De Castro J (2016) Effect of natural weathering on durability of pultruded glass fiber-reinforced bridge and building structures. J Compos Constr 20(1):04015025
24. Bai Y, Keller T, Wu C (2013) Pre-buckling and post-buckling failure at web-flange junction of pultruded GFRP beams. Mater Struct 46(7):1143–1154
25. Keller T, Haas C, Vallée T (2008) Structural concept, design, and experimental verification of a glass fiber-reinforced polymer sandwich roof structure. J Compos Constr 12(4):454–468
26. Zafari B, Mottram JT (2015) Characterization by full-size testing of pultruded frame joints for the startlink house. J Compos Constr 19(1)
27. Zhu D, Shi H, Fang H, Liu W, Qi Y, Bai Y (2018) Fiber reinforced composites sandwich panels with web reinforced wood core for building floor applications. Compos B 150(1):196–211
28. Chavan Dattatraya K, Utpat LS (2012) Pultruded FRP cooling tower—design, development and validation. Int J Eng Res Technol 1(4):1–6
29. Godat A, Légeron F, Gagné V, Marmion B (2013) Use of FRP pultruded members for electricity transmission towers. Compos Struct 105:408–421
30. Manalo A, Aravinthan T, Fam A, Benmokrane B (2017) State-of-the-art review on FRP sandwich systems for lightweight civil infrastructure. J Compos Constr 21(1):04016068
31. AASHTO (2008) Guide specifications for design of FRP pedestrian bridges, 1st edn. American Association of State Highway and Transportation Officials (AASHTO)
32. Ramaswamy GS, Eekhout M, Suresh GR (2002) Analysis, design and construction of steel space frames. Thomas Telford, UK
33. Yang X, Bai Y, Luo FJ, Zhao XL, Ding F (2016) Dynamic and fatigue performances of a large-scale space frame assembled using pultruded GFRP composites. Compos Struct 138:227–236
34. Fawzia S, Zhao XL, Al-Mahaidi R (2010) Bond–slip models for double strap joints strengthened by CFRP. Compos Struct 92(9):2137–2145

35. Xie L, Qi Y, Bai Y, Qiu C, Wang H, Fang H, Zhao XL (2019) Sandwich assemblies of composites square hollow sections and thin-walled panels in compression. Thin-Walled Struct 145:106412
36. Zhang ZJ, Bai Y, Xiao X (2018) Bonded sleeve connections for joining tubular glass fiber reinforced polymer beams and columns: an experimental and numerical study. ASCE J Compos Constr 22(4):04018019
37. Qiu C, Ding C, He X, Zhang L, Bai Y (2018) Axial performance of steel splice connection for tubular FRP column members. Compos Struct 189:498–509
38. Qiu C, Bai Y, Zhang L, Jin L (2019) Bending performance of splice connections for assembly of tubular section FRP members: Experimental and numerical study. J Compos Constr 23(5):04019040
39. Australian Standard, AS 1170.1-1981, Minimum design loads on structures—dead and live loads
40. Australia Standard. AS 1170.2-2002 Structural design actions-Wind actions
41. Australian Standard, AS4100-1998 steel structures, Sydney, 1998
42. NOHSC (1995) National code of practice for manual handling. NOHSC:2005. National Code of Practice, Canberra, Australia
43. Wu C, Zhang Z, Bai Y (2016) Connections of tubular GFRP wall studs to steel beams for building construction. Compos B 95:64–75
44. Bdeir Z (2001) Deflection-based design of fiber glass polymer (FRP) composite sheet pile wall in sandy soil. McGill University, Quebec
45. El-Reedy M (2014) Marine structural design calculations. Butterworth-Heinemann
46. Ferdous W, Bai Y, Almutairi AD, Satasivam S, Jeske J (2018) Modular assembly of water-retaining walls using GFRP hollow profiles: components and connection performances. Compos Struct 194:1–11
47. Bai Y, Keller T, Wu C (2013) Pre-buckling and post-buckling failure at web-flange junction of pultruded GFRP beams. Mater Struct 46:1143–1154
48. Borowicz DT, Bank LC (2011) Behavior of pultruded fiber-reinforced polymer beams subjected to concentrated loads in the plane of the web. J Compos Constr 15:229–238
49. Keller T, Gürtler H (2006) In-plane compression and shear performance of FRP bridge decks acting as top chord of bridge girders. Compos Struct 72:151–162
50. Turvey GJ, Zhang Y (2005) Stiffness and strength of web–flange junctions of pultruded GRP sections. P I Civil Eng-Str B 158:381–391
51. Ferdous W, Almutairi AD, Huang Y, Bai Y (2018) Short-term flexural behaviour of concrete filled pultruded GFRP cellular and tubular sections with pin-eye connections for modular retaining wall construction. Compos Struct 206:1–10

Printed by Printforce, United Kingdom